Lecture Notes in Intelligent Transportation and Infrastructure

Series Editor

Janusz Kacprzyk, Systems Research Institute, Polish Academy of Sciences, Warszawa, Poland

The series "Lecture Notes in Intelligent Transportation and Infrastructure" (LNITI) publishes new developments and advances in the various areas of intelligent transportation and infrastructure. The intent is to cover the theory, applications, and perspectives on the state-of-the-art and future developments relevant to topics such as intelligent transportation systems, smart mobility, urban logistics, smart grids, critical infrastructure, smart architecture, smart citizens, intelligent governance, smart architecture and construction design, as well as green and sustainable urban structures. The series contains monographs, conference proceedings, edited volumes, lecture notes and textbooks. Of particular value to both the contributors and the readership are the short publication timeframe and the world-wide distribution, which enable wide and rapid dissemination of high-quality research output.

More information about this series at http://www.springer.com/series/15991

Egils Ginters · Mario Arturo Ruiz Estrada ·
Miquel Angel Piera Eroles
Editors

ICTE in Transportation and Logistics 2019

Springer

Editors
Egils Ginters
Faculty of Computer Science and Information Technologies, Department of Modelling and Simulation
Riga Technical University
Riga, Latvia

Mario Arturo Ruiz Estrada
Faculty of Economics and Administration
University of Malaya
Kuala Lumpur, Malaysia

Miquel Angel Piera Eroles
Department of Telecommunications and Systems Engineering
Autonomous University of Barcelona
Bellaterra, Barcelona, Spain

ISSN 2523-3440 ISSN 2523-3459 (electronic)
Lecture Notes in Intelligent Transportation and Infrastructure
ISBN 978-3-030-39687-9 ISBN 978-3-030-39688-6 (eBook)
https://doi.org/10.1007/978-3-030-39688-6

This Springer imprint is published by the registered company Springer Nature Switzerland AG
The registered company address is: Gewerbestrasse 11, 6330 Cham, Switzerland

Preface

The development of transport and logistics services has a significant impact on the economic development of any country. The European transport industry employs more than 10 million people and accounts for about 5% of gross domestic product (GDP). In the USA, the impact of transport services on GDP is almost twice that, while the logistics in EU account for 10–15% of the cost of a finished product [1].

Like any real system, transportation and logistics are sociotechnical, characterized by the interoperability of technologies, the society, and the environment (see Fig. 1). Transport and logistics systems are characterized by their logical and physical structures, so the articles are organized into five bigger functional chapters that represent the major trends in horizontal development:

- **Business Planning and Crafting**
  Approaches and techniques for planning new policies, creating prospective transport corridors, supply chains, routes, schedules, inventory, and infrastructure development and deployment. Business planning respects the demands of the digital society by providing insights into modern data visualization trends, which must ensure proper semantic representation of information under Big Data conditions, distinguishing the essential from the insignificant;
- **Equipment Development and Diagnostics**
  The current period is characterized by a gradual change of generations of equipment and control systems. Electronic blocks are gradually being replaced by microprocessor-based controllers, devices are being diagnosed and managed remotely, they are becoming more intelligent, and equipment and systems security parameters are improving. Technologies are becoming more sociable, informative, and multifunctional;
- **Telecommunications Development**
  One of the attributes of automated management and control is telecommunication, which provides information exchange between respondents. Currently, one of the main respondents are technical devices that make up the Internet of things (IoT). The basic criterion for the management and control of technical

systems is safety. As new types of LTE, 5G, etc., develop, it is important to determine their suitability for the management and control of technical systems. Arouses interest in UWB communications possible uses;

- **Education and Training**
  Emphasizes current trends in changing teaching methods in the digital society through e-learning. New study courses appropriate to the interdisciplinary development of the industry, such as mechatronics, guided vehicles, and visual analytics, are being designed and implemented;

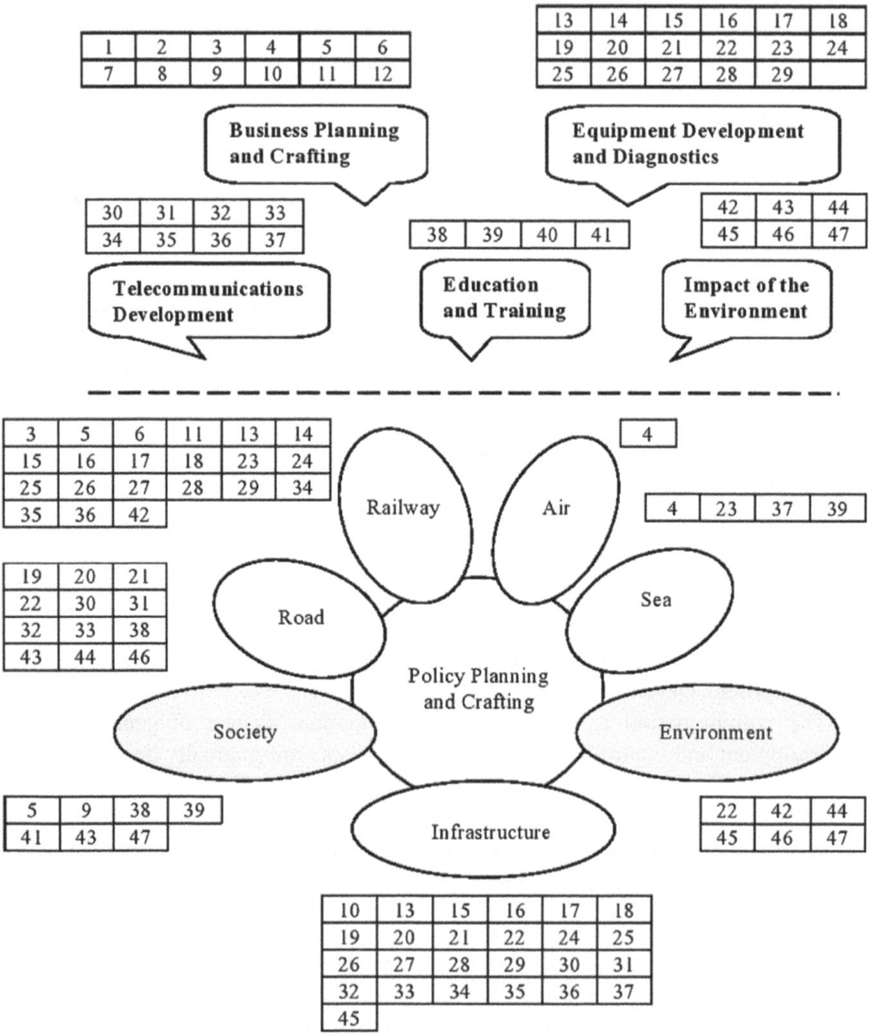

Fig. 1. Sociotechnicality paradigm in transportation and logistics.

- **Impact of the Environment**
  Technological development takes place in interaction with the society and the environment. The impact of technology must be more cautious and respect the inevitable climate change. The articles in the book look at alternative energy and green transport applications, analyze pollution and other impact factors, and look at possible transport solutions to address the consequences of natural disasters in affected areas.

The physical structure of the transport and logistics system is the environment for performing the functions defined by the logical structure (see Fig. 1). Each article can be assigned to several functions of a logical structure, as well as belonging to several groups of objects of a physical structure.

The articles in the book illustrate the impact of technologies' development on the society and the economy, as well as demonstrate the feedback that makes technologies more environmentally friendly and to comply with the society interest.

December 2019

Egils Ginters
Mario Arturo Ruiz Estrada
Miquel Angel Piera Eroles
Co-editors

The Transport Institute of Riga Technical University is the initiator of this book. Seeing new digital technologies and modern research methods emerge day by day, we consider it our duty to introduce them to as wide an audience as possible, which could contribute to a more homogeneous knowledge coverage and successful training of highly qualified professionals in transport and logistics area.

Prof. Mareks Mezitis
Head of the Transport Institute of Riga Technical University

Acknowledgements

Editors want to say sincere thanks to all the authors, as well as Dr. Thomas Ditzinger (Springer) and Mrs. Varsha Prabakaran (Springer) for her support and valuable advice in publishing the book.

The editors wish to thank the Editorial Board of Books article review, and in particular, Rosa Maria Aguilar (University of La Laguna, Spain), Krzysztof Amborski (Warsaw University of Technology, Poland), Elina Gaile-Sarkane (Riga Technical University, Latvia), Yahsi Yazicioglu (Gazi University, Turkey), Zbigniew Lukasik (Kazimierz Pulaski University of Technology and Humanities in Radom, Poland), Mareks Mezitis (Riga Technical University, Latvia), Elena Mikhailova (ITMO University of Saint Petersburg, Russia), Gaby Neumann (Technical University of Applied Sciences Wildau, Germany), Sergii Panchenko (Ukrainian State Railway Academy, Ukraine), Aleksander Sladkowski (Silesian University of Technology, Poland), Peter Sonntagbauer (cellent AG, Austria), Haris Neophytou (Interfusion Services Ltd., Cyprus), Natalya Tokmurzina (Kazakh Academy of Transport and Communications, Kazakhstan), Boyong Wang (Yantai Integrated Information Center, China), and Robert Zhang (China Bosstedia International Co., Ltd., China).

Reference

1. Transport sector economic analysis, EU Science Hub. The European Commission's science and knowledge service, https://ec.europa.eu/jrc/en/research-topic/transport-sector-economic-analysis, last accessed 2019/05/23.

Contents

Modeling Global Consumptions Trends Impact on Transport and Logistics: Scenario Analysis

Astra Auzina-Emsina(✉), Velga Ozolina, and Inguna Jurgelane-Kaldava

Faculty of Engineering Economics and Management, Riga Technical University, Kalnciema Street 6, Riga 1048, Latvia
{astra.auzina-emsina, velga.ozolina, inguna.jurgelane-kaldava}@rtu.lv

Abstract. Transport and logistics heavily depend on economic activity and consumption decisions made by the market nationally and globally. Advanced economies face new global trends caused by more eco-friendly consumption behaviour decisions, distance shopping etc. The aim of the research is to model the economic impact of consumption decisions on transport and logistics caused by global trends. The inter-industry linkages are taken into account and input-output approach and scenario analysis are applied to model the impact of detected global trends on national transports and logistics industry. One EU country, Latvia, is selected as research case, however, the method and general findings can be applied to other countries. The findings are valuable to transport and logistics companies, agencies, governments.

Keywords: Transport and logistics · Consumption modeling · Scenario analysis

1 Introduction

Rapid development of transport and its infrastructure has led to increased trade volumes both nationally and internationally, allowing transport and logistics to develop as well. On the other hand, transport and logistics heavily depend on economic activity and consumption decisions. Thus, in order to adjust to the changing situation, managers in transport and logistics have to follow recent developments and forecasts regarding economic development and new trends. They have to study not only the market they operate in, but also other markets as new trends tend to become global due to the information exchange possibilities. Moreover, global trends usually have some local peculiarities [1], which cause the necessity for adjustment.

This paper focuses on advanced economies due to the significant differences in consumption structure [2]. Global trends faced by the advanced economies are related to the technological progress, including increased speed of information flow, to the environmental concerns like plastic and other types of waste and greenhouse emissions, as well as health issues. As a result consumption decisions become more eco-friendly, many consumers seek for more healthy consumption options, use distance shopping more frequently etc. More and more people tend to participate in collaborative

E. Ginters et al. (Eds.): ICTE ToL 2019, LNITI, pp. 1–8, 2020.
https://doi.org/10.1007/978-3-030-39688-6_1

consumption, e.g. by allowing others to rent their cars or accommodation while it is not needed [3, 4]. Increasing Internet use and development of appropriate apps has facilitated the development of on-demand deliveries within a couple of hours [5]. Thus also transport and logistics companies have to adjust to these trends providing consumers more access to goods worldwide by more sustainable means.

It is evident that due to more distance shopping possibilities the share of light weight freight transport has increased [6]. To reduce negative environmental and economic impacts, the use of more environmentally friendly transport means, e.g., electric vehicles is facilitated [7] and such innovations as crowd logistics [8] are implemented. Also the impact of packaging and the possibilities to reduce or re-use it are analysed [9].

The aim of the research is to model the economic impact of consumption decisions on transport and logistics caused by global trends. While some of the new trends can be analysed using only qualitative methods, the study focuses on quantitative evaluation.

2 Literature Review

Scientific literature provides various studies on issues faced by transport and logistics. Several of them use mathematical modeling for analysis.

Mathematical programming methods provide tools for optimisation. Sarkar *et al.* [10] use mixed-integer non-linear programming (MINLP) model to minimize transportation and carbon emission costs taking into consideration container capacity, required number of containers, shipment sequence of retailers, cycle time, and remanufacturing rate. The authors conclude that without optimization, not only there would be negative environmental impact, but also economic losses. Lam and Gu [11] use bi-objective optimisation model to minimize both cost and transit time for the tactical planning of intermodal container flows. They provide analysis of trade-offs between cost and transit time. Other models, like in the study of de Keizer *et al.* [12], optimize profits within a mixed integer linear programming model. There are also more sophisticated models, which include stochastic elements [13].

Simulation models allow performing "what if" analysis. It is useful, when there is not one single optimization criterion. For example, in planning of urban freight transport [14].

There are studies, which use regression and neural network models. Binary logistic regression model is used to analyse bicycle use in Beijing, outlining facilitating and hindering factors [15]. In the study of Shafiee *et al.* [16] logistic costs are just a fraction of total costs, which are analysed in the 25-year lifecycle of the wind farms. There also are studies that examine the possibilities to model the impact of the transportation and storage industry on the economic development of national economy when the economy is relatively small economy and have also geographical location that faster also transit transport flows and business [17].

It can be concluded that transport and logistics and its impact is mostly analysed in comparatively small scale – within particular projects, supply chains or cities. This paper is filling this gap by providing a model, which can be used for the macro level analysis using one European Union (EU) economy, Latvia, as an example.

3 Methodology

As the case of the EU country is selected, then the research is based on the EU data and Eurostat classification. Symmetric input-output (IO) data in product-by-product classification is used. The latest available IO data on 2015 are used in the research. The selected modeling approach consists of analysis of trends, input-output model development, scenario elaboration, analysis of modeling results.

The input-output model includes the basic identities that ensure the equilibrium in the economy (1) and (2):

$$x = \mathrm{A}x + y \tag{1}$$

where x – output (vector), Ax – intermediate consumption (matrix) and y – final demand (vector).

Latvia is amid the majority of the EU countries that elaborate product-by-product symmetric input-output tables according to CPA classification, rather than in industry-by-industry symmetric input-output tables according to NACE classification. In the EU, 23 countries publish IO data sets according product-by-product classification, of which three countries also according product-by-product classification, however, five countries publish only according product-by-product classification. These data sets are resources and time consuming and a substantial submission delay, of more than half a year, was also observed for several EU countries by Eurostat [18]. Symmetric input-output should be mandatory published every five years in all EU countries.

In the research, the latest available Latvia's symmetric IO data on 2015 are used that are in euros as Latvia joined the Euro Area in 2014 (national currency was replaced by euro).

Therefore, A is a matrix of technical coefficients, where each element a_{ij} corresponds to the volume of intermediate input of product i products to produce a unit of output of product j. Technical coefficients represent the direct cost structure of production of product j:

$$x = \mathrm{B}\,y \tag{2}$$

where B – Leontief inverse (matrix), computed by (3):

$$\mathrm{B} = (\mathrm{I} - \mathrm{A})^{-1} \tag{3}$$

where I – identity matrix.

The use of primary factors (F) depends on the technology (technical coefficients of primary factors, showing the quantity of primary factors required per unit of output; L) and output of products x (4):

$$\mathrm{F} = \mathrm{L}x \tag{4}$$

where L – technical coefficients of primary factors (matrix).

The impact of new consumption trends and changes in consumer decisions on transport sector are modelled by the scenario analysis applying input-output model. In this research, the impact and the so-called what-if scenarios are evaluated in short term and medium term hence technological coefficients are constant. In long term, technologies and technical coefficients change as well new consumption trends arise. The what-if scenarios allow to estimate the quantitative effect of a certain (or given) change in the economy.

The research data analysis covers the whole EU, but modeling is executed on the case of one EU country - Latvia. The method and the scenarios are applicable to the EU countries and other economies facing similar research questions.

To model the impact of detected global consumptions trends impact on the transport and logistics sectors and the economy, two comparative-static scenarios are estimated taking into account the above-detected trends and assumptions in Latvia's economy:

- 10% of current households final demand expenditure on food and beverages at stores and shopping malls (i.e. retail industry) is replaced by online shopping and delivery. Distance or online shopping is an ongoing trend that results also in larger spending on land cargo transport services and warehousing and support services for transportation. At present, due to changes in national retail food and groceries market, larger shopping centres are dominant trend, but these centres are not reachable anymore by foot and hence partially food shopping is done online and additional demand for land transport and warehousing sector generated;
- 5% of current households final demand expenditure on transport vehicles (as purchase of cars) is replaced by renting and co-sharing transport vehicles when needed. Participation in collaborative consumption, e.g. by renting or borrowing the needed vehicle from others while it is not needed is an ongoing trend that results also in larger spending on land transport services. Government is more fixed in choices and linked to the existing contracts with transportation and service companies, hence only private consumption decisions are evaluated due to higher estimated likelihood to take place in near future. In long term, the government sector as well can change public consumption (including public procurement) patterns and partially participate in collaborative consumption.

The estimated scenarios give an important insight and information for market participants and analytics how the latest trends in consumption affects other sectors and the whole economy.

4 Data Analysis and Modeling Results

The results of the first scenarios estimated by input-output model claim that the impact on the economy is observable and noticeable. It results in 0.3% increase of national output on various products and services. As households consume and spend on products and services of almost all industries, only the major impacts are analysed in detail. Table 1 shows the modelled full (direct and indirect) impact results of product output changes that are larger than 0.5% (output of products with the impact below 0.5% are not included in the Table 1).

Table 1. Modeling results of the first scenario representing sectoral output change (%).

NACE/CPA CODE	Product or service	Impact (%)
H53	Postal and courier services	8.6
H52	Warehousing and support services for transportation	1.7
H49	Land transport services and transport services via pipelines	1.3
C22	Rubber and plastic products	0.6
N77	Rental and leasing services	0.5
C24	Basic metals	0.5
C20	Chemicals and chemical products	0.5

It is assumed that private households consume the same amount of food and beverages, but due to changes in purchasing style, buying online, the modelled share of products (10% relatively or accounted for 180.5 mil euros in money terms are delivered by others). It is assumed that delivery buy others (transportation companies, wholesale and retail companies, courier etc.) cost more (by 30% or accounted for 54.1 mil euro) than go by foot to local store when there are no direct transportation costs. It is assumed that transportation margin of 30% is divided into 15% land transportation (additional demand created of 27.1 mil euro), 10% warehouse services (18.0 mil euro) and 5% courier services (9.0 mil euro).

The results claim that this type of change in consumption results in considerable increase of demand for postal and courier services (8.6%), warehousing and supporting services for transportation (1.7%), and land transport services (1.3%). Other sectors have relatively smaller impact, however, as more deliveries are needed, more supporting products and services are demanded as well.

The results of the second scenarios estimated by input-output model claim that the impact on the economy is two-sided – some sectors (in this case, transport sector) have positive impact of this trend, however other sectors due to more eco-friendly and moving towards zero-waste ideas face decline in demand. It results in no national level impact as it results in 0.0% increase or decrease of total national output; however sectoral impact is very diverse. As households consume and spend on products and services of all industries, only the major impacts are analysed in detail. Table 2 shows the modelled full (direct and indirect) impact results of product output changes that are larger than 0.1% increase and −0.1% decrease (output of products with the impact between 0.1% and −0.1% are not included in the Table 2).

According to the scenario, it is assumed that private households spend on purchase of passenger cars by 5% less than before. And the 80% of saved (or un-spent) amount

Table 2. Modeling results of the second scenario representing sectoral output change (%).

NACE/CPA CODE	Product or service	Impact (%)
N77	Rental and leasing services	1.4
H49	Land transport services and transport services via pipelines	0.2
G45	Wholesale and retail trade and repair services of motor vehicles and motorcycles	0.1
C20	Chemicals and chemical products	−0.1
M69	Legal and accounting services; services of head offices; management consultancy services	−0.1
C25	Fabricated metal products, except machinery and equipment	−0.2
E37-39	Sewerage services; sewage sludge; waste collection, treatment and disposal services; materials recovery services; remediation services and other waste management services	−0.4
C24	Basic metals	−2.4
C29	Motor vehicles, trailers and semi-trailers	−7.9

of money (accounted for 8.6 mil euro) is used to buy land transport services and rental services from other members of society. It is also assumed than some part of society offer for free their vehicles or significantly reduced price; in result in short-term it is assumed that the saved 20% (or un-spent) amount of money (accounted for 2.2 mil euro) are as savings and not spend on other products or services in the modelled time period.

The transport sectoral facing such consumption trend experience positive or no changes, as society offer their vehicles to other then an increase observed in land transport services (0.2%) or rental activities (1.4%). In comparative-static scenario analysis when the impact in short term is analysed some sectors that offer materials and services to manufacture of motor vehicles (C29) experience decline in demand – as the whole sector in Latvia (output decline by −7.9%), basic metals (−2.4%), waste and metal collection services (−0.4%). It should be outlined that majority of passenger cars are imported in Latvia.

The two evaluated and modelled scenarios show that the private consumption trends can have both positive and negative impact on the economy, however, modern society are so used to transport services and the results claim that the transport and logistics can adjust to these trends if detected and duly considered.

The scenario analysis using comparative-static modeling approaches (as input-output what-if scenario analysis) is an effective tool to obtain quantitative results. The results of two modelled scenarios claim that trends of private consumption, even relatively minor trends (that are accompanied by change (increase or decrease) of 5% or 10%) results in notable impact on certain areas and sectors in the economy.

The transport and logistics sector companies, associations and policy makers should take into account the intersectoral linkages and impact size in the economy before taking significant shifts or releasing bold policy proposals.

5 Conclusion

There are several global consumption trends that transport and logistics managers have to consider to ensure sustainability. In the changing situation, managers in transport and logistics have to follow recent developments and forecasts regarding economic development and new trends.

The performed literature review indicates several trends that are transformed into two modeling scenarios' assumptions in order to obtain quantitative results on impact level and changing business environment.

Optimization and simulation methods are useful for smaller scale analysis. The research shows how input-output model can be used for modeling of transport and logistics impact on the economy at both macro level and sectoral level.

The results of the first scenario lead to the conclusion that the increasing distance and online shopping have a positive impact on transport and logistics and also on the national economy, hence the transport and logistics companies must to consider to offer the wider set of services, including also postal and courier services and also warehouse services to clients in near future. At the same time, it also indicates the changes in individual private consumption habits and it can results in downturn of existing shopping malls.

The results of the second scenario lead to the conclusion that the eco-friendly and zero-waste inspired changes in private consumption can have both positive and negative impact on transport sector. The results claim that the modeling results are sensitive, in some cases, even highly sensitive, to additional assumptions regarding the saving pattern of society (whether the un-spent (saved) sum of money is spend on other consumption purposes or increased savings or retirement fund) etc.

The findings are valuable to transport and logistics companies, agencies, governments.

References

1. Robertson, R.: Globalisation or glocalisation? J. Int. Commun. **18**(2), 191–208 (2012)
2. Furceri, D., Loungani, P., Simon, J., Wachter, S.M.: Global food prices and domestic inflation: some cross-country evidence. Oxf. Econ. Pap. **68**(3), 665–687 (2016)
3. Jiang, B., Tian, L.: Collaborative consumption: strategic and economic implications of product sharing. Manag. Sci. **64**(3), 1171–1188 (2018)
4. Fehrer, J.A., Benoit, S., Aksoy, L., Baker, T.L., Bell, S.J., Brodie, R.J., Marimuthu, M.: Future scenarios of the collaborative economy. J. Serv. Manag. **29**(5), 859–882 (2018)
5. Dablanc, L., Morganti, E., Arvidsson, N., Woxenius, J., Browne, M., Saidi, N.: The rise of on-demand 'Instant Deliveries' in European cities. Supply Chain Forum **18**(4), 203–217 (2017)
6. Mrazovic, P., Eravci, B., Larriba-Pey, J.L., Ferhatosmanoglu, H., Matskin, M.: Understanding and predicting trends in urban freight transport. In: Bilof, R. (ed.) Proceedings - 18th IEEE International Conference on Mobile Data Management - MDM 2017, pp. 124–133. CPS, Daejeon (2017)

7. Morganti, E., Dablanc, L.: Recent innovation in last mile deliveries. SpringerBriefs in Applied Sciences and Technology, pp. 27–45. Springer, Heidelberg (2014). https://doi.org/10.1007/978-3-319-09791-6_3
8. Rai, H.B., Verlinde, S., Macharis, C.: Shipping outside the box. Environmental impact and stakeholder analysis of a crowd logistics platform in Belgium. J. Clean. Prod. **202**, 806–816 (2018)
9. Meherishi, L., Narayana, S.A., Ranjani, K.S.: Sustainable packaging for supply chain management in the circular economy: a review. J. Clean. Prod. **237**, 117582 (2019)
10. Sarkar, B., Ullah, M., Kim, N.: Environmental and economic assessment of closed-loop supply chain with remanufacturing and returnable transport items. Comput. Ind. Eng. **111**, 148–163 (2017)
11. Lam, J.S.L., Gu, Y.: A market-oriented approach for intermodal network optimisation meeting cost, time and environmental requirements. Int. J. Prod. Econ. **171**, 266–274 (2016)
12. de Keizer, M., Akkerman, R., Grunow, M., Bloemhof, J.M., Haijema, R., van der Vorst, J.G.A.J.: Logistics network design for perishable products with heterogeneous quality decay. Eur. J. Oper. Res. **262**(2), 535–549 (2017)
13. Feitó-Cespón, M., Sarache, W., Piedra-Jimenez, F., Cespón-Castro, R.: Redesign of a sustainable reverse supply chain under uncertainty. A case study. J. Clean. Prod. **151**, 206–217 (2017)
14. Russo, F., Comi, A.: Urban freight transport planning towards green goals: synthetic environmental evidence from tested results. Sustainability **8**(4), 1–18 (2016)
15. Yang, M., Zacharias, J.: Potential for revival of the bicycle in Beijing. Int. J. Sustain. Transp. **10**(6), 517–527 (2016)
16. Shafiee, M., Brennan, F., Espinosa, I.A.: A parametric whole life cost model for offshore wind farms. Int. J. Life Cycle Assess. **21**(7), 961–975 (2016)
17. Jurgelane-Kaldava, I., Ozolina, V., Auzina-Emsina, A.: Modeling the influence of transportation and storage industry on the economic development of Latvia. Procedia Comput. Sci. **149**, 450–456 (2019)
18. Eurostat: Quality report on National and Regional Accounts. Publications Office of the European Union, Luxembourg (2018)

Polish State Aid for Transport Services and Infrastructure Between 2007–2016

Zofia Gródek-Szostak[1(✉)], Danuta Kajrunajtys[1], and Luis Ochoa Siguencia[2]

[1] Cracow University of Economics, Rakowicka 27, 31-510 Cracow, Poland
grodekz@uek.krakow.pl, danusia@kajrunajtys.com
[2] Akademia Wychowania Fizycznego w Katowicach, Mikołowska 72A, 40-065 Katowice, Poland
l.ochoa@awf.katowice.pl

Abstract. Financial support provided to entities of the transport sector from public funds is an important element of the implementation of the European Union's transport policy. The philosophy of the European Union's activity assumes that granting assistance to private entities from public funds should be allowed exceptionally. It is difficult to imagine the situation in order to be able to intervene in economic processes in a free manner in the conditions of a market economy without prejudice to the functioning of free competition. The article presents the results of analyses in the use of public aid in the transport sector in Poland in 2007–2016.

Keywords: Transport policy · Public aid · Market analysis

1 Introduction

Transport has accompanied humanity from the very beginnings of the development of civilization. The transport need was separated from a large group of generic instrumental needs related to the implementation of specific objectives, tasks and intentions. In contrast to the basic biological, sociological and cultural needs, they are inseparable from the production and social activity of man [1]. The importance of transport in the economy results from multiple connections and dependencies with all manifestations of the country's economic development. Transport activity contributes to the general socio-economic development of the state and enables efficient functioning of national economy departments [2, 3]. Transport is an important element of the economy of every country, and thus of the Member States of the European Union, and its efficient functioning depends to a large extent on the efficiency of management, the scale of possible benefits from the existence of the single European market, and thus the competitiveness of the Community economy on the international arena. Investment expenditures on transport and transport infrastructure in the EU reach around 1% of EU GDP. It is estimated that over 10 million people work in the EU transport sector, which is 4.5% of all employees, and that it generates 4.6% of gross domestic product. (GDP). Goods and services related to transport consume an average of 13.2% of the

E. Ginters et al. (Eds.): ICTE ToL 2019, LNITI, pp. 9–14, 2020.
https://doi.org/10.1007/978-3-030-39688-6_2

household budget [4]. The aim of the article is analysis the use of public aid in the transport sector in Poland in 2007–2016.

2 Research Approach and Methods

The publication analyses theoretical and empirical achievements regarding the financial support provided to entities of the transport sector from public funds. The subject of the analysis were changes (their intensity and direction) co-financing of selected modes of transport with public funds. The analysis covered the years 2007–2016. Study of the use of public funds in the following fields of transport: rail, onshore, pipeline, maritime, inland, air, intermodal and transport support activities. The division used resulted from the availability of source materials. The analysis was quantitative in nature, to which basic statistical measures were applied: percentage share and indicators formulated for the needs of the present research.

The subject matter is in the area of research interests of authors regarding the international activity of innovative enterprises [5].

3 Public Aid in the Transport Sector

Financial support provided to entities of the transport sector from public funds is an important element of the implementation of the European Union's transport policy. The philosophy of the European Union's activity assumes that granting assistance to private entities from public funds should be allowed exceptionally. It is difficult to imagine the situation that in the conditions of a market economy one could intervene in economic processes in a free manner, without prejudice to the functioning of free competition.

The legal regulations in force in the European Community and Poland regarding the admissibility and supervision of state aid are subordinated to the implementation of general principles that underlie the legal order of the European Union in this area. These principles are: transparency, proportionality, consistency.

There is no legal definition of state aid in the legal system of the European Union. The basic legal source defining the notion of state aid is Article 87 of the Treaty establishing the European Community (TEC). Although, based on the provision of Article 87 of the EC Treaty, one can infer what public aid is, it is not a definition of state aid. Article 87 should be treated only as a very general guideline for member states to identify public aid. The premise of state intervention or by means of state funds is understood in a special way according to the state aid law. The answer to the question about the scope of this premise boils down to determining the circle of potential aid donors. Descent from the state or from state sources has support granted from the resources of the public administration body - governmental and local government, regardless of its position in the hierarchy of state organs.

For the purposes of reporting and reporting, public aid instruments in accordance with the guidelines of the European Commission are compiled in five groups [6]:

- Subsidies and tax breaks;
- Capital and investment subsidies;
- So-called soft crediting;
- Sureties and loan guarantees;
- Others, that is forms of help that are difficult to allocate to a total of four previous groups.

In national legislation, forms of public aid have been listed exhaustively in the Regulation on the detailed method of calculating the value of state aid granted in various forms [7], where 15 types of aid have been shown. On the other hand, a more detailed list of forms of public aid is provided in the Regulation on reports on public aid granted [8], where as many as 44 examples of state aid have been identified.

4 Public Aid for the Transport Sector in Poland

Public aid in Poland is monitored by the Office of Competition and Consumer Protection (UOKiK). The UOKiK collected information is presented in annual reports. The subject of the analysis were changes [their intensity and direction] co-financing of selected modes of transport with public funds. The analysis covered the years 2007–2016. Study of the use of public funds in the following fields of transport: rail, onshore, pipeline, maritime, inland, air, intermodal and transport support activities (see Fig. 1).

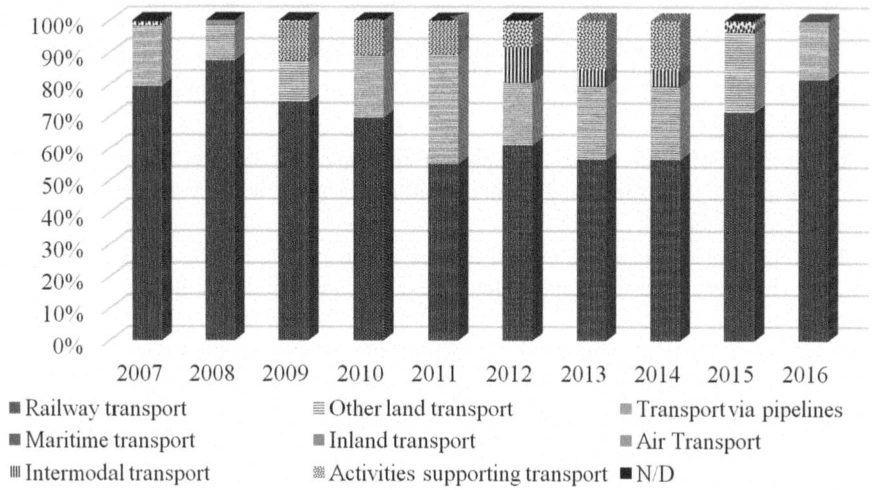

Fig. 1. The structure of public aid for individual transport modes in 2007–2016.

The division used resulted from the availability of source materials. The analysis was quantitative in nature, to which basic statistical measures were applied: percentage share and indicators formulated for the needs of the present research. The text volume does not allow to build a more complex and extended model with a list of factors that should be analysed.

In the analysed period, air transport benefited from public aid in 2008 by Poland Zloty (PLN) 43.6 million (1.5% of aid granted in 2008), in 2009: PLN 0.001 million and 2010: 0.01 (in both years it is below the promise of total help). Sea transport in 2009 used PLN 0.01 million of aid, in 2014: PLN 03 million, in 2015 and 2016, PLN 0.2 million. No pipeline transport was installed, which in the analysed period did not use public aid. As can be seen in Figs. 1, 2 and 3, the largest share in the consumption of subsidies was recorded in the railway transport and the land transport shortly followed.

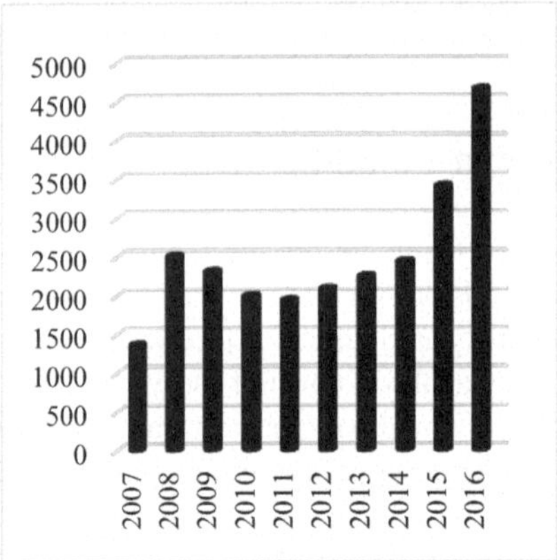

Fig. 2. The level of co-financing for rail transport in 2007–2016 in PLN million.

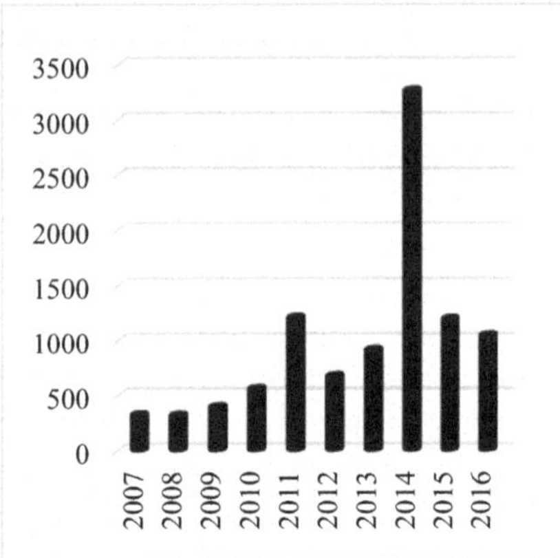

Fig. 3. Level of co-financing of land transport remaining in 2007–2016 in PLN million.

In the case of both forms of transport (railway and land-based), the level of involvement of public funds supporting economic activity and creating development opportunities is generally visible.

The financing of railway transport and land transport remained subject to further analysis. The level of co-financing of both these transport modes in the analysed period is shown in Figs. 2 and 3.

The figures illustrating the level of co-financing (in PLN million) was imposed on the dynamics of changes in this co-financing level (see Figs. 4 and 5). Observing the dynamics of changes in the level of co-financing, very large (and multidirectional) changes can be seen in both areas of transport. The involvement of public funds changes radically from year to year.

The co-financing level was related to the number of kilometres of railway lines and public roads with a hard surface. This does not mean that all these expenditures were used for investments and changes in this infrastructure. The reference has been accepted for research taking into account the fact that this infrastructure predominantly determines the success of the implementation of basic transport tasks.

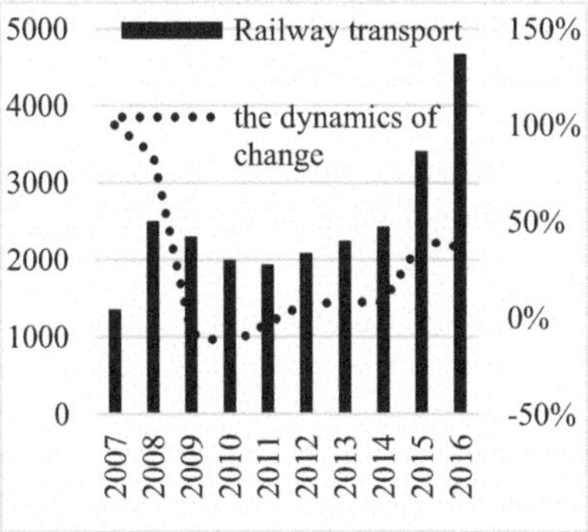

Fig. 4. Dynamics of the rail transport co-financing level in 2007–2016 in PLN million.

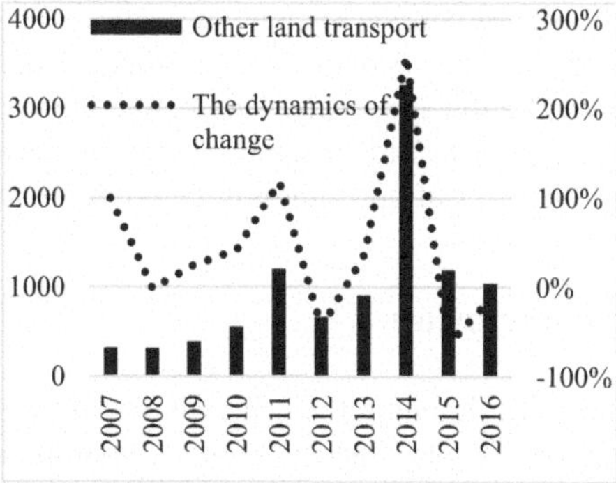

Fig. 5. Dynamics of the level of co-financing for land transport remaining in 2007–2016 in PLN million.

This infrastructure is used in both cases to transport people as well as physical goods (goods, materials or production machines). The index of public capital expenditures per one km of respectively railway lines and public roads with a hard surface is shown in Figs. 6 and 7.

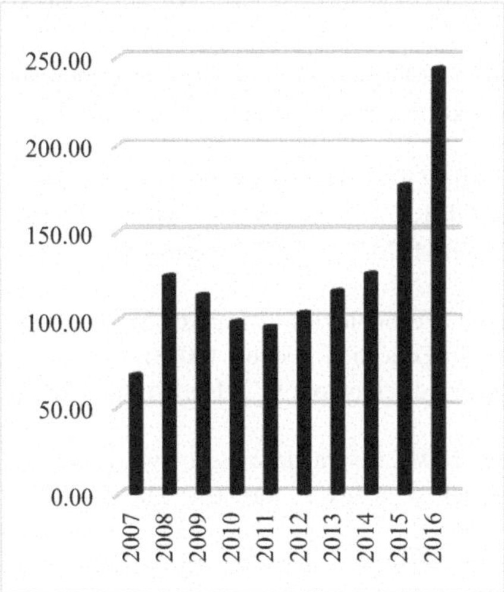

Fig. 6. Co-financing level of 1 km of railway lines (PLN thousand) in 2007–2016.

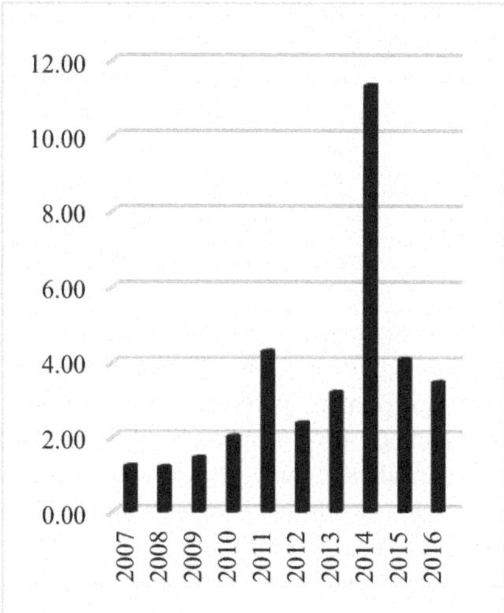

Fig. 7. Co-financing level of 1 km of public roads with hard surface (PLN thousand) in 2007–2016.

The statistical co-financing of one kilometre of respectively hard-line rail and public roads was different in individual years. As can be seen in Figs. 6 and 7, expenditures per one km tend to increase. This is a good prediction. It seems that one of the factors that cannot be quantified is the increase in the experience of the staff responsible for preparing and conducting projects.

A complete further analysis should allow to formulate an answer to the question why the level of consumption of public funds in individual years was so varied.

5 Conclusion

Public aid, although it is generally not allowed due to distortions of competitiveness, is common and plays an important role in the economic development of the country. In the transport sector, due to its nature, it is of particular importance. It is worth paying attention to the disproportions in the level of co-financing in individual years. These investments generate market demand for specialists implementing works covered by these activities. They cover many regions of the country, generating periodic high demand for labour.

References

1. Rydzkowski, W., Wojewodzka-Krol, K.: Transport, 39. Wydawnictwo Naukowe PWN, Warszawa (2005)
2. Grzywacz, W., Burnewicz, J.: Ekonomika transportu, 41. Wydawnictwo Naukowe PWN, Warszawa (1989)
3. Kopiec, A., Ochoa Siguencia, L., Grodek Szostak, Z., Marzano, G.: Transport infrastructures expenditures and costs analysis: the case of Poland. Procedia Comput. Sci. **149**, 508–514 (2019)
4. Commission Press release of 28 March 2011, MEMO/11/197. Transport 2050: the major challenges, the key measures. Commission Press Release of 28 March 2011, MEMO/11/197. https://ec.europa.eu/commission/presscorner/detail/en/MEMO_11_197. Accessed 19 Nov 2019
5. Ochoa Siguencia, L.: Contemporary Information Technologies. Business Management. Publishing House of the Research and Innovation in Education Institute (2018)
6. Raport o pomocy publicznej w Polsce udzielonej przedsiebiorcom w 2014r. Urzad Ochrony Konkurencji i Konsumentow, Warszawa (2015)
7. Rozporzadzenie Rady Ministrow z dnia 11 sierpnia 2004r. w sprawie szczegolowego sposobu obliczania wartosci pomocy publicznej udzielanej w roznych formach, Dz. U. 2004, No. 194.1983.3 (2004)
8. Rozporzadzenie Rady Ministrow z dnia 7 sierpnia 2015 r. zmieniajace rozporzadzenie w sprawie sprawozdan o udzielonej pomocy publicznej, informacji o nieudzieleniu takiej pomocy oraz sprawozdan o zaleglosciach przedsiebiorcow we wplatach swiadczen naleznych na rzecz sektora finansow publicznych, Dz. U. 2015, Poz. 1474 (2015)

Efficiency of Routes in Middle Easter Transport Corridor

Zhaken Kuanyshbaev(✉), Muratbek Arpabekov, Tynys Suleimenov, and Aidana Ybray

Eurasian National University named after L.N. Gumilyov, Nur-Sultan, Republic of Kazakhstan
lev-49-07@mail.ru, arpabekov_m@mail.ru, maina_96@mail.ru

Abstract. The article presents the calculations for comparison and the choice of the most effective route in the organization of international transport. The southern and northern routes on the railways of the Republic of Kazakhstan are presented when cotton is transported in 20-foot containers from China to the station of the same name and the port of Bandar-Abbas of the Islamic Republic of Iran.

Keywords: Railway transport corridors · Tariffs for freight transportation · Costs assessment

1 Introduction

For shipment by rail transport presented cargo is raw cotton. As is known, the People's Republic of China is one of the largest producers and suppliers of raw cotton to the world market. For many centuries cotton has been one of the main textile cultures of China. Twenty-four of China's 35 provinces grow cotton and about 300 million people participate in the cultivation and processing, over 10 million people of which are involved in the industrial processing of cotton itself. In November 2017, the cotton harvest in China has ended. According to the latest data from the China Cotton Association (CCA), cotton production in the country this season was around 7.5 million tons. Import for January-November 2007 is amounted to 2,14 million tons. Transportation is carried out by rail. The cargo is packaged and placed in a 20-foot containers. The transportation routes are shown in Figs. 1 and 2. It should be noted that one of the routes passes along the southern border of the Republic of Kazakhstan and then follows the railways of the Republic of Uzbekistan, the Republic of Turkmenistan and the Islamic Republic of Iran. The second route follows the northern border of the Republic of Kazakhstan and then follows the railways of the Republic of Turkmenistan and the Islamic Republic of Iran. The length of the southern route is around 9276 km, the length of the northern route is 6016 km. Type of rolling stock - platform and 20-foot containers, where two 20-foot containers are located on the platform. The cargo is raw cotton with tariff group 2 and cargo code according to UTSNC (Uniform tariff and statistical nomenclature of cargo) is 031013. The cargo code for HCN (Harmonized cargo nomenclature) is 52010010. The leadership of the Islamic Republic of Iran

E. Ginters et al. (Eds.): ICTE ToL 2019, LNITI, pp. 15–19, 2020.
https://doi.org/10.1007/978-3-030-39688-6_3

attaches great importance to the development of transit routes and an increases the volume of transit of goods through the country. Iran seeks to fulfill the mission of a bridge to transport goods from east to west and from north to south of the continent. Through the railways of Iran with a length of about 12 thousand kilometers are going important transport corridors. Currently, 20% of all transit transportation accounted for the railway.

Due to its geographical position, Iran can actively cooperate with European countries in the field of freight traffic in the Persian Gulf. Twenty years ago, after the opening of the North-South railway corridor through the Sarakhs station, interaction with the railways of Turkmenistan, Uzbekistan and Kazakhstan began. The new route through the Inche-Burun border crossing reduces the transport of goods by more than 400 km. Geographical features of Kazakhstan (wide territory, lack of access to the sea, uneven distribution of human settlements and natural resources) make it's economy one of the most cargo-intensive in the world, causing a high dependence on the transport system. For comparison, freight calculations for two routes of international freight traffic are presented (see Figs. 1 and 2). The aim of the article is charge assessment for cargoes delivery respecting efficiency and responsiveness of supply chain in both directions.

The first route is "southern": stations Dostyk – Sary-Agash – Keles – Khodzha-davlet – Farab - Sarakhs (Turkmenistan) – Sarakhs (Iran) – Bandar-Abbas (Iran).

Fig. 1. Middle Eastern Corridor Route "southern": Dostyk – Sary-Agash – Keles – Khodzha-davlet – Farab - Sarakhs (Turkmenistan) – Sarakhs (Iran) – Bandar-Abbas (Iran).

The second route is "northern": st. Dostyk – Aktau – Bolashak (Kazakhstan) – Serhetiyaka (Turkmenistan) – Akyaila (Iran) – Bandar-Abbas (Iran).

Fig. 2. Middle Eastern Corridor Route "northern": Dostyk – Aktau – Bolashak (Kazakhstan) – Serhetiyaka (Turkmenistan) – Akyaila (Iran) – Bandar-Abbas (Iran).

2 The Freight Charge Assessment

2.1 The Southern Route Option

The freight charge on the first "southern" route by railways of the countries participating in international freight traffic would be following. Freight on the route Dostyk station – Sary-Agash station (Kazakhstan) in national currency is amounted to 1096331,56 tenge (see Fig. 3).

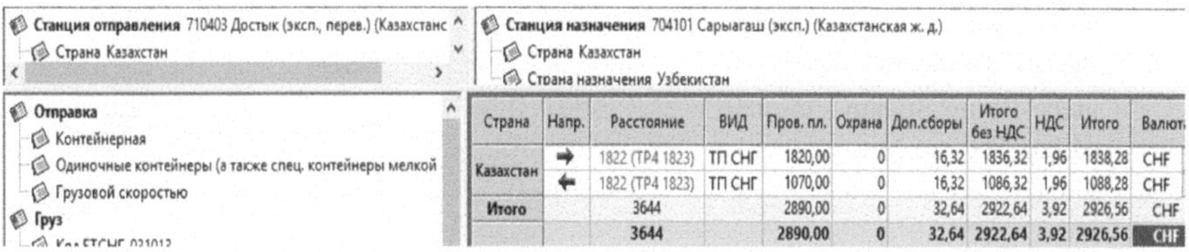
Станция отправления 710403 Достык (эксп., перев.) (Казахстанс
Страна Казахстан
Станция назначения 704101 Сарыагаш (эксп.) (Казахстанская ж. д.)
Страна Казахстан
Страна назначения Узбекистан
Отправка
Контейнерная
Одиночные контейнеры (а также спец. контейнеры мелкой
Грузовой скоростью
Груз

Страна	Напр.	Расстояние	ВИД	Пров. пл.	Охрана	Доп.сборы	Итого без НДС	НДС	Итого	Валют
Казахстан	→	1822 (ТР4 1823)	ТП СНГ	1820,00	0	16,32	1836,32	1,96	1838,28	CHF
	←	1822 (ТР4 1823)	ТП СНГ	1070,00	0	16,32	1086,32	1,96	1088,28	CHF
Итого		3644		2890,00	0	32,64	2922,64	3,92	2926,56	CHF
		3644		2890,00	0	32,64	2922,64	3,92	2926,56	CHF

Fig. 3. Determination of the charge on the route Dostyk – st.Sary-Agash (Kazakhstan).

However, freight on the route Dostyk station – Sary-Agash station (Kazakhstan) is amounted to 2926,56 ChF, but the fare on the route of Keles – Khodzhadavlet stations (Uzbekistan) is around USD 1967,48 and on the route of Keles - Khojadavlet stations (Uzbekistan) it is 1987,15 ChF. The fare on the route of st.Farab - st.Sarakhs (Turkmenistan) is around 1108,10 ChF for the same operating conditions.

Between Sarakhs and Bandar-Abbas stations (Iran), the fare is determined according to the Unified Transit Tariff method. The tariff distance between these stations is 1615 km. For a 20-foot loaded container the charge is 1107,00 ChF, and for an empty one - 554,00 ChF. Then, the total amount of freight paid by railways of the Islamic Republic of Iran will be 1661,00 ChF. The performed calculations can be presented in tabular form (see Table 1).

Table 1. Definition of the charge on the south route Dostyk st. – Bandar-Abbas st.

No.	Destination stations	Tariff distance, km	Freight charge	Freight charge, ChF
1	Dostyk station - Sary-Agash station, Kazakhstan	1822,0	1096331,56 tenge	2926,56
2	Keles station - Khojadavlet station, Uzbekistan	732,0	1967,48 USD	1987,15
3	Farap station - Sarakhs station, Turkmenistan	469,0	–	1108,10
4	Sarakhs station - Bandar Abbas station, Iran	1615,0	–	1661,0
5	Total	4638,0	–	7682,81

2.2 The Northern Route Option

The charge on the “northern” route by railways of the countries participating in international freight traffic can be determined as following (see Table 2).

Table 2. Definition of the charge on the north route Dostyk st. – Bandar-Abbas st.

No.	Destination stations	Tariff distance, km	Freight charge	Freight charge, ChF
1	Dostyk station - Bolashak station, Kazakhstan	3408,0	695493,00 tenge	1856.56
2	Serkhetyaka station - Akyaila station, Turkmenistan	700,0	1967,48 USD	1720.90
3	Inche-Borun Station - Bandar-Abbas Station, Iran	1908,0	–	1971.0
4	Total	6016,0	–	5547.56

The route follows the railways of the Republic of Kazakhstan from Dostyk station to Bolashak station. The tariff distance of the route is 3408,00 km. The route goes through the border stations Serkhetyaka and Akyaila exp. (Turkmenistan). By rail of the Islamic Republic of Iran, the route follows from Inche-Borun station to Bandar-Abbas station (port). The tariff distance of the route by railways of the Republic of Iran is 1908,00 km. Between the stations Inche-Borun and Bandar-Abbas, the fare is determined by the method of the Unified Transit Tariff. For a 20-foot loaded container, the charge is 1314,00ChF, and for an empty one - 657,00ChF. Then, the total amount of freight paid by railways of the Islamic Republic of Iran is around 1971,00ChF.

3 Conclusion

A comparison of the options of the routes of the Middle Eastern Corridor shows that the length of the southern route is 4638 km, the fare on the route is 7682,81ChF, but the length of the northern route is 6016 km, and the fare on the route is 5547,56ChF. Calculations show that the effectiveness of the northern route on the railways of the Republic of Kazakhstan is beyond doubt. Based on the above, it could be advisable to organize the international rail transport using the defined above northern route on the railways of the Republic of Kazakhstan. The economic effect from the use of the northern route $\boldsymbol{E_{north}}$ is 2135,25ChF. In the calculation of one million tons of cargo carried the total economic effect $\boldsymbol{E_{total}}$ will be around 53381250ChF (1):

$$E_{total} = E_{north} \times C_{total}/W_{2cont} \quad (1)$$

where, E_{north} = 2135,25 - the economic effect of using the northern route, ChF; C_{total} - million tons of cargo transported, tons; W_{2cont} = 40 (2 containers) weight of carload dispatch, tons. A number of provisions of the scientific article were set out in the period from 2014 to the present [1]. Moreover, when organizing international transportation in the system of the International Freight Transport, a model of crediting was proposed forfaiting [2, 3]. A method of determining the main components, in particular, its active part – locomotives, is proposed, which will make it possible to adjust the system of market relations [4].

Acknowledgements. The authors express the gratitude to the staff of the department “Organization of Transportation” Eurasian National University named after L.N. Gumilev (Nur-Sultan) and colleagues from Riga Technical University (Riga).

References

1. Kuanyshbayev, Zh., Arpabekov, M.: Projection of logistic schemes in intermodal communications. J. Appl. Econ. Sci. **12**(4), (2018)
2. Kuanyshbayev, Z., Ybray, A.: Structuring cash flows based on the model of forfaiting for the grain transportation. Sci. World **2**(5), 40–45 (2019)
3. Kuanyshbayev, Zh.: Using the FORFAITING model in the transportation process. Cerificate. Ministry of Justice of the Kazakhstan, No. 2584 of, 8 August 2018
4. Kuanyshbayev, Zh., Suyunbayev, Sh., et al.: A study of locomotive components in intermodal and unimodal transportation. Sci. World **1**(6), 43–50 (2019)

Digitalization in Maritime Industry: Prospects and Pitfalls

Viktorija Babica[1(✉)], Deniss Sceulovs[1], and Elvira Rustenova[2]

[1] Institute of Business Engineering and Management, Riga Technical University, Kalnciema Street 6-408, Riga 1048, Latvia
viktorija.babich76@gmail.com, deniss.sceulovs@rtu.lv
[2] West-Kazakhstan Agrarian-Technical University named after Zhangir Khan, Zhangir Khan Street 51, 090009 Uralsk, Kazakhstan
elvira02@mail.ru

Abstract. Roughly ninety percent of world trade is carried by the international shipping industry which remains the most cost-efficient means for the global transportation of goods. Obstinate culture of the industry and barriers shipping must overcome in order to adopt digitalization are the reason why the sector has lagged behind in the process of digital transformation (DX). Shipping is in transition to acquire prospects of digital transformation with respect to optimized cargo handling, improved maritime procurement and logistics processes, enhanced efficiency, safety and minimizing environmental impacts. Based on a literature review the current study provides an overview of methods of digital technology implementation in maritime industry, discusses accompanying challenge areas and presents trends for it surmounting. Main result of the study is identifying and grouping of factors which characterize digitalization process in shipping industry.

Keywords: Maritime transport · Digital transformation · Digitalization assessment

1 Introduction

Waterborne transport is the key logistical unit in the world trade, it provides low-cost and efficient transportation services. According to the United Nations Conference on Trade and Development (UNCTAD) the annual average growth rate of maritime trade is estimated 3.4 per cent for the period 2019–2024 [1], but the same time the shortage of officers is increasing rapidly [2]. In early 2019, the total world fleet stood at 95,402 ships and reached 1.97 billion dead-weight tons of capacity [3].

International Maritime Organization's (IMO) increased attention towards sustainability, disruptive technologies, which are transforming industries around the globe, trade policy crosscurrents, geopolitics and sanctions, environmental restrictions, fuel economics - all these aspects are challenging the conservative culture of the maritime industry. Despite the great opportunities provided by digital technologies brought by the Industry 4.0, digitalization of the backbone of world commerce - shipping industry - significantly lags behind major industries. Nevertheless, shipping is in transition to

E. Ginters et al. (Eds.): ICTE ToL 2019, LNITI, pp. 20–27, 2020.
https://doi.org/10.1007/978-3-030-39688-6_4

acquire prospects of the digital transformation (DX) with respect to optimized cargo handling, improved maritime procurement and logistics processes, enhanced efficiency, safety and minimizing environmental impacts. Trends in shipping and port operations' DX were categorized according to the selected domains of digital transformation: autonomous vehicles and robotics, artificial intelligence, blockchain and big data, virtual reality, internet of things and digital security [4]. These major digital technology innovations are among the most studied ones in the frames of maritime industry.

The study based on a systematic literature review, demonstrates the potential of maritime industry – shipping and port management – to adjust digital transformation. The present paper illustrates the prospects and pitfalls arising from the necessity to develop maritime industry to the same digital level as other industries involved in maritime logistics and reasons the need of maritime industry for DX. The results of the study may benefit scholars and practitioners involved in the development of maritime digital technologies, the study suggests areas of further research and stands out the main trends in digital transformation of the industry.

2 Digital Transformation of Maritime Industry

Digital technologies can significantly transform areas of maritime administrative, logistics, shipping and port operations' functions. Those make possible to optimize processes, create new values and transform logistic chains.

The word "digitalization" in the present paper means the use of digital technologies to change a business model and provide new revenue and value-producing opportunities [5]. Digital transformation of maritime industry is meant as implementation disruptive digital technologies to standard procedures, measurement and control of operations, and the creation of digital communication models [6]. According to UNCTAD effects of digitalization on waterborne transport can be divided into the following three stages: optimization, extension and transformation [7]. Those shortly can be described as: improving existing processes (ex. smart logistics, port-call optimization), creating a foundation for creating new business opportunities (ex. use of blockchain technology for bunkering, cargo tracking) and reinvention of existing models.

2.1 The Present of Maritime Industry

The maritime industry is a dynamic, complex and safety-oriented system, actors of which and technological tools interact, oriented towards a common goal of running their operations efficiently and safely [8]. Since the maritime industry deals with cross-border trade and transportation, the process involves multiple languages and institutions, and the process of digitization and standardization is slow. Groenfeldt [9] argues that up to 50% of maritime transportation costs in the Global container shipping industry are due to the time spent on paperwork.

Table 1 provides summary of drivers relevant to DX acquisition in shipping and seaports. Further are discussed the major changes in shipping and seaport management brought by DX.

2.2 "Smart" Shipping

Autonomous Vehicles and Robotics. Autonomous vehicles are found in all transport sectors. Ship automation and autonomy, digital connectivity, traffic automation are making shipping 4.0 with unmanned vehicle and integrated logistics transport system possible. The term "intelligent ships" is becoming trendy among parties of shipping industry, the expression of which is remotely controlled and autonomous ships [10]. The safety and efficiency of sea transport have a key impact on sustainable development. International shipping law requires ocean-going vessels to be equipped with crew, [11] that means, that fully autonomous, unmanned ships will not be allowed in inter-national waters, but since 2017 International Convention for the Safety of Life at Sea (SOLAS), has been working on policies to allow waterborne transport with no captain or crew to operate [12]. Unmanned vessels up to date are used by the marine researchers and by the defense units. Those sail only along precisely defined shipping lanes [13] and do not extend 20 m in length.

Maritime Internet of Things (IoT). Maritime IoT provides gain of efficiency, simplifies management complex transport and supply chain systems, including both technical-operational and co-ordination efficiency. As the commercial shipping is at the fledging stage of adoption policies for autonomous ships and is already implementing e-navigation, usability of maritime IoT is increasing [14]. The maritime IoT includes also satellite technologies and Telematics, which, in the maritime sector, have the potential to considerably improve navigation, safety, remote monitoring and maintenance, communication and environmental efficiency. Those provide real-time transmission of data from a device back to a receiver; IoT represents the broader concept connecting physical objects to facilitate real-time communication between sensors, applications and people [15].

2.3 Digital Seaports Operations

The concept of "smart port" became widespread, expressing automation of port processes and the application of new intelligent technologies for the processing of goods and ships in ports [16]. The concept of smart port, aims to adopt modern information technologies to enable a better planning and management within and between units of maritime industry, it includes systems that support tools for handling cargo, managing traffic, dealing with custom, providing conditions for safety, and monitoring energy use.

Artificial Intelligence (AI). Digital technologies in port are able to process information coming from other ports, ships, weather services, cargo shippers and other sources, and inform continuously the ships going to the port, what should be their optimal speed and other seagoing characteristics so that they can reach the port in exactly right time and should not wait in order. Information management and digital transformation cover the biggest share in the current port logistics evolution [6].

Smart sensors easier track, operate, and maintain the physical infrastructure and facilities. Those are inbuilt in quay walls and bridges to transmit real-time data about operating conditions of berths and other infrastructure [17].

2.4 Pitfalls in Adopting DX

The European Union lacks a clear strategy towards digitalization in the maritime industry [18]. Innovations are accompanied by the fear of unknown and uncertainty with the respect to the risks ahead. Specialists of maritime industry argue that risks and safety in shipping will remain at the same only with different condition, that risk of human error is being shifted on technologies.

The major challenge for the industry to adopt digitalization is the mindset, stakeholders are loyal to the traditions. Conservativism of the units involved in the maritime industry brings up trust issues. Transparency of transactions and processes, which is possible with digital technologies should be supported and encouraged by the stakeholders. As for blockchain it has become the new mainstream in all industries and maritime sector as well, brining much visibility and efficiency into shipping and logistics [19].

Unmanned vessels do not overcome need for digital literacy, only with a combination of new skills, seafarers will continuously build their competence on technical and commercial capabilities within the digital sphere. Competent and highly skilled seafarers are needed to monitor and guide vessels using AI or the machine learning. Ship operators aim to reduce operational expenses by removing crews from vessels, but they will still need remote crew.

Reliance in information technology (IT) software and communications systems elevate cyber risk for modernized ships, which rigged with sensors, networks to monitor performance and maintenance. Cyber-incidents may threaten control of vessels' steerage. Cyber-attack in June 2017 temporarily paralyzed terminal operations in several countries, it created huge problems to the world's biggest carrier – Maersk - its vessels stood still at sea and its 76 port terminals around the world ground to a halt. The organization suffered financial losses loss of revenue, IT restoration costs and extraordinary costs related to operations [20].

Ellingsen and Aasland [21] argue that the potential value added and issues related to adopting new advanced technologies can be reviewed in two dimensions: tangible and intangible, and internal and external, mentioning such as new knowledge and skillsets, lack of expertise, lack of financial means to invest, lack of confidence, speed of technological development exceeds strategy development. The paradox of DX strategy is that it aims to reach better efficiency by disruption, when standardization is the process of making an innovation efficient; it aims to rise competitive advantage, same time efficient DX adoption is possible only by cluster collaboration. The main challenges of maritime DX according to the available scientific researches and maritime organization's publication are defined in Table 1.

Table 1. Accordance of maritime industry needs to DX indicators, prospects and pitfalls.

Needs	Prospects	Indicator	Pitfalls
Effective communication between stakeholders and units; information integration; accurate information sharing	A unified system of ports and ships, in which components are constantly in communication	Digital Security, Blockchain; IoT	The ultimate complexity of the system; Safety issues, internal disturbances; Inefficient energy consumption; Developing countries lagging [22]
Effective logistic management; Minimizing environmental damage	Optimal timing of arrival and departure from port; reduction of waiting time on the anchorages; minimization of capacity of port warehouses; reduce carbon-dioxide emissions	Port Community Systems, IoT	Need to coordinate simultaneously the personal needs and interests of many stakeholders [13, 23, 24]; Entrepreneurs are not ready to contribute to the system [25]
Minimizing fuel consumption	Cost-efficiency; Reduce negative externalities and streamline the transport of international trade.	Blockchain, AI	Unauthorized malicious intrusion into system [21]
Relevant and up-to-date information	Remote monitoring of offshore assets; reduced intuitive approach; Control of complex objects and systems	AI; IoT	The imperfection of the leading global economic model and human qualities; [26]
Increased safety of navigation; enhanced maritime security	Standardized and harmonized communication technologies between ships and shore	IoT	Cyber-attacks, digital technology trust [8, 27]
Effective teamwork, work efficiency	Prediction of behavioral patterns	AI	Technological error, but human [28]
Prototyping	Digital shipyard	Digital twin, VR	Financial assets [29]
Traceability and transparency	Easing paperwork, tracking and tracing cargo, customs clearance, and managing maritime shipping operators	Blockchain application	Related cost-benefit analyses are unclear, mid- and long-term assessments of returns on investment are unclear [30, 31]
Reduce time-to-market and reduce cost-of-non-quality	Supply chain optimization		Complexity and big data; Policy and regulatory issues;

3 Conclusion

Digital technology can be a significant enabler of growth and innovation or disruption. Digital technology is a means, not an end, the success of digital transformation is not only in the adoption of modern technologies and advanced methods, but also supporting administrative policy and digital literacy. This study has set out some of the trends, opportunities, challenges and preconditions for successful shipping and seaports sector digitalization. DX of maritime industry operations can lead to a decline in accidents at ports and at sea, reduce pollution and emissions and minimize waiting time on the anchorages; Digitalization and the automation of activities in the shipping industry could lead to better cost-efficient in maritime operations.

Strong facilitators of the digitalization are investments into technology and cooperation's for promoting information sharing and a better coordination and collaboration, often regarded as an obstacle in highly competitive environments.

The barriers for successful adaptation of smart shipping is that there is no "one-size-fits-all" solution, the developing countries are faced to extreme their abilities to maintain the level of other parties involved. Findings show that while developing countries have not yet fully implemented electronic systems, paperless systems of developed economies do not work efficiently due to harmonization issues and fragmentation of different services and agencies. Automatization offers a solution to many issues facing the maritime industry, cybersecurity remains the key challenge and predicting and fighting cyberattacks is crucial for every shipping company. One of the major issues regarding unmanned vessels is the lack of clarity of the current policies. Ambiguity of legal status, the financial feasibility, the interaction with manned vessels, and very traditional mindset are all factors that have slowed down the process of developing operational unmanned vessels. To enable faster DX of maritime industry could help international standardization and collaboration with governmental institutions. The present study benefits the target audience by confronting the needs of the maritime industry with the possibilities provided by digital transformation, and eliciting out the main barriers of obtaining the benefits.

Further research could be done on analysis of DX possibilities drawn from other industries such as aviation in order to improve process chain, coordination and adaptation of technologies. DX application can be evaluated from a perspective of technology acceptance model.

References

1. United Nations Conference on Trade and Development: Review of Maritime Transport 2018. https://unctad.org/en/PublicationsLibrary/rmt2018_en.pdf. Accessed 20 Oct 2019
2. International Chamber of Shipping: Global Supply and Demand for Seafarers. http://www.ics-shipping.org/shipping-facts/shipping-and-world-trade/global-supply-and-demand-for-seafarers. Accessed 27 Oct 2019
3. United Nations: Review of Maritime Transport 2019. United Nations Publication, USA (2019)
4. ECMAR: Maritime Technology Challenges 2030. ECMAR, Brussels (2018)

5. Gartner. https://www.gartner.com/en/information-technology/glossary/digitalization. Accessed 25 Oct 2019
6. Heilig, L., Lalla-Ruiz, E., Voß, S.: Digital transformation in maritime ports: analysis and a game theoretic framework. Netnomics **18**, 227–254 (2017)
7. United Nations Conference on Trade and Development: Review of Maritime Transport 2017. https://unctad.org/en/PublicationsLibrary/rmt2018_en.pdf. Accessed 20 Oct 2019
8. Costa, N.A., Lundh, M., MacKinnon, S.N.: Identifying gaps, opportunities and user needs for future e-navigation technology and information exchange. In: Advances in Human Aspects of Transportation. Advances in Intelligent Systems and Computing, vol. 597. Springer, Cham (2017)
9. Yang, C.: Maritime shipping digitalization. Blockchain-based technology applications, future improvements, and intention to use. Transp. Res. Part E **131**, 108–117 (2019)
10. Cowan, D.: https://www.raconteur.net/finance/autonomous-ships. Accessed 15 Oct 2019
11. Matthews, C.: http://theconversation.com/unmanned-ghost-ships-are-coming-83324. Accessed 19 Oct 2019
12. IMO. http://www.imo.org/. Accessed 14 Oct 2019
13. Alop, A.: The main challenges and barriers to the successful smart shipping. Int. J. Mar. Navig. Saf. Sea Transp. **13**(3), 521–528 (2019)
14. IMO, e-navigation: http://www.imo.org/en/OurWork/Safety/Navigation/Pages/eNavigation.aspx. Accessed 12 Oct 2019
15. Intrnational Transport Forum. Information Sharing for Efficient Maritime Logistics. https://www.itf-oecd.org/sites/default/files/docs/information-sharing-maritime-logistics.pdf. Accessed 11 Oct 2019
16. Berns, S., Dickson, R., Dragt, J., Vonck, I.: Smart Ports. Point of View. Deloitte Port Services, The Netherlands (2017)
17. Riedl, J., Delenclos, F., Rasmussen, A.: To get smart, ports go digital (2018)
18. Heilig, L., Voß, S.: Information systems in seaports: a categorization and overview. Inf. Technol. Manag. **18**(3), 179–201 (2017)
19. Safety4sea. https://safety4sea.com. Accessed 21 Oct 2019
20. Maersk Cyber Attack. https://safety4sea.com/cm-maersk-line-surviving-from-a-cyber-attack. Accessed 13 Oct 2019
21. Ellingsen, O., Aasland, K.E.: Digitalizing the maritime industry. A case study of technology acquisition and enabling advanced manufacturing technology. J. Eng. Technol. Manag. **54**, 12–27 (2019)
22. Bartolacci, M.R., LeBlanc, L.J., Kayikci, Y., Grossman, T.A.: Optimization modeling for logistics. Options and implementations. J. Bus. Logist. **33**, 118–127 (2012)
23. Aydogdu, Y.V., Aksoy, S.: A study on quantitative benefits of port community systems. Marit. Policy Manag. **42**, 1–10 (2015)
24. Min, H., Lim, Y.-K., Park, J.-W.: Supply chain analytics for enhancing the maritime security. Int. J. Logist. Syst. **28**, 164–179 (2017)
25. Flin, R., O'Conner, P., Crichton, M.: Safety at the Sharp End. A Guide to Non-technical Skills. Ashgate Publishing Limited, Aldershot (2008)
26. Vries, L.: Success factors for navigational assistance: a complementary ship-shore perspective. In: Proceedings of the Human Factors and Ergonomics Society. Europe Chapter 2014 Annual Conference, pp. 175–186, Lisbon (2015)
27. Safety at Sea. https://safetyatsea.net/news/2018/maritime-sector-unprepared-for-major-threats-report-warns/. Accessed 15 Oct 2019
28. Siemens PLM: Newport News Shipbuilding launches the digital shipyard. https://www.youtube.com/watch?v=93za-vO_ffs. Accessed 14 Oct 2019

29. Taylor, N., Human, C., Kruger, K., Bekker, A., Basson, A.: Comparison of digital twin development in manufacturing and maritime domains. Studies in Computational Intelligence, pp. 158–170 (2019)
30. Meersman, H., van de Voorde, E., Vanelslander, T.: Port competitiveness now and in the future. What are the issues and challenges? Res. Transp. Bus. Manag. **19**, 1–3 (2016)
31. Lee, P.T.-W., Kwon, O.K., Ruan, X.: Sustainability challenges in maritime transport and logistics industry and its way ahead. Sustainability **11**, 1331 (2019)
32. Kenyon, G.N., et al.: Improving the return on investment in ports: opportunities in data management. Marit. Econ. Logist. **19**(2) (2017)
33. Lieber, A.: Trust in Trade: Announcing a new blockchain partner. https://www.ibm.com/blogs/blockchain/author/aaron-lieber/. Accessed 14 Oct 2019

Optimization of Transport Logistical Flows Between Railway and Urban Passenger Transportation Systems

Mareks Mezitis[1(✉)], Ruslans Muhitovs[1], Muratbek Arpabekov[2], Zura Sansyzbajeva[2], Baglan Togizbayeva[2], and Alimzhan Assiltayev[2]

[1] Riga Technical University, Azenes Street 12-316, Riga LV–1048, Latvia
{mareks.mezitis, ruslans.muhitovs}@rtu.lv
[2] L.N. Gumilyov Eurasian National University, Satpayev Street 2, Nur-Sultan 010000, Kazakhstan
arpabekov_m@mail.ru, zura_astana@mail.ru, baglan099@mail.ru

Abstract. The aim of the article is the optimization of load on transport corridor and the artery of the city by organizing the interaction of railway and urban transport, taking as an example the city route of Nur-Sultan [2]. In the result, the model has been worked out and the effectiveness of including a railway corridor with rail buses running along into the urban network of passenger transport has been proved. To improve the quality of the services provided in the article, it is additionally proposed to open a combined railway bus route and calculate its cost [3, 4]. The maximum network flow was calculated for the given initial conditions. Subsequently, by modeling the inclusion of additional branches that are other transport corridors into the algorithm, the possibility of increasing the maximum flow with a simultaneous improvement in the quality of urban passenger transport services was shown.

Keywords: Heterogeneous urban transport · Route planning · Efficiency assessment

1 Introduction

In conditions when in the market of passenger transport services along with municipal transport, separate carriers of private form of ownership act independently of each other, the issues of improving the management of urban passenger transport become particularly relevant.

In the transport networks of large cities there are many routes, and it is rarely possible to determine the time of arrival, knowing the time of departure, and vice versa. There is a problem with the implementation of the algorithm for finding the shortest path, provided that the public transport timetable is known [1, 8]. In discrete mathematics, there is a number of optimization algorithms on a graph, in particular, algorithms for finding the shortest distances, but in the classical it is implemented on static weights, i.e. arc weight does not change over time. In practice, if a graph is a model of the movement of public transport, the travel time may be different. This is due to the

E. Ginters et al. (Eds.): ICTE ToL 2019, LNITI, pp. 28–34, 2020.
https://doi.org/10.1007/978-3-030-39688-6_5

fact that the weight of the arc depends not only on the speed of movement, but also on the waiting time and the parking of the route. The problem of finding the shortest path is not a new issue [7, 9]. There are a number of algorithms for solving such problems. In network planning, various problems of optimizing graphs with random weights are considered [6]. Operations research provides algorithms for finding the maximum dynamic flow. In this article, authors consider the problem of finding the optimal path on graphs, where the capacity of arcs changes dynamically and depends not only on time, but also on control, under the action of which movement occurs [10].

2 Transport Network as a Graph and the Ford-Fulkerson Algorithm

2.1 Nur-Sultan Transport Network

The article considers the transport problem of Nur-Sultan city (see Fig. 1). The city has transport network with main components: network of railways, roads, specific lines, etc.). At the same time, restrictions are imposed on individual components – their maximum permissible load.

It is necessary to determine the maximum possible number of passengers, vehicles, etc. which can be transported on this network. It is also required to determine how to organize the transportation of this maximum possible quantity. Often it is a problem of maximizing the flow of a product between two given network nodes, provided that the flow along each arc is restricted. An obvious example is the flow of urban transport.

Fig. 1. A map of Nur-Sultan city.

It is possible to show the transport network as a directed graph G = (E, U) with a set of vertices E and a set of arcs U (see Fig. 2). The vertices E mean transport nodes, where different kinds of transport have connections, but arcs U mean the transport corridors or ways with the number that shows each corridor maximum possible flow.

For this graph the following conditions are met:

- There is one and only one vertex X_0, which no arc from the set U enters;
- There is one and only one vertex X_N, from which no arc comes;
- Each arc $u \in U$ is associated with an integer c (u) $\geq$ 0, called the bandwidth of the arc.

X_0 is called the input (source) of the network, X_N – is the output (drain), and c (u) is the bandwidth of the arc u. Figure 2 shows the graph of a transportation network.

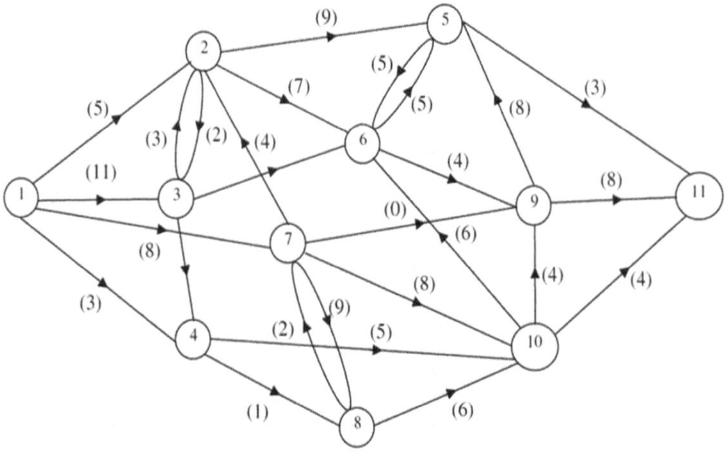

Fig. 2. Transportation network graph.

A problem in the transport network appears when the transportation request exceeds the capabilities of individual transport corridors. To search for combined transport corridors, the article proposes to use the Ford-Fulkerson method. Based on the calculation results, it will be possible to quickly change the network configuration and connect other transport arteries of the city to solve peak loads.

2.2 The Ford-Fulkerson Algorithm

The Ford-Fulkerson algorithm is a maximum flow theorem of a graph – the value of the maximum flow is equal to the size of the minimum cut. Adequacy: any flow between the vertices t and s is less than or equal to the value of any section. Let some flow and some section be given. The value of this stream is the sum of the "passenger" values carried along all possible paths from the top of t to s. Each such path must have a common edge with a given section. Since for each edge of the section it is impossible to transfer "passengers" more than its carrying capacity, therefore the sum of all cargoes is less than or equal to the sum of all throughput capabilities of the edges of this section.

The Ford-Fulkerson algorithm for finding the maximum flow in a graph is based on this theorem, which is also proof of the necessity of this theorem, which means that it is constructive. The Ford-Fulkerson algorithm solves the problem of finding the maximum flow in the transport network.

The Ford-Fulkerson algorithm is defined as: for a given transport network, the maximum flow value is equal to the minimum cut throughput [5]:

$$max\ \varphi_{X_N} = \min C\left(U_{A}^{-}\right), where\, AI\, X_0 \notin \mathrm{A}, \mathrm{X}_N \in A \quad (1)$$

where: φ_{X_N} – maximum flow in vertex N; $C(U_A^-)$ – cut throughput; A – vertex set; N – vertex; X_0 – start vertex; X_N – end vertex. And to calculate flow is possible to use this formula:

$$(\varphi_{X_0} = \sum\nolimits_{u \epsilon U_{Xi}^{+}} \varphi(u)\ and\ \varphi_{X_N} \sum\nolimits_{u \epsilon U_{Xi}^{-}} \varphi(u)) \rightarrow (\varphi_{X_0} = \varphi_{X_N}) \quad (2)$$

where: φ_{X_0} – maximum flow in start vertex; φ_{X_N} – maximum flow in end vertex; $u \epsilon U_{Xi}^{-}$ – many arcs entering X_i; $u \epsilon U_{Xi}^{+}$ – many arcs going out to X_i.

2.3 Calculation of Routes

The article describes the Ford-Fulkerson algorithm relating to a specific problem. It is assumed that there is the need to find the maximum flow of the transport network, a graphical image of which is shown in Fig. 3.

The source node is, for example, vertex 1, and the drain node is vertex 6 [5]. The task is to construct the maximum flow (F) between these vertices. The initial value of F is zero. It is obvious that the data structure for the description of F is a matrix of the same type as the matrix C, in which the carrying capacities of the arcs are defined.

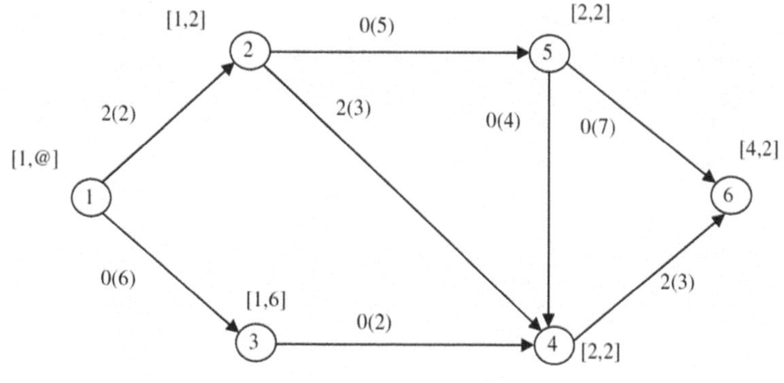

Fig. 3. An example of one of the graphs.

- First interaction;
  The first step. Let us assign the label [1, @] to the vertex 1. Consider arcs whose origin is vertex 1 – arcs (1, 2) and (1, 3). Vertices 2 and 3 are not marked, therefore it is required to assign labels to them, for the 2nd one – [1, 2] and the 3rd one – [1, 6]. The tags description is the following: the first digit is the number of the vertex from which the stream is coming; the second digit is the numerical value of the stream that can be passed along this arc.
  The second step. Select the marked but not viewed vertex. The first in the corresponding data structure recorded vertex 2. Consider the arcs for which it is the beginning – arcs (2, 4) and (2, 5). Vertices 4 and 5 are unmarked. Give them the labels [2, 2] and [2, 2]. So, in the second step vertex 2 is viewed, vertices 3, 4, 5 are marked, but not viewed, other vertices are not marked.
  The third step. Choose vertex 3. Consider an arc (3, 4). Vertex 4 is marked. Proceed to the next vertex, the fourth, the corresponding arc - (4, 6). Vertex 6 is not marked. Label it [4, 2]. Having reached the top of the runoff, thereby finding a path (a sequence of arcs), the flow through which can be increased. Information about this path contains vertex labels. In this case, the path $1 \to 2 \to 4 \to 6$. The maximum possible flow that can be passed along the arcs of this path is determined by the second digit of the drain top-bar label that is 2. The flow in the network became equal to 2.
- Second interaction;
  The first step. Let us assign the label [1, @] to the vertex 1. Consider arcs whose beginning is labelled vertex 1. These are arcs (1, 2) and (1, 3). Vertex 2 cannot be marked because the arc throughput (1, 2) has been exhausted. To top 3 need assign a label (1, 6).
  The second step. Select the marked but not viewed vertex. This is vertex 3. Repeat the action. As a result, vertex 4 receives the label [3, 2].
  The third step. Select vertex 4, only it is marked and not viewed. Vertex 6 is assigned the label [4, 1]. At the previous iteration, two units of throughput of this arc have been consumed, only one remained. Top-flow is reached. An increasing flow path is found. This is the path $1 \to 3 \to 4 \to 6$, on which a single stream can be transmitted. The resulting stream in the network is three. The change in the network state on this iteration is shown in Fig. 4 [5].

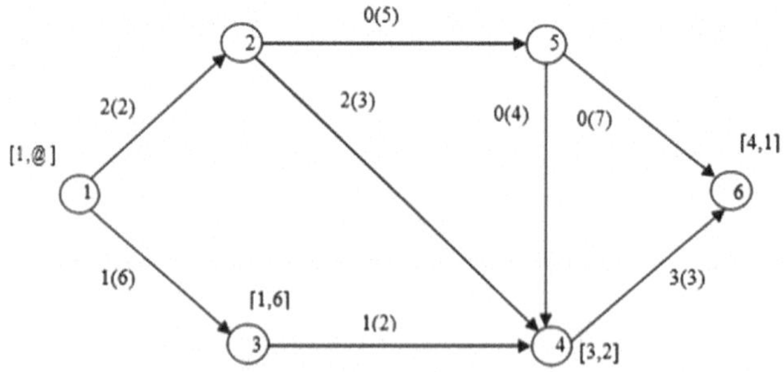

Fig. 4. Changing the interaction.

Thus, in the presence of powerful computing systems, it is possible to calculate the maximum flow from start point to end point in transport network for all possible routes, which has been done by the authors of the article. To enhance the calculations, software module has been written in the DELPHI environment.

Based on the results of calculations, and in order to eliminate the overload of the transport network of the Nur-Sultan city, it was proposed to launch a rail bus along the railway line. The route of this bus (13,73 km) is shown in Fig. 5.

Fig. 5. Route of a rail bus along the railway line.

Thus, the authors have proved the effectiveness of the interaction of urban and railway transport in solving transportation load during peak hours.

3 Conclusion

In the article, a thorough analysis of the assessment of the quality of urban passenger transport services has been conducted. The result of this analysis is the development of a strategy to improve the quality of urban passenger transport services with a reduction in operating financial costs while increasing passenger flow.

The article simulates the transport network of the city of Nur-Sultan from the standpoint of the Ford-Fulkerson algorithm: nodal points are defined, corridor capacities (graph branches) are designated and initial and final points (source and drain) are designated as examples.

The maximum network flow has been calculated for the given initial conditions. Subsequently, by modeling the inclusion of additional branches that are other transport corridors into the algorithm, the possibility of increasing the maximum flow with a simultaneous improvement in the quality of urban passenger transport services is shown.

In order to automate the calculation process, a software model was developed in the Delphi environment that allows to simulate various changes in both graph nodes and branches.

Most well-known methodologies for calculating urban transport networks do not allow the use of railway infrastructure as an integral part of urban transport, but offer solutions as a separate mode of transport. The proposed approach for enhancing urban transport networks can be used in route planning and can solve logistical issues with the integration of railway networks in urban transport networks.

The study conducted in this article shows that for the calculation of a real city network with many start and end points, as well as with changing values of the maxi-mum flows of each edge (ways) the decision depends on many important and stochastic factors, therefore further research would be aimed to introduction of the artificial neural networks that can give better recommendations on changing traffic flows for the city in real time.

References

1. Gudkov, V.A., Velmozhin, A.V.: Theory of transport processes and systems, pp. 1–167. Transport, Moscow (1998)
2. Assiltayev, A.: Optimization of logistical flows in the urban passenger transportation system. Master Thesis, pp. 1–75 (2019)
3. Shander, O., Shumyk, D., Shander, Y., Ischuka, O.: Improving the technology of freight car fleet management of operator company. Proc. Comput. Sci. **149**, 50–56 (2019)
4. Ischuka, O.: Main directions of improvement of the method for calculation of idle time of cars at technical stations. Proc. Comput. Sci. **149**, 475–482 (2019)
5. Dorofeeva, V.I.: Algoritmi na grafah I ih oprilozhenija. Uchebnoje posobije, pp. 9–13. Oryol State University, Orel (2002)
6. Romanova, A., Vygnanov, A., Vygnanova, M., Sokolova, E., Eiduks, J.: Problems of the formation of a single transport space on sections of international transport corridors. Proc. Comput. Sci. **149**, 537–541 (2019)
7. McKinnon, A., Button, K., Nijkamp, P.: Transport Logistics, pp. 1–680. Elgar, Germany (2002)
8. Prokhorchenko, A., Parkhomenko, L., Kyman, A., Matsiuk, V., Stepanova, J.: Improvement of the technology of accelerated passage of low-capacity car traffic on the basis of scheduling of grouped trains of operational purpose. Proc. Comput. Sci. **149**, 86–94 (2019)
9. Martin, H.: Warehousing and Transportation Logistics, pp. 1–680. Transport, Hamburg (2008)
10. Guido, G., Klaus, N.: Modelling Public Transport Passenger Flows in the Era of Intelligent Transport Systems, pp. 1–522. Springer, Switzerland (2016)

Improving the Quality of Railway Operation to Extend the Services of Transportation of Cargo of Small Volume

Oleksandr Lavrukhin[1], Anna Kravets[1(✉)], Larysa Parkhomenko[1], Daria Kulova[1], Artem Panchenko[2], and Natalja Ivaskovska[3]

[1] Ukrainian State University of Railway Transport, Kharkiv 61001, Ukraine
tilavalval@gmail.com, anya.obuchova@gmail.com, d.coolava@gmail.com, l.o.parkhomenko@ukr.net
[2] V. N. Karazin, Kharkiv National University, Kharkiv 61022, Ukraine
artempanchenkotop@gmal.com
[3] Institute of Transport, Riga Technical University, Azenes Street 12, Riga 1048, Latvia
natali5000@gmail.com

Abstract. In the article some ways to improve the quality of the railway operation and to satisfy the demand for the transportation of cargos of small volume using the existing infrastructure of the railway and other modes of transport are considered. Possible schemes of passenger service and delivery of their freight luggage are considered in combination with the principles of logistics and interoperability to extend the servicing options of the railway. The most complete servicing scheme involves comprehensive customer service, either as a passenger or as a customer of the door-to-door cargo delivery service. The propositions for convenient implementation of the servicing and delivery process include: creation and use of a software product that will be implemented as a client-server application, development and introduction of a single shipping document which will not require any re-registration when the cargo is transferred to another mode of transport involved in the delivery.

Keywords: Railway transport · Operations planning · Shipping unification

1 Introduction

Railway transport is one of the important sectors of the Ukrainian economy and a leading industry in the national road transport complex, which transports 82% of freight and almost 50% of total number passengers transported by all modes of transport.

Today, a considerable number of passengers need new services, the main ones being the delivery of luggage and freight baggage to stations and points of destinations. Road transport is competing with rail transport for medium distances. Therefore, the purpose of this article is considering ways to improve the quality of operation of the railway and to satisfy the demand of the population for transportation of small volumes of freight using the available railway infrastructure and other modes of transport.

E. Ginters et al. (Eds.): ICTE ToL 2019, LNITI, pp. 35–43, 2020.
https://doi.org/10.1007/978-3-030-39688-6_6

2 Analysis of Published Data

In [1], the organization of small freight transportation in China was analyzed to find that there is no effective pricing method for this type of transportation and for this reason the stations providing such services have low profits. To solve this problem, a model was created with the objective of maximizing profit for the station and the method of selecting the optimal route of transportation.

In [2], the advantages and disadvantages of charter transportation of small volumes of freight by road are considered. The bottlenecks in the organization of the transportation process and pricing for this service were identified, and the ways of their elimination using the Rubin Stein-Starr method were proposed.

In [3], a model to optimize the use of both freight and passenger planes for freight transportation is presented. This linear programming model, implemented using ILOG, allows finding the optimal route for freight by combining transportation in freight and passenger airplanes using heuristic methods.

The conducted analysis suggests that the issue of transportation of small volumes of freight is rather urgent. Transportation of such freight requires finding the best way of delivery, by combining both different modes of transport and freight and passenger traffic. In order to promote further competitiveness of the railway in the transport services market and to balance the interests of Ukrzaliznytsia and those of passengers and customers, measures should be taken to include the rail transport into the logistics chain when small volumes of freight are transported.

3 Service and Development of Transportation of Small Freight Volumes

The need of people for small-scale freight transportation has increased significantly. Thus, with the existing demand for transportation, the economic profitability of delivering freight depending on its volume negate all advantages of rail transport over other types of transport.

Therefore, directions should be considered that would separate and give each mode of transport its own niche for provision of services taking into account individual advantages and disadvantages. The first direction involves renewal and increase in the attractiveness of the Luggage and Freight Baggage Transportation service.

At the stations open for luggage operations, passengers can check in their luggage before boarding for an extra fee to be transported as freight baggage. That is, the existing system of luggage delivery may be considered. In this case, a separate mail and luggage wagon is used for the freight transportation.

However, let us consider other possible passenger service and baggage delivery schemes that extend the range of railway services.

If a passenger travels by train intending to change and further travel by air, then the luggage transportation scheme can be extended with two options:

- Scheme 1 – passenger checks in the luggage at the departure station – passenger travels (luggage delivery) – passenger is at the transfer point (transfer of luggage to the road transport) – passenger travels (luggage delivery by road) – passenger receives luggage in the airport;

- Scheme 2 – Passenger checks in luggage at the departure station – travel of the passenger (luggage delivery) – passenger is at the changeover point (transfer of luggage to the road transport) –passenger travels (luggage delivery by road) – passenger is at the airport (transfer of luggage to the air transport).

It is proposed to re-equip one compartment in a passenger wagon and adjust part of the premises in high-speed wagons for the placement of freight baggage to transport small freight batches and rationalize the use of freight space in rolling stock.

As optional services, the customer can also be offered maximum convenience in servicing: early delivery of their luggage to the luggage department of the departure station by the service office of the station, as well as the transfer of the customer's luggage from the airport in the country of destination to their temporary stay, for example, to a hotel (see Fig. 1).

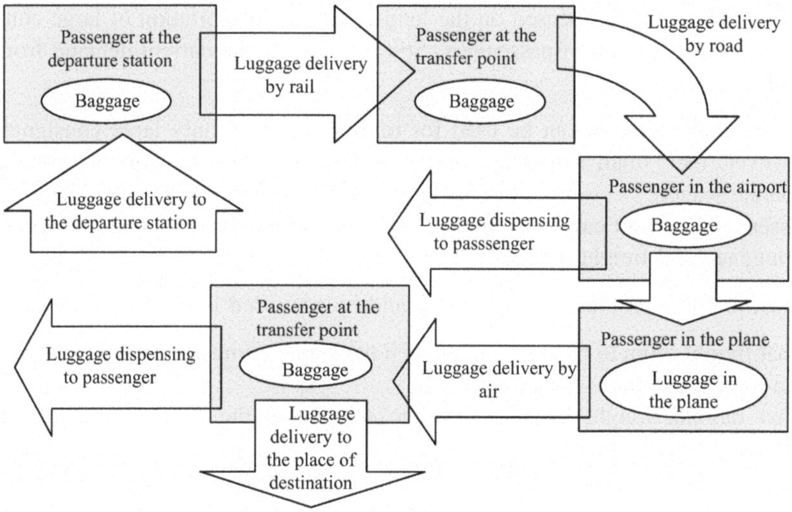

Fig. 1. Door-to-door luggage delivery with more than one mode of transport.

In addition, when this type of service is implemented, a single baggage document should be used to avoid multiple check-in to transfer the luggage to another mode of transport. However, it should contain more detailed information:

- The owner of the luggage;
- The place of delivery of the freight baggage;
- Overall dimensions and weight of the freight baggage or packaging;
- Route of travel, indicating all points of transfer of luggage between modes of transport.

In order to maintain quality and enhance the attractiveness of the service, it is very important that the entire process from the delivery of the luggage by the passenger until the receipt at the point destination is as simple and convenient as possible.

Now any individual or legal entity can send luggage by rail: there are no restrictions, provided the consignment does not include things banned for transportation and luggage is properly packed [4].

In general, the entire procedure is similar to that performed by other freight carriers (UkrPoshta, Nova Poshta, Meest Express, etc.). However, only a few dozen luggage offices operate at the Ukrainian railways. Most of them are located at the stations of Kyiv, Kharkiv, Dnipro, Lviv, Ivano-Frankivsk, Odessa-Main. By comparison, Nova Poshta network has more than 4,500 offices. In 5 months of 2019, the company delivered over 81 million parcels, which is 23% more than in the same period last year [5]. Meest Express also has a well-developed office network. However, unlike Nova Poshta, it is more focused on international shipments and of door-to-door deliveries with company couriers [6], i.e. taking into account the experience of other carriers, it can be stated that there is a demand for the shipment of small volumes and dimensions.

Railway transport is focused on the long distance transportation of large consignments and transportation of passengers. Still, in view of the current demand from the population:

- Freight transportation can be used for transportation of only large consignments, however, the "small shipment" option should be returned, without providing a separate wagon;
- Passenger transport can be used more actively and efficiently for the transportation of luggage and freight baggage.

Thus, to solve problems, questions should be answered in the following order:

- What freight room to choose: re-equipped passenger compartment; re-equipped part of a high-speed train; baggage wagon;
- What rational freight baggage stowage scheme in the freight room to choose seeking its maximum load;
- Calculation of the transportation cost of freight baggage depending on what scheme the freight will be sent: from the departure station to the destination; from the departure station to the airport terminal; from the departure station to the point of destination of the passenger abroad.

4 The Problem of Substantiation of the Quantitative Composition of Freight Consignments

Delivery of freight baggage has limitation of time when the baggage can be accepted, as the departure depends on the time of departure of trains.

In a specified term which, for example, begins 2 days and ends one hour before the departure of the train, a luggage office operator (LOO) accepts a request or freight baggage for transportation. The *min* consignments size becomes known as soon as the first request is received, and the *max* size is known only after when the time for luggage receipt expires. LOO processes the requests and plans the formation of collective freight consignments in a freight room of size i with the following main characteristics: D_i, W_i are load capacity and freight capacity of the freight room, respectively. LOO should justify the quantitative composition of freight baggage consignments of different customers and choose a room among the available ones to maximize the utilization rate of its freight capacity ($W_i \to max$) and/or load capacity ($D_i \to max$).

The statement of the problem that LOO has to solve in order to choose the rational loading of the freight room is specified as follows.

LOO receives z number of requests ($z = \overline{1, Z}$) for delivery of k freight baggage ($k = \overline{1, K}$; $K = K^H \cup K^L$): some pieces of freight baggage are "heavy" ($k = \overline{1, (k-1)}$; $k \in K^H$), their specific loading volume is: $\overline{\upsilon_1^H}$, $\overline{\upsilon_2^H}$, …, $\overline{\upsilon_{k-1}^H}$, m³/kg; and some pieces of freight baggage are "light" ($k = \overline{k, K}$; $k \in K^L$), their specific loading volume is: $\overline{\upsilon_k^L}$, …, $\overline{\upsilon_{K^L}^L}$, m³/kg.

To choose the freight room, LOO should determine the number of units in the freight baggage consignment. In this case, the objective function maximizes the number of all freight baggage units in loading the freight room by volume (1) with limitations (2)–(4), which takes into account the total weight of freight baggage and load capacity:

$$P = \sum_{k=1}^{K=K^H \cup K^L} (M_k \cdot \overline{\upsilon_k}) = \sum_{k=1}^{K=K^H} (M_k^H \cdot \overline{\upsilon_k^H}) + \sum_{k=1}^{K=K^L} (M_k^L \cdot \overline{\upsilon_k^L}) \to \max \tag{1}$$

limitations:

$$\sum_{k=1}^{K=K^H} M_k^H + \sum_{k=1}^{K=K^L} M_k^L \le D_i \quad (i = \overline{1, I}) \tag{2}$$

$$\sum_{k=1}^{K=K^H \cup K^L} (M_k \cdot \overline{\upsilon_k}) = \sum_{k=1}^{K=K^H} (M_k^H \cdot \overline{\upsilon_k^H}) + \sum_{k=1}^{K=K^L} (M_k^L \cdot \overline{\upsilon_k^L}) \le W_i \ldots i = \overline{1, I} \tag{3}$$

$$M_k^H \ge 0; \quad M_k^L \ge 0 \quad (k = \overline{1, K}; \quad K = K^H \cup K^L \tag{4}$$

where K, M_k are parameters of freight baggage in the batch, by their number and weight, respectively, classified into "heavy" (K^H, M_k^H) and "light" (K^L, M_k^L), kg.

Taking into account all checked-in luggage, LOO maximizes the utilization rate (α_W^i) (5) of the freight capacity of the freight room with limitations (6)–(8):

$$\alpha_W^i = \frac{\sum_{k=1}^{K=K^H \cup K^L} (M_k \cdot \overline{\upsilon_k})}{W^i} = \frac{\sum_{k=1}^{K=K^H} (M_k^H \cdot \overline{\upsilon_k^H}) + \sum_{k=1}^{K=K^L} (M_k^L \cdot \overline{\upsilon_k^L})}{W^i} \to 1 \tag{5}$$

limitations

$$\frac{\sum_{k=1}^{K=K^H} M_k^H + \sum_{k=1}^{K=K^L} M_k^L}{D^i} \leq 1; \tag{6}$$

$$\frac{\sum_{k=1}^{K=K^H} (M_k^H \cdot \overline{\upsilon_k^H}) + \sum_{k=1}^{K=K^L} (M_k^L \cdot \overline{\upsilon_k^L})}{W^i} \leq 1 \tag{7}$$

$$M_k^H \geq 0; \quad M_k^L \geq 0 \quad (k = \overline{1,K}; K = K^H \cup K^L) \tag{8}$$

Given the spatial limitations of the freight room, the recommended maximum utilization rate of freight capacity (α_W^i) is up to 0.8.

According to this indicator, LOO chooses a freight room, namely:

- Using only a re-equipped freight room in a passenger wagon as a freight room;
- Using a freight baggage wagon, too.

The next question is the issue of the rational placement of freight baggage in the chosen freight room. The space of the room is limited and has certain size and volume. In addition, freight baggage is considered as a freight unit without a hard container, that is, it cannot be placed in tiers. Freight baggage is stowed on shelves, which also limits the stacking options in the freight room.

Taking into account the variability of possible solutions, the solution of the "knapsack problem" which belongs to combinatorial optimization problems can be used to solve the problem of rational and secure placement of freight baggage in the freight space [7]. The optimal knapsack loading or the knapsack problem is extremely relevant; knapsack problem solving algorithms are used in economics, computer science, mathematics, bioengineering [8] and logistics. The application of algorithms for solving the knapsack problem is further expanded since the knapsack problem is NP-complete, i.e., many tasks of the same class can be reduced to it. Then the second problem statement will be as follows: there is a set of k_i freight baggage units, each with two parameters: volume V and freight baggage processing cost C. There is a freight room of certain capacity The task is to form a freight baggage consignment of the maximum total volume without exceeding the weight limit of the room.

Subject to the specific conditions under consideration, let us consider one of the types of the "knapsack problem" – the "*multi-dimensional knapsack problem*" [8]. In this case, p_j resources are added to the weight and volume of the luggage in the problem situation, and each freight baggage consumes a given amount of each resource.

The problem statement for LOO to determine the minimum cost of transportation of the declared luggage consignment with rational loading of the freight room is specified as follows. LOO has information about k number of freight baggage units to be taken from the a-th point of departure from d ($a = [1...d]$) and b_y number of freight baggage units to be delivered to the p-th point of destination ($p = [1...s]$). The number of points of destination (s) depends on the number of stations of destination indicated in the transport documents and the selected freight baggage transportation scheme (station – station; station – airport; station – airport – hotel).

At the same time, when freight baggage is loaded into a freight room at the i-th point, a number of resources should be taken into account ($j = [1 \ldots m]$) which limit the general conditions for performing this process, namely.

The total cost of transportation of k freight baggage (9) will be:

$$C = \sum_{i=1}^{d} \sum_{p=1}^{s} C_{ip} \cdot k_{ip} \tag{9}$$

The objective function minimizes the total cost of luggage handling, including transportation, and has the form (10), (11):

$$C = \sum_{i=1}^{d} \sum_{=1}^{s} C_p \cdot k_p \rightarrow min \tag{10}$$

$$\begin{gathered} \sum_{i=1}^{d} \sum_{p=1}^{s} p_{ipj} \cdot k_i \leq P_j, j = 1, 2, \ldots m \\ k_i \geq 0 \end{gathered} \tag{11}$$

where C_{ip} is the cost of transporting a unit of freight baggage from the i-th point to the p-th, which depends on the distance of transportation and the cost of handling of the k-th unit of freight baggage; P_j is the total consumption of each p-th resource.

The total cost of delivery of a unit of luggage is the sum of individual costs of all operations with freight baggage, including other modes of transport, and tends to a minimum. In addition, the set of costs that determine the total cost of shipping a unit of luggage depends on the delivery scheme chosen by the customer.

5 Interaction of the Customer and the Luggage Office Operator

In order to introduce the interaction between the customer and the luggage office operator, a software product as a client-server application should be developed which will consist of two main parts: client and server. In Fig. 2 is shown an example of the interface of the client part of the "Fast luggage" application with the main pieces of information that the client must provide in order to receive the complete freight delivery service.

The customer enters all information which LOO requires and upon completion of the registration receives an electronic document, which can be used to check in the freight baggage for transportation.

When the freight baggage arrives to the luggage office, the employees verify the correspondence of the information in the luggage document and the actual parameters

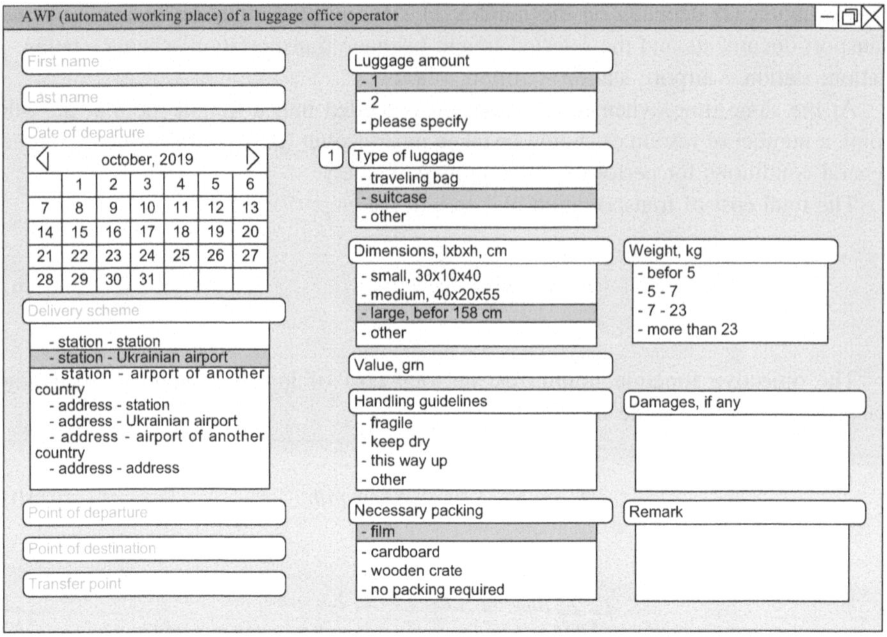

Fig. 2. Example of the interface of the "Fast luggage" application.

of the luggage: dimensions, weight, damage, if any. However, loading of luggage is planned, first of all, according to the data specified by the client at registration.

6 Conclusion

The article considers the necessity of providing an additional type of service to passengers of the Ukrainian railway: handling and delivery of small volume luggage. The stages of practical implementation of quality delivery are given. Different possible options of freight rooms are proposed depending on the size of the batch of freight baggage and the route of shipment. The size and weight of individual articles and freight room should be taken into account for rational formation of a batch. Implementing door-to-door freight delivery is proposed in order to take advantage of the opportunities of different modes of transport and to strengthen the ties between them.

In further studies, more detailed investigation of various possible options of delivery of freight baggage, their practical implementation, development, as well as implementation of information support for this type of service, including the formation of a single baggage document, is planned.

References

1. Sun, X., Li, X., Cheng, H.: A pricing method for small-size cargo express service in long distance highway transportation. J. Transp. Syst. Eng. Inf. Technol. **13**(3), 115–120 (2013)
2. Sun, X., Cheng, R.: Chartered pricing for small-size cargo express service in long distance highway passenger transportation station. J. Transp. Syst. Eng. Inf. Technol. **17**(4), 13–18 (2017)
3. Nahum, O., Hadas, Y., Kalish, A.: A combined freight and passenger planes cargo allocation model. Transp. Res. Procedia **37**, 354–361 (2019)
4. Rules of carriage of passengers, luggage, cargo and mail by rail of Ukraine. Order of the Ministry of Transport and Communications of Ukraine, no. 1196, 27 December 2006
5. 81 million parcels, 3400 branches, 78% - NPS: Nova Poshta sums up 5 months of 2019: official site of Nova Poshta. https://novaposhta.ua/en/news/rubric/2/id/6098. Accessed 21 Jan 2019
6. In June, the branch network of Meest increased! Official site of "Meest Express". https://www.meest-express.com.ua/ua/pro-nas/novyny/u-chervni-merezha-viddilen-meest-zbilshylas/. Accessed 21 Jan 2019
7. Solution of classical optimization problems. https://znatock.org/s11117t1.html. Accessed 21 Jan 2019
8. Goos, P., Syafitri, U., Sartono, B., Vazquez, A.: A combined freight and passenger planes cargo allocation model. Transp. Res. Procedia **37**, 354–361 (2019)

Improving Inventory Management Under Uncertainty of Data - Optimization Problem in Logistics

Zbigniew Lukasik and Malgorzata Gorska(✉)

Kazimierz Pulaski University of Technology and Humanities,
26-600 Radom, Poland
{z.lukasik,m.gorska}@uthrad.pl

Abstract. An optimization problem in inventory management is considered, taking into account uncertainty of model parameters. The aim is to support decision making on the basis of an analysis of costs and risks. Options for the implementation of the model and its optimization are presented. A first approach is devoted to the stationary variant of the problem. For a case study, it is shown that optimized rule-based control improves the profit considerably.

Keywords: Inventory management · Logistics · Modeling

1 Introduction

An optimization problem in inventory management is considered, taking into account uncertainty of model parameters. The aim is to support decision making on the basis of an analysis of costs and risks. Options for the implementation of the model and its optimization will be presented.

The aim of the paper is twofold. First, a model problem for storage management under random order activity is formulated, with particular focus on risks connected with shortage of material and overflow of capacity. Second, a parameter optimization problem for a parametric decision rule on the basis of a sufficiently high number of instances of the random process is performed to find a compromise between the two considered risks.

The techniques applied to the presented logistics problem make use of special properties of the formulated model of inventory management. In particular, monotonicity, smoothness, positivity and convexity of certain functions are essential for the feasibility and efficiency of certain solution methods [6, 16].

An extension of first analysis will concern multi-product transport and storage. In that case instead of a scalar quantity Q, a vectorized inventory balance model will combine delivery and storage of several goods, including competition for space and synergetic effects due to common exploitation of resources. The research is intended to support management in actual assessment of options for the logistic operation of their firms.

E. Ginters et al. (Eds.): ICTE ToL 2019, LNITI, pp. 44–52, 2020.
https://doi.org/10.1007/978-3-030-39688-6_7

2 Structure of Costs and Revenue

A factory processing a selection of raw materials into a set of products is considered. The materials have to be purchased at certain costs, the products are sold leading to a profit depending on the difference between income from sales and the costs of production and materials. Since market prices of goods are variable, depending on time, but also on size and regularity of orders, it is reasonable to place orders for material at suitable times and in such amounts to avoid the necessity of purchases, when prices are high. However, this requires the possibility of storage – which comes at a cost. Likewise, products may be sold immediately or stored until a later time, when a better price may be obtained. Advantages and disadvantages of maintaining storage at the factory as opposed to ordering supplies just in time are summarized into a scalar objective function. To this end, one needs to formulate assumptions as to the structure of prices, including random variations.

A major contribution to the objective is the cost of supplies. Two types: ordinary and express, are considered. Ordinary, regular orders are assumed to have a nearly constant price level. In the case of temporary shortages due to high demand of a product extra orders may turn out more expensive.

At given a constant price c, this would lead to a cost of acquisition of a quantity Q of:

$$f(Q) = c \cdot Q \quad (1)$$

More generally, allowing for contributions independent of the amount purchased, the cost may be given with fixed and variable prices c_0, c_v as const function:

$$f(Q) = c_0 + c_v \cdot Q \quad (2)$$

Taking into account discount for larger purchases, this may be further refined to:

$$f(Q) = c_0(Q) + c_v(Q) \cdot Q \quad (3)$$

with a monotonously decreasing specific prize c_v and an increasing absolute contribution c_0. The latter may increase stepwise, e.g. due to the cost for a box, palette, container etc.

A main issue is that in a continuous production, deliveries come in installments, i.e. the total quantity Q is split into a sum of smaller positive portions $Q_i, Q_i > 0$, $i = 1, 2, 3 \ldots m$,

$$Q = Q_1 + Q_2 + Q_3 + \ldots + Q_m = \sum\nolimits_{i=1}^{m} Q_i \quad (4)$$

If one calculates the total cost F as the sum of the costs of the contributions,

$$F = f(Q_1) + f(Q_2) + f(Q_3) + \ldots + f(Q_m) = \sum\nolimits_{i=1}^{m} f(Q_i) \quad (5)$$

then it turns out that in general, F is not at all a function of Q, i.e., it cannot be calculated without knowing the split of the total into its fractions. More but smaller installments result in higher costs, the lowest possible cost being obtained in the case of a single shipment.

Practically, a single delivery is not a feasible option, and that for multiple reasons. For one, material is needed all the time, from the beginning of the considered time span, but the amount needed to meet orders from customers is not known in advance. Second, there are constraints in the form of financial and physical restrictions on the size of orders from the supplier. That means that material is ordered periodically at times t_i, adapting quantities Q_i to the demand from clients, the storage potential and the financial status of the firm.

The second main contribution to the objective is the revenue generated from the sales of products, driven by customer demand. As with the supplies, also the total sales split usually a set of orders placed by customers and dealt with by the producer [5]. Denoting the amount of the product P, sold during the considered period, the total is split into a sum:

$$P = P_1 + P_2 + P_3 + \ldots + P_n = \sum\nolimits_{j=1}^{n} P_j \tag{6}$$

The shipments are made and invoiced at a different set of time instants, say $\tau_1, \tau_2, \tau_3, \ldots \tau_n$, the clients will be charged according to an analogous price scheme of the form:

$$g(P) = d_0(P) + d_v(P) \cdot P \tag{7}$$

so that the total inflow of profit can be expressed as:

$$G = g(P_1) + g(P_2) + g(P_3) + \ldots + g(P_m) = \sum\nolimits_{i=1}^{m} g(P_i) \tag{8}$$

Again, the total income G is not a deterministic function of the total amount of the sold good. The random distribution of said amount P into separate orders influences the income G.

Next, the cost of production has to be considered as negative contribution to the objective. Classically, the demand of resources, and hence the cost, is calculated in a linear way by means of technological coefficients [7, 8]. More precisely, however, the cost of production depends on both the sequences concerning orders by customers and from suppliers via a nonlinear cost-function:

$$K = K(\tau, P, t, Q) \tag{9}$$

Here the symbols τ, P, t, and Q denote sequences, i.e. vectors rather than single scalar quantities. The total cost K comprised fixed coasts like for machines, equipment, but also labor.

The objective function is completed by terms representing the cost of storage S and of capital [12, 13]. It is postulated here that in each moment of time, there is a certain nonnegative state of stock of materials and products. This way production can be kept up if a shipment of material should be delayed or should orders of products be placed that exceed the processing capacity. With time elapsing, this state has to be updated at each of the instants t_i – the stock increases by Q_i, likewise with τ_j and P_j, when the reserve of the product drops. Consequently, a price for storage has to be taken into account. Again, the structure comprises a fixed component and a variable one, the first independent on the amount stored, the latter increasing with the stored volume. Thus the storage cost function $h(S)$ with nonnegative coefficients e_0 and e_v is considered:

$$h(S) = e_0 + e_v S \tag{10}$$

The amount of storage S is complementary to the financial status of the firm. Goods in the magazine are frozen capital, leading to costs in terms of interest. The details of the evolution of the financial status can be calculated from the two order sequences and the price schemes for materials, goods, storage and production. Denoting the financial balance by B, taking into account costs and income (1) through (10), at time t one obtains:

$$B(t) = B_0 - \sum\nolimits_{t_i \le t} f(Q_i) + \sum\nolimits_{\tau_i \le t} g(P_i) - H(t) - K(t) \tag{11}$$

where B_0 is the initial balance at start.

Now, the objective is to ensure a possibly high gain in the assets of the firm, considering (a) the change in the balance B and (b) the storage S, i.e.:

$$\varphi = B(t_{final}) - B(t_{initial}) + c(S(t_{final}) - S(t_{initial})) \tag{12}$$

The increments in (12) depend on the initial state, decision variables and the random order activity of clients.

In the next section, the above factors are implemented into a simulation model. Its behavior under selected strategies – considering ordering of material and storage management – are studied in dependence of a scalar decision variable. Maximization of the expected value of the random gain φ, as expressed by (12), is performed next in a case study.

3 Simulation Model

Assuming a random sequence of incoming orders, characterized by their number and by their size, the generated levels of storage and financial status are calculated. For the purpose of the presented analysis, Julia language scripts were developed. The choice of the simulation environment was based on availability and convenience.

The level of storage, at which an order for a new shipment of material is issued, is the key decision variable v for optimization of the gain [9, 14]. Comparison of selected schemes demonstrates, the risk of unexpected surges in demand, thus of running out of material, the need of expensive express orders of material, see [10, 11]. Also interesting research were described in [3, 4], where simulations and IT systems were well described in the field of logistics.

The aim of the simulation is to find a trade-off between the conflicting goals: low cost and low risk [1]. For a case study a period of time of 100 working days is simulated. The production level, according to the average number of incoming orders, amounts to 10 units per day, which make a total of 1000 produced and sold units over the observation time. As an example, the following settings are chosen (basic data and assumptions of simulation):

- Supply: 100 units each ten days, starting from day 2 (3, 4, 5);
- Sale: according to incoming orders, no delay in production;
- Storage: initial level 25, no delay after supplies;
- Account: initial level 200, payments without delay;
- Prices: supply 3.6, sale 4.8, fixed costs 10.0 per day.

In a first reference scenario, the daily orders are constant, equal to the average a = 10. The difference between the price for the resource and the finished product of 4.8–3.6 yields a turnover of 12 currency units per day. After allowing for the cost of constant 10 units per day, a profit of 2 currency units per day is left, which gives an expected total gain of 200 for the whole period.

In Fig. 1, the top part shows the incoming orders of customers, the second row corresponds to the material in storage, while the last part of the diagram gives the bank balance as function of time.

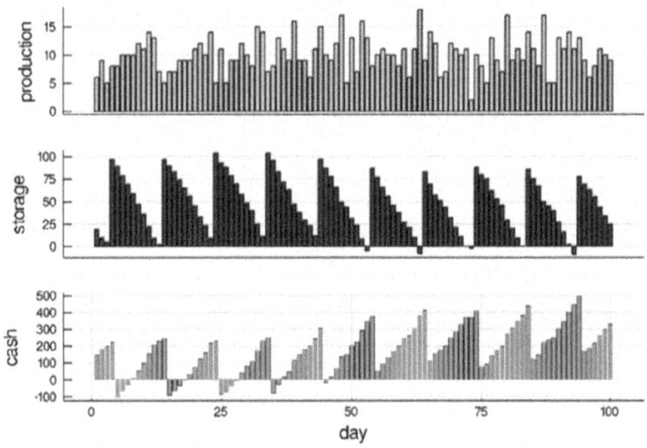

Fig. 1. A random sequence of orders and its consequences.

As shown, the random demand results in a negative financial status in the initial stage of operation of the firm. Further, frequently a shortage of material threatens the continuity of the production process.

In fact, in general, this would be impossible to maintain. However, if it will be assumed (a) that the bank allows overdraft, and (b) shortage of material can be handled by express shipments, the simulation may considered valid.

In practice, the bank would charge the firm for the overdraft at an interest rate $p > 0$, also the express shipments would come at an increased price, either per shipment, or per unit, or both. Those extra costs are not incorporated in the cash balance in Fig. 1.

The next figure (see Fig. 2) shows that a time shift of the supply arrangement, together with down payment instead of upfront payments, can avoid a negative balance in the bank account. Therefore the shortage of material is worse.

A larger storage and/or a better financial status at the start would help with both problems. On the other side, earlier shipments and a higher initial storage might lead to an overflow of the magazine. After a period of low business, the level in the magazine might exceed its capacity, raw material may be lost or has to be sold with losses.

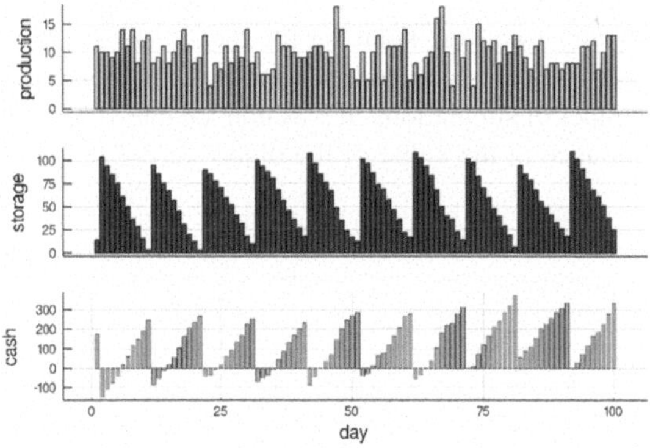

Fig. 2. Alternative random sequence of orders with different supply rule.

Figure 3 represents a better initial situation, the considered firm has more supplies in storage and more cash in reserve. Now, however, the level of 150 in the magazine is regularly exceeded. Assume this to be the capacity of the storage room, then the extra supplies may be considered as losses, which diminishes profit.

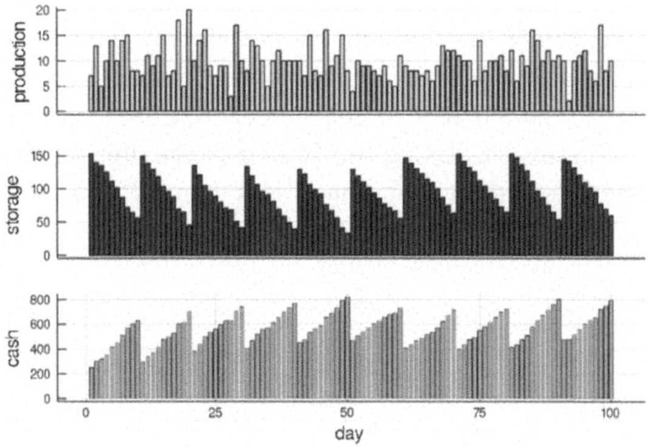

Fig. 3. Higher starting levels.

In order to address the issues of a negative bank balance, shortage of material and overflow of the magazine's capacity, a flexible order strategy is proposed and tested [2]. First of all, a supply is ordered at a given level of the storage, rather than on a given date [15]. It is assumed that shipment takes 3 days, payment is due on arrival. Next, if storage is at zero, an express order of 2 units at a price of 3 units in regular shipment is placed. Finally, if the balance is negative, the firm is charged 0.05% of the saldo per day.

The success of the firm in terms of gain in the balance over the period of time under consideration is measured. This gain φ is measured as the balance in the account plus the worth of the supplies in the magazine (12).

Of course, the ideal value is 200, which is possible to obtain in a uniform flow of in pouring orders. Higher fluctuation may lead to the need for express deliveries or to losses of storage. Simulation of 10000 runs for each set of parameters allows to evaluate different settings (see Fig. 4) taking the mean gain as objective function. So, for the purpose of optimization, the random variable φ is replaced by its mean value over the assumed number of simulations. It was noticed that 10000 simulations ensured a sufficiently low variance in the value of the objective.

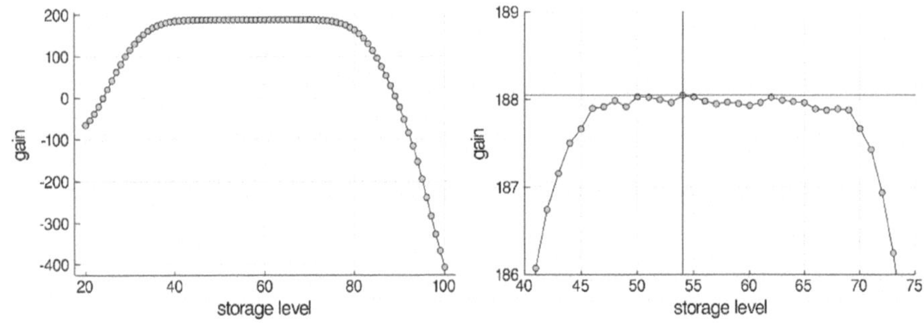

Fig. 4. Left: objective function – gain over 100 days as function of threshold for storage. Right: zoom of objective function – maximum of 188.058 is attained at 54.

It should be noticed that the optimum is not sharply defined, the (still random) objective exhibits a flat plateau starting from around 45, reaching up to 65. The results depend further on other parameters of the model, like maximum capacity, price of express orders, interest rates etc. Using SimJulia package, the simulation was repeated 100 times and obtained a standard deviation of less than 0.752, which is less than 0.5% of the maximum value of the considered quantity.

4 Conclusion

The article concerns a model problem for designing decision rules for economic processes under random influence. This should be of interest for researchers in the fields of logistics and managers confronted with the necessity to make decisions on the basis of uncertain data.

The considered model of a small or medium enterprise shows advantages and disadvantages of selected strategies of storage management. Relying on external logistics – ordering materials just in time – competes with a local in-house storage management. A quantitative analysis on the basis of simulations allows to decide about the preference, the possible gains and risks.

The model depends on many factors, in particular on the temporal distribution of the demand for the considered product. Adaptation of the model to actual market data is subject of ongoing research.

The proposed approach demonstrates rigid schedules of regular orders of raw materials are not suitable in a realistic free market situation. On the other hand, rule based operation always depend on the choice of parameters, here the thresholds at which orders are executed. It is shown that optimal choices can be taken on the basis of repeated simulation. The presented approach is intended as a framework for more sophisticated models of the logistics around the operation of a firm, the developed tools are easy to modify and apply and computationally highly efficient.

The presented research is a first step in an attempt to identify external factors influencing the random variables governing the considered process. Next, basing on forecasts concerning mentioned factors, further optimization of the decision making will be undertaken.

References

1. Baldick, R.: Applied Optimization: Formulation and Algorithms for Engineering Systems. Cambridge University Press, Cambridge (2006)
2. Christopher, M.: Logistyka i zarządzanie łańcuchem dostaw, strategie obniżki kosztów i poprawy poziomu usług, 2nd edn. PCDL, Warszawa (2000)
3. Ciszewski, T., Wojciechowski, J.: Logistyczne zastosowania systemów informacyjnych. Logistyka **3**, 349–352 (2012)
4. Ciszewski, T., Kornaszewski, M., Olczykowski, Z.: Information and IT systems in logistics. Podniková ekonomika a manažment **4**, 30–34 (2010)
5. Diaz, R., Bailey, M., Kumar, S.: Analyzing a lost-sale stochastic inventory model with Markov-modulated demands: a simulation-based optimization study. J. Manuf. Syst. **38**, 1–12 (2016)
6. Gołembska, E.: Kompendium wiedzy o logistyce, 4th edn. PWN, Warszawa (2019)
7. Krajewska, R., Łukasik, Z.: Model procesu zaopatrzenia elektrociepłowni w węgiel opałowy. Logistyka **1**, 256–264 (2016)
8. Krajewska, R., Łukasik, Z.: Zapewnienie płynności dostaw paliw stałych do elektrociepłowni poprzez sprawne zarządzanie ryzykiem operacyjnym. Autobusy: Technika, Eksploatacja, Systemy transportowe **3**, 2221–2230 (2013)

9. Krzyżaniak, S.: Podstawy zarządzania zapasami w przykładach, 4th edn. Instytut logistyki i magazynowania, Poznań (2008)
10. Kuśmińska-Fijałkowska, A., Łukasik, Z.: Koordynowanie działań w organizacji w odniesieniu do systemu zarządzania jakością. Logistyka **3**, 3570–3576 (2014)
11. Łukasik, Z., Nowakowski, W., Kuśmińska-Fijałkowska, A.: Zarządzanie bezpieczeństwem infrastruktury krytycznej. Logistyka **4**, 758–763 (2014)
12. Milewski, D.: Problemy optymalizacji i modelowania w uzupełnianiu zapasów. Zeszyty Naukowe Uniwersytetu Szczecińskiego. Problemy Transportu i Logistyki **17**, 141–150 (2012)
13. Nowakowski, T.: Problematyka niezawodności procesów logistycznych. Proces magazynowania. Logist. Transp. **4**(1), 63–71 (2007)
14. Sabo, A., Trzaskalik, T.: Optymalizacja poziomu zapasów z uwzględnieniem prognoz sprzedaży przy wykorzystaniu modeli adaptacyjnych prognozowania. Studia Ekonomiczne/Uniwersytet Ekonomiczny w Katowicach **96**, 393–407 (2011)
15. Sarjusz-Wolski, Z.: Sterowanie zapasami w przedsiębiorstwie. Polskie Wydawnictwo Ekonomiczne, Warszawa (2000)
16. Zamkowska, S., Zagożdżon, B.: Systemy logistyczne w obsłudze przedsiębiorstw. Wydawnictwo Politechniki Radomskiej, Radom (2010)

Architectural Approach to the Digital Logistics System Design

Alexey Nekrasov[1(✉)], Anna Sinitsyna[2], and Mark Styskin[1]

[1] Moscow Automobile and Road Construction State Technical University, Moscow, Russia
tehnologistic@mail.ru, 328587@bk.ru

[2] Russian University of Transport, Moscow, Russia
acc-lgkr@mail.ru

Abstract. Management of transport and logistics resources in the context of digital logistics is always at the center of research and business practice. This problem becomes especially important in the context of transition to Industry 4.0, when high demands are set to the infrastructure both in terms of costs and on the part of the strategic approach, taking into account the sustainability criteria and the response time to situations of fault. The article describes a comprehensive approach to life cycle management of transport and logistics infrastructure, based on the concept of enterprise architecture and logistics engineering. Coordinated interaction of enterprises within the life cycle stages "design – operation - operational control - verification and corrective actions - control system verification" is accepted as the basis of highly effective architecture of integrated TLS (ITLS), which is especially important in the conditions of digital transformation. The life cycle model forms the framework of a new generation of ITLS operating within supply chains on the principles of self-regulation of transport and logistics links (elements) with common resources that provide highly effective interaction with customers on the basis of IT-service. As a project approach based on logistics engineering, the use of a library of software procedures and models with an open description of it-services interfaces with a common database is considered. A systematic approach to service management allows businesses to manage risks, improve customer relationships, optimize costs and accelerate the flow of all business processes.

Keywords: Digital logistic · System service architecture · Logistics system life-cycle

1 Introduction

Prospects for the development of digital technologies are determined by the implementation of the concept of Industry 4.0 and the formation of new principles, models and architecture of engineering to ensure the digital transformation of the transport industry and related markets interacting with it. The end-to-end penetration of technologies into all sectors of the economy, both as digital (intangible) assets in the form of new business models and in the form of the industrial Internet of things leads to the formation of large arrays of economically significant industry and inter-industry data.

E. Ginters et al. (Eds.): ICTE ToL 2019, LNITI, pp. 53–60, 2020.
https://doi.org/10.1007/978-3-030-39688-6_8

As a result of research at the end of the twentieth century, an integrated approach to the formation and functioning of transport and logistics infrastructure on the basis of system architecture and its life cycle was formed. The openness of standards and models allows to increase and modernize the tools of it service, while the integration of logistics processes will be provided with digital data coming from the integrated infrastructure of transport and logistics systems (ITLS) based on the methods of system engineering.

2 Methods of Problem Analysis and Materials for Research

Analysis of the structural elements of transport infrastructure and logistics practices at individual transport facilities showed that one of the key areas of development along with digital technologies is the application of the methodology of system engineering and the development of enterprise architecture. Digitalization of the transport and logistics sector transforms this market on the basis of innovative tools related to the management of life cycle processes and land transport infrastructure [1–4].

Management of transport and logistics resources in digital logistics is always the focus of research and business practice [5, 6]. This issue is of particular importance in the context of the transition to Industry 4.0, when there are high demands on infrastructure both in terms of costs and from the strategic approach, taking into account the criteria of sustainability and response time to failure situations.

ITLS digital data models based on production infrastructure systems and best practice standards form the information model system architecture across the asset lifecycle. Digital transformation is the transformation of relevant business processes in the new economy, or the combination of operational and information technologies.

3 Theory and Practical Application

Traditional (functional) logistics, solving the problems of placing production, warehouses, consumers and modeling the supply chain in order to optimize costs and lead times, in modern economic conditions, involves a reorientation of processes to adaptive digital methods. A new approach was required, based on the system mechanism of integrated management not only of the supply chain, but also of processes at a separate enterprise on the basis of interaction of intellectual and Internet technologies (such as "Internet of things").

Used systems, tools and complexes often have such features as: multi-aspect, multi-structure, uncertainty of functioning, hierarchy, redundancy of elements and connections, multivariate implementation of functions and processes and mobility of components. This all largely depends on the level of sustainability of logistics processes that ensure the effectiveness of the main elements of the systems (products) throughout the life cycle of both the products and processes and assets (resources) [7, 8].

Standards focused on engineering approaches to the description of models form an effective architecture of ITLS – a new area of activity of GERAM [9]. GERAM is considered as a generalized standard architecture of a complex organizational and technical object and the methodology required to design and maintain an integrated enterprise or extended enterprise (network).

The basic element of the methodology is a common standard architecture of complex organizational and technical objects (SOTO), which combines both the description of business processes and the technology to support these business processes. The private model category (PEMs) uses processes and characteristics common to many infrastructure objects, including workflows and infrastructure components.

Proactive management of SOTO infrastructure, in contrast to the traditionally used in practice reactive management focused on rapid response and subsequent prevention of incidents, involves the prevention of incidents by creating in the appropriate monitoring and management system fundamentally new predictive and proactive capabilities based on the concept of system (integrated) modeling [10, 11].

Highly efficient operation of transport is directly related to the structure (processes) of the enterprises themselves, as well as the equipment and production assets used, which are included in the various stages of the life cycles of the systems. All this increases the complexity of organizational and technical objects, which include transport and transport and logistics systems.

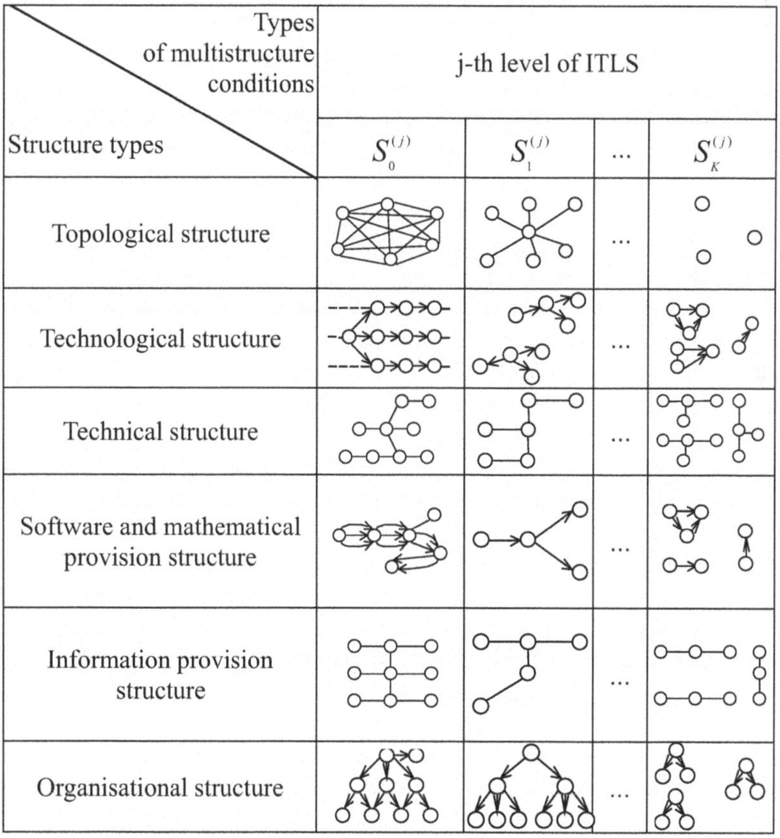

Fig. 1. Structure types in structure dynamics of ITLS.

Methodology, technological solutions and tools of organizational interaction on the basis of proactive management should be based on a new generation of ITLS, which combine subsystems of transport and logistics services, supply chains and provide stability and customer focus at a more efficient level. This is one of the priorities of cyber-physical systems envisaged by the Industry 4.0 strategy [12].

Possible variants of SOTO States based on the methodology of structural dynamics are presented in the Fig. 1. Types of structures - correspond to the transformed elements of the 4D model, which assumes the description of not individual subsystems, but the whole state of the whole object [1].

System integration of information infrastructure, digital data and IT services, organizational structures, products (transport work) is realized within the framework of an integrated customer focused transport system, using the principles of adaptation of the interaction of its elements in the life cycle process.

4 Results

Functioning of ITLS should be considered at such stages of life cycle as: planning (creation), operation (transportation), updating (improvement). The life cycle is the evolution of an ITL system from conception to discontinuation. In the course of the study, the ITLS architecture was formed, providing digital transformation of any type of enterprise.

It is based on the system interaction of such stages of the life cycle as "planning" - "operation" (application) – write-off (recycling). This approach has created conditions for the practical application of the GERAM methodology based on the methods of logistics engineering and project management [13].

The completion of the project approach using logistics engineering is the use of a library of software procedures and models with an open description of it-service interfaces with a common database of the ITIL library [14]. Initially, ITIL was organized by 10 main components of it services management, each of which corresponds to 38 processes. The model consists of a set of interrelated recommended practices, fundamental principles and procedures that together provide complete guidance to ensure the efficiency and reliability of IT solutions and services [15]. Metrics should be designed according to customer requirements and tested for feasibility. The monitoring ensures that the data obtained corresponds to the specified threshold values in order to correct the problems.

Based on the recommended model practices, it is proposed to use 12 categories of metrics for the monitoring system of the object under consideration (TLS) that describe the entire cycle of logistics activity and form the architecture of the monitoring system. The goals and objectives of the use of metrics were related to the coordination of IT services with the tasks of logistic activities and the improvement of processes (see Fig. 2).

Each metric describes the TLS process of a transport enterprise. An automated process monitoring system is considered as a multi-level adaptive management system focused on intelligent technologies. It covers the entire life cycle of processes in TLS and is based on the use of a special unit implemented at the program level - the

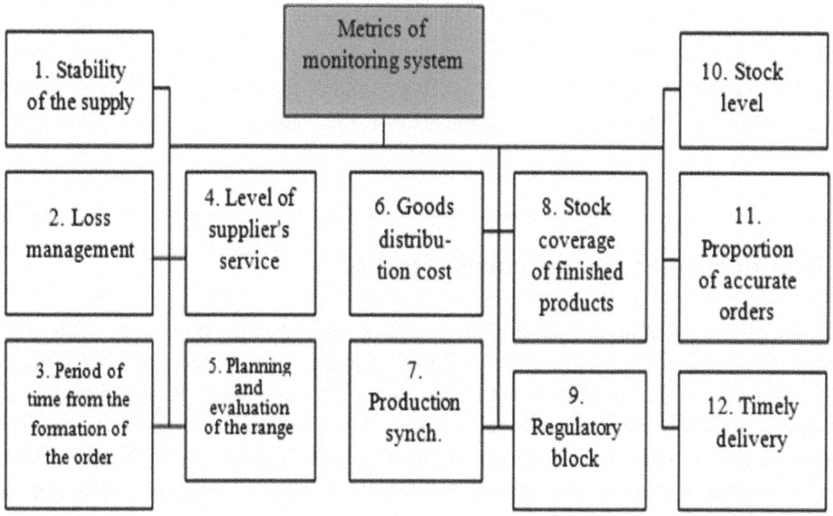

Fig. 2. The architecture of the metrics in the monitoring system of logistic activities.

neurologistical module. The neurologistical module forms, on the basis of the classified incident received, the possible consequences - the forecast - and provides a response to identified incidents in the logistics functional cycle.

Thus, by processing the obtained data through the neurologistical module, users receive a warning signal indicating critical values for the control effect on the process. In case of assigning the value "1", the risk of going beyond the forecast limits is considered critical, "0" - within acceptable limits.

The description of the formal neuron of the hierarchical neural network used for prediction can be represented by the following Eqs. (1) and (2):

$$\alpha_i = arctg\rho_i \tag{1}$$

$$\rho_i = \sum\nolimits_{j=1}^{N} \alpha_j x_{ij} + A_i \tag{2}$$

where A_i - external input signals; α_i - the signal of the i-th neuron in the output vector of the neural network, α_j - the input signals from other blocks; x_{ij} - the weights of interneuron connections.

For each metric from 1 to 12, its value is calculated. The result class "1" means that in this situation the functional cycle did not fail, the result class "2" - the failure occurred. After obtaining a predictive score using the neurological module, the result is predicted for the following case. Based on the use of the neurologistical module (algorithm) during operation, limit values for 12 metrics were calculated experimentally, calculated on the basis of average statistical data, and using an automated enterprise functioning system (see Table 1).

Table 1. Quantitative metrics.

No p/p	1	2	3	4	5	6	7	8	9	10	11	12
Metric values	0 «0»	10% «0»	3,5 days «0»	85% «1»	4 out of 5 points «0»	7,5% «0»	86% «1»	13,5 days «1»	0 «0»	12% «0»	62% «0»	89% «0»

5 Significance of the Research Results

Methods of digital transformation and logistics engineering, integrated with digital technologies, in the XXI century are developing mainly as the interaction of processes of TL objects at different stages of the life cycle. This approach is global and creates new opportunities for the digital transformation of the architecture of the transport and logistics system in the implementation of the concept of Industry 4.0. The main driving force in global transport systems is the need for highly efficient production and transport processes based not only on digital technologies, but also on methods of logistics engineering focused on the enterprise architecture.

The relevance of the study is also associated with overcoming disintegration in the design and operation of TLS throughout the entire life cycle within the framework of a holistic architecture. The solution of this problem in the scientific and applied plan is of particular importance when conducting analytical work with digital data and IT services. An integrated approach was used as the main method, which allowed us to combine business processes and digital technologies into a single architecture, which form the basis for creating a digital service platform, as well as individual elements of the theory of neural networks. The use of innovative technologies of neural networks with metrics, as a means of measuring and evaluating, made it possible to identify predicted inconsistencies and possible failures at an early stage of their formation.

Currently, architectural constructions of systems are able to control complex objects, the scale of which varies from robots and big data, to intelligent transport and logistics systems and complexes. The described approach leads to the need for introducing fundamentally new models and methods for constructing and monitoring the architecture of digital transport and logistics infrastructure and IT service. The application of an integrated approach contains methodological provisions, system development and recommendations in the field of logistics engineering and digital transformation of TLS, previously applied only in the IT field.

6 Conclusion

Among the key factors of a balanced system of strategic business models and technologies, it is necessary to highlight the management of the full life cycle of both the infrastructure of enterprises and products, services and resources based on the architecture of the enterprise, embedded in system engineering and its applied discipline – logistics engineering.

The use of a standard and modified set of metrics in the monitoring system allows you to compare your own metrics with the best practices of other organizations that apply similar criteria. Implementing the ITIL model allows you to step-by-step evaluate process improvements and determine how much time and at what stages of the life cycle changes are required to achieve a set level of change.

Architectural approach to logistics engineering allows to maintain the functioning and transformation of the entire enterprise system throughout its life cycle at an effective level. It consists of a set of related organizational and technical concepts and principles that combine it service and business objectives. The logistic architecture is presented using methods, including monitoring, corresponding to the thematic groups of descriptions, and includes not only functional architecture based on the GERAM description methodology, but also behavioral and temporal architecture.

One of the main ideas of a proactive life cycle management system is the effective use of a consistent representation of the system and its environment in inherently incompatible organizational and technical systems of the «extended organization» supported by a digital network.

References

1. Trofimenko, Yu., Nekrasov, A., Sinitsyna, A., Atyev, K.: Modeling principles of the digital infrastructure of it services in sustainable low carbon transport systems. Int. J. Eng. Technol. **7**(2.28), 386–389 (2018)
2. Kupriyanovsky, V., Namiot, D., Sinyagov, S.: Cyber-physical systems as the basis of the digital economy. Int. J. Open Inf. Technol. **4**(2), 19–29 (2016)
3. Nekrasov, A., Sokolov, B., Ataev, K.: Life cycle management system (transformation into a digital infrastructure). Publishing House: Technical Polygraph Center, Moscow (2017)
4. Mirotin, L., Nekrasov, A., Gudkov, V.: Increase in the efficiency of freight transportations on the basis of the creation of a stable transport-logistical system of a modular type for high-speed handling and delivery of goods. Monograph: Technical Polygraph Center, Moscow (2013)
5. Sokolov, B., Yusupov, R.: Neo-cybernetics-opportunities and prospects for development. In: Report at the General Plenary Session of the 5th Scientific Conference «Management and Information Technologies» (UIT-2008), St. Petersburg, p. 10. (2008)
6. Shvab, K.: Fourth Industrial Revolution. Publishing house «E», Moscow (2017)
7. Nekrasov, A., Sinitsyna, A.: Logistic engineering as a tool for integrating logistics systems. J. Logist. **12**, 40–45 (2016)
8. Oykhman, E., Popov, E.: Business reengineering. Organizational reengineering and information technology. Finance and Statistics, Moscow (1997)
9. Standard ISO 15704 – 2008: Industrial Automated Systems: Requirements for enterprise standard architectures and methodologies. Standardinform, Moscow (2010)
10. Okhtilev, M., Sokolov, B., Yusupov, R.: Theoretical and technological foundations of the concept of proactive monitoring and management of complex objects. Sci. Tech. Appl. J. «Izvestia SFU. Technical science». Section V. Systems and control points (1), 162–174 (2015). http://izv-tn.tti.sfedu.ru/?page_id=1576. Accessed 09 Oct 2019
11. Yusupov, R., Sokolov, B.: Problems of the development of cybernetics and computer science at the present stage. In: Collection « Cybernetics and Computer Science » , pp. 6–21. Publishing House of SPbSPU, St. Petersburg (2006)

12. Meyer, C., Davis, S.: Living Organization. «Good Book», Moscow (2007)
13. Don Taylor, G.: Systems Engineering Handbooks, Manuals, etc. Logistics Engineering Handbooks, Manuals, etc. New York (2008)
14. ITIL® Service Strategy. https://www.best-management-practice.com. Accessed 09 Oct 2019
15. Van Bon, J., Kemmerling, G., Pondman, D.: IT Service Management: Introduction. IT Expert, Moscow (2003)

The Spread of Supply Chain's Consumers' Emotions as Function of Their Social Network Structure

Mara Pudane[1(✉)], Bernard Brooks[2], and Michael A. Radin[2]

[1] Department of Artificial Intelligence and Systems Engineering, Riga Technical University, Kalku Street 1, Riga 1658, Latvia
mara.pudane@rtu.lv

[2] Rochester Institute of Technology, Rochester, NY 14623, USA
{bpbsma,michael.radin}@rit.edu

Abstract. Modeling and simulation of complex sociotechnical systems such as supply chains and consumer markets include not only strictly the rational but the social components as well. Consumers are impacted by various parameters such as others' opinions, rumors, overall mood of the community etc. Computer simulations are often used to observe how such phenomena form and change in time. The systems theory principles state that the system's dynamics cannot be researched independently from the system's structure, i.e., it is necessary to investigate the media over which the system operates. This leads to the inference that next to the observable behavior of the model, the structural composition must be considered. The paper investigates how different types of network structures impact the dynamics of contagion processes, in particular, the emotional state dynamics. The paper applies graph theory and network metrics to show that the structure parameters have an impact on the dynamics of the contagion.

Keywords: Emotional state dynamics · Sociotechnical systems modeling · Supply chains

1 Introduction

From semantical point of view, complex sociotechnical systems are impacted by many parameters. To run such a system in simulation models, the system is often simplified, otherwise it would be nearly impossible to draw conclusions from the results [1]. The social dimension is one of the first dimensions that is omitted from the models, although multiple sources (see, e.g. [2]) have shown the importance of social aspect on individuals behaviour. The social processes are also related to emotional components such as moods and attitudes [3]: in one word, affects.

A modern market is one of the most obvious examples of sociotechnical systems, yet research in supply chains often ignores the user's needs. Social networks and overall frequency of communication has made the modern market an extremely dynamic place, which means that supply chains also must be adaptable. Modeling social aspects might contribute to one of the most promising directions of supply chain research, namely, the research of optimization of the supply chain.

E. Ginters et al. (Eds.): ICTE ToL 2019, LNITI, pp. 61–68, 2020.
https://doi.org/10.1007/978-3-030-39688-6_9

The primary goal of this paper is to see how different structure of consumer groups might impact the emotional contagion. Research on emotion contagion in crowds has been done by several researchers; in these cases the network is either defined arbitrarily or based on physical proximity [4]. However, it is clear that contagion processes differ depending on the network – this can be observed in different agent-based models, e.g., where the spreading of the disease are impacted by the human interaction [5]. This paper explores how emotion dynamics change based on different group structures and generalizes the results to find a predictor of such a change.

The paper is organized as follows: Sect. 2 describes the related works from the perspective of sociotechnical systems, research on systems structure, as well as the structure formalization. Section 3 renders the model and its implementation, while Sect. 4 provides the description of the experiments and results. Section 5 contains conclusions and future work.

2 Related Works

The related works of this topic are divided into two parts: general modeling of the systems' structure and the consumer modeling in sociotechnical systems.

2.1 Modeling Sociotechnical Systems

The structure of various systems' as well as its impact on the systems' performance and other parameters have been researched in multiple application areas such as organizational theory, information flow management [6] and technical systems analysis [7]. Overall, most of these studies either directly or indirectly belong to the body of research devoted to general systems theory. General systems theory is a basic science which "… deals, on an abstract level, with general properties of systems, regardless of physical form or domain of application" [8]. According to this theory, any system (either physical, organizational, or abstract) consists of the structure, i.e., a static part and processes that may change systems structure over time, i.e., the dynamic part [8]. It is widely accepted that the system's behavior and functioning cannot be explained and researched by understanding just one of these components, instead, it is crucial to analyze the entire system [7, 8]. Multiple approaches have been proposed based on this philosophy. One of such examples is the structural modeling which seeks to formally measure the complexity of technical systems by analyzing the corresponding graphs (see e.g. [7]). In organizational theory, such approaches as Viable Systems Model are aimed at mapping the functionality of organization on the physical structure [6]. A system's analysis, approach widely used in the development of IT systems has also derived from this area [8].

With growing complexity of the technical and socio-technical systems, it becomes increasingly important to find methods and approaches that allow to research complex systems. Systems modeling is one of the methods for systems' analysis, which provides a way to re-play multiple scenarios on a system [8].

Systems' modeling can be performed by using multiple tools and is researched by the area of modeling and simulation. There are multiple modeling approaches that allow to simulate and possibly predict the behavior of the system; these approaches can be widely divided into two groups: equation-based modeling, which is used for fluid dynamics and other technical processes, and the agent-based modeling that is mostly used for simulating complex systems [1]. The main difference between these two approaches lies in the overall way of approaching a problem – the equation-based modeling focuses on creating an umbrella model for the problem, i.e., uses top-down approach. On the contrary, agent-based modeling primarily defines behavior and interactions of individual components in the system, thus implementing bottom-up approach. By applying this method it becomes possible to model heterogeneous elements. Moreover, it enables discovering unintuitive, i.e., emergent phenomena. These are the primary reasons for using the agent-based modeling for simulation of social systems. Since this paper is predominantly concerned with researching social systems, we will use the agent-based modeling.

2.2 Consumer Modeling in Sociotechnical Systems

Due to above mentioned pros, the agent-based models are often used in modeling consumer behaviour. Such models' examples include transportation network modeling [9] which also includes retailers (in this case, the retailer can be considered a customer) demand for the transportation, however, these do not include the interaction between customers' or end-users' impact on the network. Similarly, modeling fluctuations in the e-music market [10] considers the consumer characteristics, but not the interaction between consumers.

Opinion formation and general dynamics of opinion formation is researched in multiple cases; one such example is [11] which analyses such behaviors in an abstract case. [5] examines how policies of vaccination are impacted by social influences of opinion. The rumor propagation is another social phenomena that has been modelled. In [12], a model of negative rumor transmission was developed which can be directly applied to how customers make decisions in a network.

Currently, the market is greatly influenced by social media. The structure of social media is a scale-free graph as confirmed by multiple sources (see e.g. [13, 14]). On the other hand, scale-free structure does not correspond to typical family-based market. These differences lead to the conclusion that structural predictors of social changes in the network would enable better change management within a supply chain.

3 The Simulation Model

The definition of the agent-based modeling in broad terms consists of defining the single agent's behavior (inspired by [15], we call it micro-level) and defining the interaction among the agents (i.e., macrolevel). Defining interactions among the agents includes two parts; defining with whom the agents will interact and second of all, how they interact.

3.1 Micro-level

The emotions semantically differ among themselves based on various parameters that can be found in the psychology research such as arousal, positivity, energy etc. A formalization is needed to implement these parameters in a computer program. This is not a simple task, especially, if one wants the model to correspond to actual emotions. Anger due to its visible consequences is an emotion that is much researched in psychology thus allowing defining the needed parameters for the model. More details regarding semantic grounds for parameter calculation can be found in [16].

Overall, the agents' general behaviour consists of getting anger from other agents, processing it and expressing the anger again to the connected agents, thus inducing them in other agents. Within the single agent intensity of anger dynamics is calculated based on three intertwining functions. There is an activation function that determines the relationship between the objective irritation strength and the subjective anger intensity level, a decay function that determines how fast emotion will pass and finally, the threshold function that determines if the agent will "catch" emotion from other agent. The whole agent model in detail is published in [16].

3.2 Macro-level: How the Agents Interact

In agent-based models, the agent interaction design consists of two parts: (1) deciding with whom they will interact and (2) deciding how they will interact. The "how" part is semantically related to emotion mirroring, i.e., people tend to mirror emotions of those around them, thus "catching" emotions [17].

Regarding the agent interaction frequency, Poisson distribution was chosen [18]. This discrete probability distribution allows the calculation of the probability with which an event will happen within the given time span. Since the agent's affective dynamics are running in real-time, it is not needed to split seconds and thus the discrete distribution works well. At any interaction time, agent passes his emotional state (in this case, anger) to other agents which receive it in their input.

Once the network is created, one node is chosen as the initial flash point, that person is annoyed by an outside stimulus. The network is directed so that the interaction among two people only happens one way – this would correspond to a single interaction among two people as opposed to conversation.

3.3 Macro-level: With Whom the Agents Interact

There are multiple ways to formalize the systems' structure such as class diagrams, ERD diagrams or concept maps. What all these examples have in common, they are all based on graphs. Graph theory allows modeling systems' structure as relations among the elements. We commenced with several highly regular structures: a ribbon graph, family graph and toroidal graph (see Fig. 1). Each of the graph types has equal number of nodes and edges as well as local degrees of the nodes.

The chosen regular networks all have the property that every node is structurally equivalent, that is, the nodes all have the same degree, same clustering coefficient, same average path length and are connected to each other in the same structure. Thus the

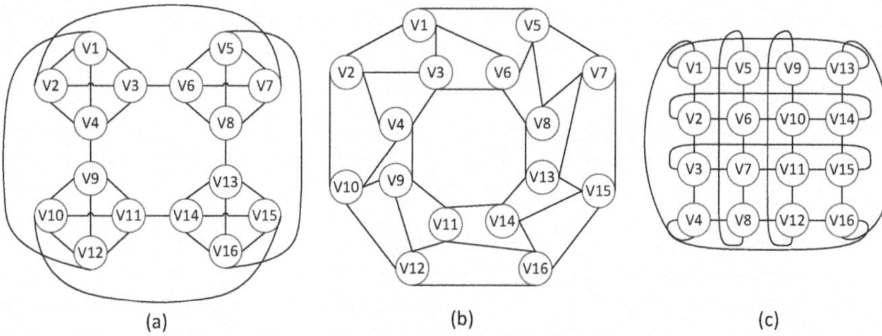

Fig. 1. (a) a family graph; (b) a ribbon graph; (c) a toroidal graph. All three graphs gave the same number of nodes and edges, but are structured differently.

experiment can choose any of the nodes to be the initial infected node and the results will be identical.

Since the human group will most likely not be that regular, we did the process of rewiring of each graph. The rewiring was done in a way that each edge had a varying probability of being changed. This means that at each step the graphs were becoming less regular and increasingly random. For example, if each edge had a 0.1 probability of being rewired, we say that 10% of the graph is rewired.

As the rewiring is random, 10 instances of each graph in each degree of rewiring was generated (at 10%, 20%, etc.), in total creating a data set of 303 graph instances.

4 The Experiments on the Created Structures

The experiment consists of two parts: producing the graph structure and performing the emotion dynamics' measurements on these structures.

4.1 The Measurement of the Experiments

Graph theory methods enable calculating different types of measurements to obtain the complexity of the system. During the experiments, four measures were initially chosen: degree distribution, average clustering coefficient, as well as two measures based on the path lengths in the graph: closeness and betweenness, as taken from [19].

Degree distribution is a number of connections that the vertices have. However, this measurement was dropped since it did not differ for various structures, i.e., it could not be used to characterize any of the three structures.

The clustering coefficient of an individual measures the proportion of existing connections in comparison to the total possible connections between their neighbors that might exist. A person with 4 friends would have 6 possible connections between those 4 friends. If only 2 of those possible 6 connections existed that person would have a clustering coefficient of 1/3. Clustering coefficient is a measure of how tight or cliquish a group of people are and contagions often have trouble penetrating groups with high clustering coefficients [20].

It is hypothesized that networks that have a high average path length will have a slower spread of the contagion. The figure shows all three measurements over rewiring for the ribbon graph. All the measurements for different structures can be seen in Fig. 2.

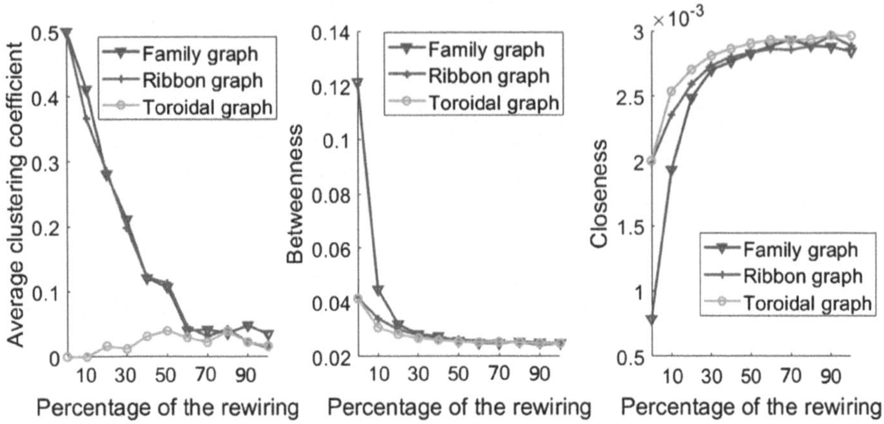

Fig. 2. The change of the graph measurements based on various percentage of rewiring.

4.2 The Impact of the Structure on Emotion Dynamics

Two measures were calculated based on the simulation results:

- Maximum average intensity of the model;
- The moment at which the highest emotion was achieved.

The first measure corresponds to the emotion level which the consumer market can acquire and works as a predictor of how strong the consequences of events leading to emotional states will be. The moment at which the highest emotion is achieved shows the delay in the network and depicts how soon the consequences of events can be expected.

On each of the structures a simulation was run and the mean for each percentage of rewiring was calculated. It is interesting to note that only after about 20%–25% rewiring of the graphs, the graphs lose their properties. It was determined by observing the results of dynamics simulation that the best predictor for the delay of dynamics in the graph is betweenness – it means that the larger the betweenness, then the higher emotion will be achieved.

On the other hand, for the delay in the network, i.e., the time when the highest emotion will be achieved, the best predictor is inverted closeness, i.e., the further the paths in the network, the longer it takes for the emotion to achieve the highest point (Fig. 3).

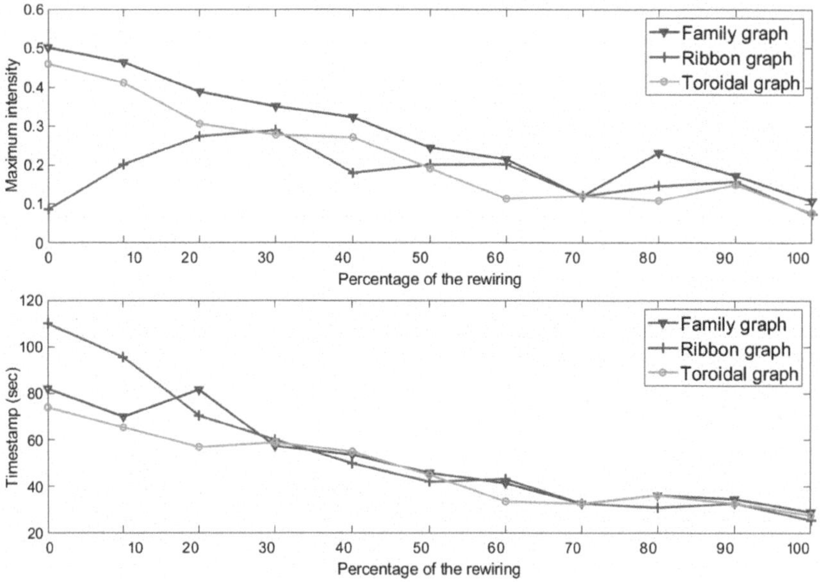

Fig. 3. The results of emotion dynamics in the graph structures.

5 Conclusion

The paper examines the impact of emotion dynamics on various systems' structures. It was concluded that the best predictor of the emotion dynamics are characteristics that are associated with paths, in particular, closeness and betweenness.

As in the scale-free graphs the graph path measurements depend on the size of the network [21], and increase logarithmically as the size of the network grows. It can be concluded that the consumer emotion dynamics matter much more for those suppliers whose customers' average path length is shorter. On the other hand, for those suppliers, whose network is large, it may take a while for the changes to take effect. Note that this is a directed graph, so in the future it might be useful to obtain the results from non-directed graph which would correspond to conversing, as well as simulating various event effect mitigating strategies.

Acknowledgement. This work has been supported by the European Council Seventh Framework Program FLAG-ERA project "Rethinking Robotics for the Robot Companion of the Future" (RoboCom++).

References

1. Helbing, D., Balietti, S.: How to do agent-based simulations in the future: from modeling social mechanisms to emergent phenomena and interactive systems design. In: Social Self-Organization, pp. 25–70. Springer, Berlin (2012)

2. Altaweela, M., Alessab, L.N., Kliskeyb, A.D.: Social influence and decision-making: evaluating agent networks in village responses to change in freshwater. J. Artif. Soc. Soc. Simul. **13**(1), 15 (2010)
3. van Kleef, G.A.: The emerging view of emotion as social information. Soc. Pers. Psychol. Compass **4**(5), 331–343 (2010)
4. Bosse, T., Duell, R., Memon, Z.A., et al.: Agent-based modeling of emotion contagion in groups. Cogn. Comput. **7**, 111 (2015)
5. Xue, J., et al.: Simulation analysis of immunization policy diffusion in social network with ABM approach. In: Proceedings of 2016 International Conference on Behavioral, Economic, Socio-Cultural Computing (2017)
6. Beer, S.: Diagnosing the Systems for Organizations. Wiley, Hoboken (1985)
7. Zeltmate, I.: Development of intelligent system for structural modelling of complex systems. Ph.D. thesis. Rīga, RTU, (2012)
8. Skyttner, L.: General Systems Theory: An Introduction. Macmillan Education, London (1996)
9. Sopha, B.M., Siagian, A., Asih, A.M.S.: Simulating dynamic vehicle routing problem using agent-based modeling and simulation. In: Processing of 2016 IEEE International Conference on Industrial Engineering and Engineering Management (2016)
10. Schuster, S., Gilbert, N.: Agent based simulation for modelling the distribution of online music. In: First International Conference on Automated Production of Cross Media Content for Multi-Channel Distribution (AXMEDIS 2005) (2005)
11. Ono, K., Harao, M.: Multi-agent based modeling and simulation of consensus for mations in arguments. Int. J. Agent-Oriented Softw. Eng. **1**(2), 193–204 (2007)
12. Brooks, B., Difonzo, N., Ross, D.: The GBN-dialogue model of outgroup-negative rumor transmission: group membership, belief, and novelty. Nonlinear Dyn. Psychol. Life Sci. **17**, 269–293 (2013)
13. Aparicio, S., Villazon, J., Alvarez, G.: A model for scale-free networks: application to Twitter. Entropy **17**, 5848–5867 (2015)
14. Barabasi, A.L., Albert, R.: Emergence of scaling in random networks. Science **286**, 509–512 (1999)
15. Wooldridge, M.: An Introduction to MultiAgent Systems. Wiley, Chichester (2009)
16. Pudāne, M., Brooks, B., Houston, R., Radin, M.: Agent based model of anger contagion and its correlations with personality and interaction frequency. Int. J. Educ. Inf. Technol. **12**, 7–12 (2018)
17. Barsade, S.G.: The ripple effect: emotional contagion and its influence on group behavior. Adm. Sci. Q. **47**, 644–675 (2002)
18. Ahrens, J.H., Dieter, U.: Computer generation of poisson deviates from modified normal distributions. ACM Trans. Math. Softw. **8**(2), 163–179 (1982)
19. Harris, J.M., et al.: Combinatorics and Graph Theory, vol. 2. Springer, New York (2008)
20. Brooks, B., Shahmohamad, H.: Simulating the evolution of strategic business alliances with a dynamic graph theoretical Monte Carlo model. Far East J. Appl. Math. **63**, 15–23 (2012)
21. Cohen, R., Havlin, S.: Scale-free networks are ultrasmall. Phys. Rev. Lett. **90**(5), 058701 (2003)

Roadside Services Facilities Planning Approach for Public Roads in the Republic of Kazakhstan

Timur Sultanov[1(✉)], Tynys Suleimenov[1], Gulnar Tlepiyeva[1], and Abdy Kiyalbaev[2]

[1] Eurasian National University, Satpaeva Street 2, 010008 Nur-Sultan, Kazakhstan
rumit78@mail.ru, STB2007@yandex.ru, tlepievagm@yandex.ru

[2] L.B. Goncharov Kazakh Road Academy, Rayymbek Street 415, 050061 Almaty, Kazakhstan
abdi-ki@mail.ru

Abstract. The content of the article is aimed at planning the location of service facilities on a public roads on the international transit corridors of Kazakhstan. The basic principles of creating a roadside infrastructure with the growth of motorization, freight and passenger traffic in transport are analyzed. The growth of transit traffic increases the efficiency of using the reserves of transportation capacity of national transport systems, stimulates their reproduction and improvement. The consequence of the growth of transit traffic can increase the budget revenues, revenues of transport organizations and, accordingly, their effective development. The globalization of the economy and the accompanying processes of development of foreign economic relations require a new approach to the development of transport, the redistribution of freight and passenger flows. The growth of foreign trade requires a qualitative revision of the share of the transport component, but typical approaches to the formation this component must be revised.

Keywords: Transport logistics · Roadside infrastructure · Service integrator

1 Introduction

The aim of the study is to develop a methodology for selecting types of roadside service facilities on the roadside strip of public roads, taking into account the views of road users. Today, the quality of customer services is one of the most important factors of consumer choice. The good ratio of price and quality of service inures the basis of constant demand, and, this, ensures profit and a stable position of the company on the market. The considering the service activities of the company, special attention is paid to factors shaping consumer interest, as well as to the main aspects of the service culture.

In the conditions of the Republic of Kazakhstan, in order to achieve the required effect, it is necessary to solve a whole range of tasks to increase the level of transport

E. Ginters et al. (Eds.): ICTE ToL 2019, LNITI, pp. 69–74, 2020.
https://doi.org/10.1007/978-3-030-39688-6_10

services provided to cargo owners and passengers. The importance and relevance of services is confirmed by the specific role of this field in the GDP of developed countries, which accounts to 70–80%. Therefore, transport services, both in goods and passengers transportation, are related to material services. It is logical that their main volume accounts to international transit roads [1].

The problem of roadside infrastructure is the funding this type of activity in the transport logistics system. Currently, in Kazakhstan, financing of the construction of infrastructure facilities for road transport and the road industry is carried out both at the expense of state budgetary allocations and at the expense of the private sector.

The concept of development of the road sector of the Republic of Kazakhstan for 2001–2008, approved by the Decree of the Government of the Republic of Kazakhstan dated May 29, 2001 No. 726 and the Law of the Republic of Kazakhstan "On Public-Private Partnership" dated 321.10.2015 No. 379-V, brought to the country's road sector, states the mission to develop a new program for the development of the infrastructure complex through two sources of financing. For objects included in the road complex - at the expense of the state budget, and for the second group of objects - at the expense of private entrepreneurs [2].

2 Roadside Infrastructure Planning Approach

2.1 Business Integrator

Based on the foregoing, the organizational concept of the technical level provides for the appropriate location of service facilities on the roadside of public roads. In the functional and supporting parts of this system, the main role played by the business integrator. At that, the ultimate goal of creating business processes is to increase the road safety and the quality of business processes of the integrated service for drivers and passengers taking business and tourist trips by ergonomically positioning service facilities on the roadside zone and making innovative organizational and economic decisions in the research area.

In the theory of economics, management is the conscious influence of a person on objects and processes, as well as on the participation of people in them, carried out with the aim of giving a certain direction to economic activity and obtaining the desired results. In the twentieth century, a science was formed that studies the general principles of management, regardless of where and in relation to which objects management takes place. The part of cybernetics that studies management processes in the transport economy is called transport and economic cybernetics [3].

Thus, when creating service facilities on the roadside territory of international transit corridors, the business integrator has three main objectives:

- Economic analysis, which consists of an initial study, research of economic and transport processes in the area of influence of the studied road, the conditions for their passage in retrospect, in the past period, in the establishment of stable trends, the identification of problems. This analysis precedes the development of control decisions and impacts, serves as their starting point and justification;

- Forecasting, i.e. events, hypotheses, scenarios, models of economic processes that may take place in the future. At the same time, before formation, the business integrator comprehensively studies the processes of goods and passenger traffic on this road, organizes analyses to study the technical and transport and operational indicators of the road separately for economic distances, collects information on weather, climate, relief, historical and resort places etc.;
- According to the information gathered, the activities engaged in a roadside service business are planned. Planning is a development process, building a plan, a way of future actions, determining the path of development of an economic entity.

2.2 The Methodology

The methodological basement is following:

- A systematic approach to select parameters of service objects;
- The basic principles of the ISO 2001 standard, considering the location of roadside service facilities as an integrator of private business, i.e. systems of business processes in the framework of public-private partnerships;
- A project-integrated approach that considers the business processes of roadside service management as a set of integration vectors in horizontal and vertical projections at the conceptual-organizational, technological and financial levels;
- An organizational concept of a technical level, providing the optimal location of service facilities on the roadside of public roads, which plays the role of a business integrator in the functional and supporting parts of the system.

The ultimate goal of research is to increase the safety of transportation and the quality of business processes for the comprehensive service of drivers and passengers making business and leisure trips by ergonomically positioning service facilities on the roadside zone and making innovative organizational and economic decisions in the field of research.

The innovation lies in the fact that the hypothesis of an integrated project approach to the development of business processes as an integrated quality management system (IMSC) based on the ISO 2001 standard has been put forward and investigated. The integration process is considered on the example of roadside infrastructure on a national highway “Almaty – Ust-Kamenogorsk”. The results of the processes when compared may be equal, may have minor differences or completely differ. If the processes are combined according to the attribute “result”, then the inputs to the processes of different subsystems should be the same. The functional diagram of the process is developed on the basis of the requirements of standards in the field of management, therefore, it is necessary to compare the requirements of international standards. During combining processes, the attribute “mechanism” is also compared, in particular the competence of the human resources involved in the process. Based on literature analysis, the following method is selected to identify 8 homogeneous processes and to develop a list of processes for each of the subsystems of the IMSC. The following processes are the infrastructure components of roadside service management:

- Process 1. The process of medical care for passengers and drivers;
- Process 2. The process of maintenance and repair of automobiles;
- Process 3. The process of providing fuel and lubricants;
- Process 4. The process of organizing food points;
- Process 5. The process of organizing the rest of drivers and passengers;
- Process 6. The process of information support;
- Process 7. The process of evacuation support of vehicles;
- Process 8. The process of training (personnel management).

To collect the service types in one complex, a homogeneity matrix of these processes is provided, combining the services into one object. Thus, the structural model of a business integrator in transport logistics is seen as a complex service business. As a result of the comparison of processes from different subsystems, a matrix of process homogeneity is constructed (see Table 1). The intersection of processes from different subsystems is indicated in the matrix where, based on the three presented selection rules, a positive result is obtained [4].

Table 1. Process homogeneity matrix.

Subsystem 1 / Subsystem 2	Process 1	Process 2	Process 3	Process 4	…	Process *n*
Process 1						
Process 2						
Process 3						
…..						
Process m						

In that way, a list of processes of an integrated service management system is developed, which includes heterogeneous processes (types of services) from different subsystems, as well as homogeneous processes [5]. The processes of the second level should include roadside trade in consumer goods, etc. Vertical integration is supposed and to be considered as the sum of heterogeneous processes, united at the organizational, legal, financial and technological levels. The processes listed above, in accordance with the recommendations of [6], can hypothetically be attributed to the main processes, which in their development will be formed into horizontally integrated processes (HIP). HIPs are processes that are technologically homogeneous throughout the investigated transport route. In the first version (see Fig. 1), the following processes are included in the roadside service complex under consideration: “Process 1-1” provides fuel and lubricants at 2 points; The “2-2 Process” also provides technical services at 2 points; “Process 3-3” - 3 power points; “Process 4-4 - 4 minimarkets; “Process 5-5” - 1 medical center; “Process 6-6” - no services; “Process 7-7” - no services; “Process

8-8" - 1 point for the rest of drivers and passengers. The completed services are deployed in another embodiment (see Fig. 2). For example, for the "Process 1-1" the number of fuel and lubrication points is 3; for "Process 2-2" the number of maintenance points is 1; for "Process 3-3" the number of points of food points - 2; for "Process 4-4" the number of minimarket points is 3, but for other processes, i.e. 4-4, 6-6, 7-7 and 8-8 the number of points of provision with a medical center, information, evacuation and organization of rest for drivers and passengers is 1 unit for each.

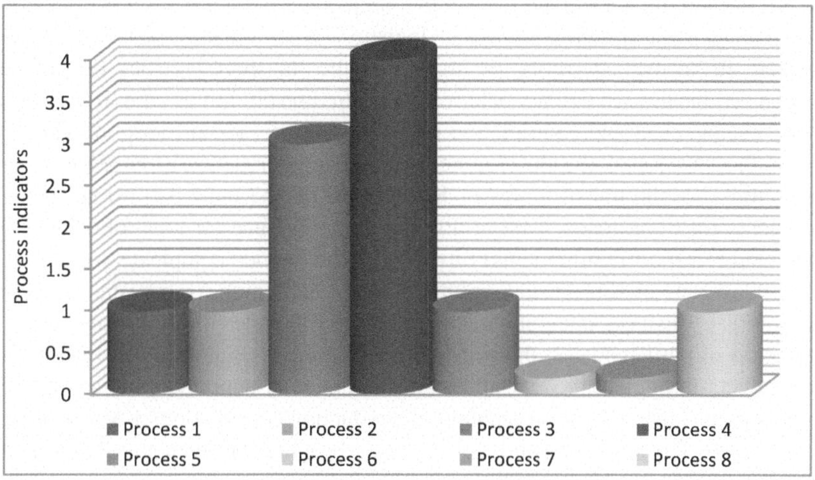

Fig. 1. The designed road section without accounting support and an evacuation point.

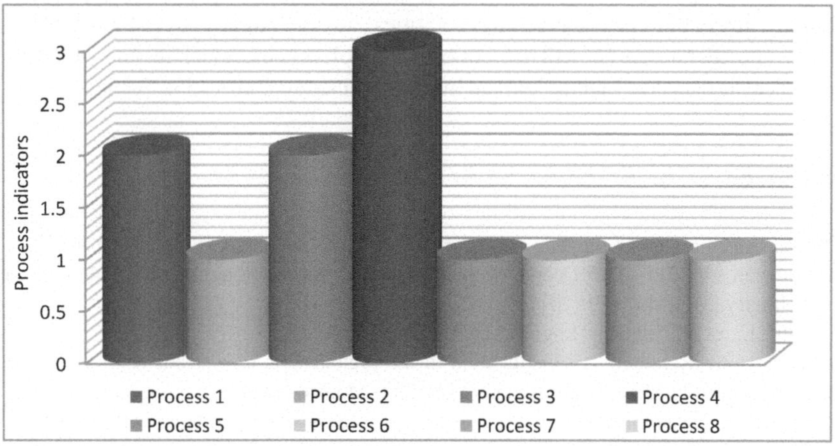

Fig. 2. The designed road section with a complete set of business processes.

3 Conclusion

The considered options for service processes are considered at a specific address of the studied road. This makes it possible to accurately assess in terms of optimizing services on a given road. Thus, the task for the business integrator is to compare profitability and clarify the correspondence of the offered service packages to the demand of road users. The initial data includes indicators obtained on field studies and correspond to the consumer demand of road users (drivers, passengers, shippers, residents of highway areas, etc.). These data adequately vary the completeness of each service object in the service management system of transport logistics. When creating service facilities in the system of roadside territory of highways, the transport logistics of small and medium-sized businesses will not achieve the desired result. In this case, it is necessary to create management and coordination of business processes in transport. This is the basis for the success of the development of transport logistics in creating roadside infrastructure. So, the transit routes of Kazakhstan for the most part do not meet the requirements for trunk roads. First of all, this is the absence (or non-compliance with international transport requirements) of modern road service complexes located in the near-highway zone of Kazakhstan's public roads, functioning for convenient and safe traffic flow [7]. Business integrator requirements set includes the following functions: coordination of participants in the transport chain in overcoming joints; creating synergistic effects through cargo consolidation; commercial support of settlements with customers in the business plans preparation; conducting consultations when performing work related to the implementation of the state program between the Public-Private Partnership and the protection of the interests of the parties, etc.

The results of the work are designed for entrepreneurs engaged in services, in determining the type of service objects and choosing their appropriate location on the roadside of roads. The calculation results were used to draw up a feasibility study for designing the deployment of service facilities on the Almaty – Ust-Kamenogorsk highway. Further research will be aimed to involving the mechanisms of artificial intelligence in decision making enhancing the efficiency and responsiveness.

References

1. On the strategic plan of the Ministry of Transport and Communications of the Republic of Kazakhstan for 2011–2015 Decree of the Government of the Republic of Kazakhstan dated 11 February 2011, no. 129 (2011)
2. Mirotin, L., et al.: Efficient Logistics, p. 311, Moscow (2017)
3. Kiyalbaev, A., Kiyalbay, A.: Justification of the organization of roadside service infrastructure. In: Scientific Monograph, p. 232. MAADO, Moscow (2017)
4. Reizberg, B.: Economics course. In: Textbook, p. 720. INFRA-M, Moscow (2015)
5. Kolinichenko, B., et al.: Optimization of complex technical systems and processes. In: Tutorial, p. 100. MADI, Moscow (2014)
6. Sultanov, T., et al.: Development of transit potential in conditions of integration of the Republic of Kazakhstan into the world transport system. Procedia Comput. Sci. **149**, 430–435 (2019)
7. Smirnov, S.: Transport complex of Kazakhstan: state, problems, opportunities. Transit Econ. **3**, 10–15 (2014)

Development of a Train Schedule in Terms of Implementing the Transport Strategy

Olena Malakhova and Grygorii Sikonenko(✉)

Ukrainian State University of Railway Transport, Feuerbach Sq. 7, Kharkiv 610050, Ukraine
{malakhova, sikonenko}@kart.edu.ua, gregsik79@gmail.com

Abstract. In accordance with the Transport Strategy of Ukraine for the period up to 2030 and the State Target Program for the Reform of the Railway Transport for 2010–2019 it is necessary to implement the Association of Agreement between Ukraine, on the one hand, and the European Union, the European Atomic Energy Community and their member states, on the other hand. In this regard, the Ministry of Infrastructure has developed plans for the implementation of some acts of the EU legislation on rail transport, including the implementation of Directive 91/440/EU and 2001/14/EU and their respective roadmaps. These acts provide separation in the management of infrastructure and the organization of transportation on the basis of non-discriminatory access. To develop train schedules taking into account practical throughput, it is advisable to apply the principles of risk management theory.

Keywords: Railway Transport · Strategic planning · Risk management

1 Introduction

In modern conditions of the passage of trains on a schedule (exact observance of technological standards) on Ukrainian railways, some reasons interfere. Thus, the normative schedule is an idealized model with minimization of intervals on the stations and between trains, section travel time. This affects the accuracy and reliability of the schedules negatively.

To improve the quality of the schedule fulfillment, supplement of the technological standards with a reserve of time is possible. The reserve is understood as [1] the amount of time for which the train movement process can be delayed in accordance with the schedule without consequences for the following steps in the sequence of execution of the train schedule. This time interval can be used in situations when there is a risk of train delay relative to the time set by the schedule.

In addition, the operating time of the movement is calculated on the basis of the traction characteristics of the locomotives. However, the physical wear of part of the train locomotives does not allow following section travel time. This affects the section speed negatively and leads to the creation of primary delays.

Organizational reasons include erroneous actions of employees of various services, the inability to receive trains at the station in a timely manner, and failures in the movement of the following trains along the section (secondary delays).

E. Ginters et al. (Eds.): ICTE ToL 2019, LNITI, pp. 75–81, 2020.
https://doi.org/10.1007/978-3-030-39688-6_11

European scientists prove the advisability of using buffer time in their studies. In particular, Schittenhelm [2] in his studies describes the use of buffer time in train traffic using various planning strategies on the Danish railways, Kroon, Dekker [3] in their work analyzed train delays on Dutch, Swedish and German railways and defined threshold values for delays for various categories of trains in these countries. Researches done by Cerreto, Nielsen, Harrod [4, 5] are devoted to the analysis of train delays on Danish railways, their division by type and the analyses of evolution of train delays. All studies are aimed at improving the quality of scheduling trains, reducing the time for the transportation of passengers and goods and the efficient use of rolling stock. Regulatory documents do not provide for buffering (compensation) time on the railways of Ukraine when trains move along the section [6]. However, the studies of European scientists cannot be completely applied for use on the railways of Ukraine, since most of them are focused on the use of trains that are not used on the railways of Ukraine when constructing cyclic schedules.

This work is aimed at improving the construction of train schedules taking into account the risk of conflict situations, as well as reducing the duration of primary and secondary train delays.

This work is aimed at improving the construction of train movement schedules taking into account the risk of conflict situations, as well as reducing the duration of primary and secondary delays of the trains.

2 Problem Statement

To understand the theory of train delays in a railway section, it is important to determine the main components in the actual time of train movement in the section. The actual time of train movement in the section consists of scheduled time and delays (see Fig. 1).

Scheduled travel times may consist of:

- Time of the train section;
- Parking time at stations (technical operations, boarding/disembarking of passengers);
- Time reserve.

The main travel time is defined as the minimum time required to complete this section on the basis of the technical characteristics of the rolling stock and the maximum permissible speed. Parking time for technical operations is allocated for overtaking, crossing, boarding/disembarking passengers. The reserve (compensation time) is defined as the period of time added to the main travel time or time associated with the movement. This is the time necessary to restore the movement of trains on a schedule, which helps to prevent the delays of train movement in the section. Its duration may vary between trains. The amount of reserves is established by current regulations.

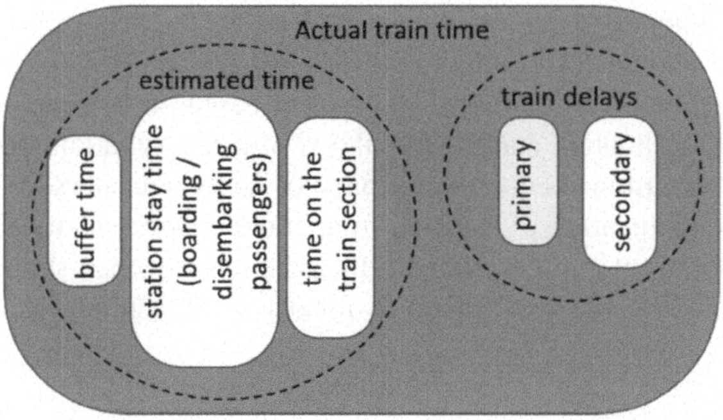

Fig. 1. Time of train movement in the section.

On EU railways, the time lost is defined as the difference between the main travel time and the run time in the section. In lost time, delay time, buffer time and station stay time are taken into account.

Failure in the train schedule - errors, malfunctions or conditions within the railway network that affect the normal movement of trains, in which deviations of the train time from the standard occur.

Failures can occur both due to external factors, as well as due to malfunctions of individual units.

Delays caused by failures are called major delays. Secondary delays occur for trains departing after the main delay ("domino effect"). For the site as a whole, a significant portion of the delays is secondary. The total time of the secondary delays for the railway section can be reduced by including buffer time in the main train time.

3 The Method of Risk Assessment in Schedule Development

When developing a train schedule, attention should be paid to emerging conflict situations, and as a result, the emerging risks of schedule failure. In risk management, it is customary to distinguish several key stages:

- Identification of risk and assessment of the likelihood of its implementation and the extent of the consequences, determination of the maximum possible loss;
- The choice of methods and tools for managing identified risks;
- Development of a risk strategy in order to reduce the likelihood of risk implementation and minimize possible negative consequences;
- Implementation of the risk strategy;
- Assessment of the results achieved and adjustment of the risk strategy.

A key stage in risk management is considered to be the stage of choosing methods and tools for risk management.

An important step in the risk management process is the assessment of the magnitude of the risk, which is compared with estimates of the magnitude of other risks to determine the priority of risk management measures, as well as to allocate resources.

The greatest amount of allocated resources or the highest priority of risk management measures should correspond to the risks with the highest value estimate. Risks should be assessed using a qualitative assessment based on two parameters: the consequence of the risk situation and the likelihood of a risk situation (see Table 1). The consequences are determined either on the basis of safety indicators, process efficiency, or on the basis of indicators of damage from traffic safety violations. If the consequences of a risky event can be different, then the consequence should be used with maximum severity. The risk assessment R_z for one event can be defined as:

$$R_z = P_{RS}\gamma_z \tag{1}$$

where P_{RS} - assessment of the probability of a risk event; γ_z - graph fill factor (see Table 1).

Risks located in the red zone (on the right) ($R_z \geq 2,5$) high risk require immediate management. These high-risk risks should be reduced, at least, to the level of medium and low risk by introducing control measures or taking safety measures in emergency situations (corrective measures). Risks located in the medium and low risk zone should be manageable, and measures to manage these risks should be constantly monitored for their effectiveness. Risks located in the zone of low risk - the green zone (on the left) - ($R_z < 0,5$), are considered acceptable and do not require additional control measures.

Table 1. Risk matrix.

		Fill factor				
Probability appear	High (5)	0,5	1,25	2,5	3,75	4,5
	Medium (4)	0,4	1	2	3	3,6
	Probable (3)	0,3	0,75	1,5	2,25	2,7
	Unlikely (2)	0,2	0,5	1	1,5	1,8
	Almost unbelievable (1)	0,1	0,25	0,5	0,75	0,9
		Minor (0,1)	Moderate (0,25)	Significant (0,5)	high (0,75)	Critical (0,9)

As an evaluation of probable damage, you can use the value of the time of non-fulfillment of the traffic schedule for primary and secondary delays. As an indicator of frequency, mathematical expectation is used (see Table 2). The damage is calculated as:

$$U = \sum\nolimits_{t_o}^{t} t_z \Delta\beta_x \tag{2}$$

where t_z – is the total time of stops in the movement schedule for primary and secondary delays; $\Delta\beta_x$ - change in the coefficient of the speed on the section.

In this case, the coefficient of the section speed with a non-packet schedule is calculated by the formula [7]:

$$\beta_x = 1 - \frac{(0.75 + 0.3\gamma_z)(N_f + 1.2\Delta N_p)t_s + (1 - 0.7\Delta)N_p t_0}{1440}, \tag{3}$$

where N_f – is the number of freight trains; Δ – is the ratio of the travel time of a pair of passenger trains to the travel time of a pair of freight trains; N_p – the number of passenger trains; t_s – is the time of crossing; t_0 – is the time of overtaking.

Table 2. Mathematical expectation of conflict emergence.

Probability of occurrence	Description	Mathematical expectation of conflict emergence
1	It will happen no more than once a year	0,05
2	May occur once a month	0,15
3	May occur several times per month	0,3
4	May occur several times a day	0,4
5	Expected several times a day	0,8

To assess risks when developing traffic schedules for the whole site and for direction, taking into account operational circumstances, the application of this approach may be insufficient, since risk has to be taken into account only as a static parameter [8]. Thus, the risks of fulfilling the train schedule for the direction are a time integral with a changing upper limit on the generalized risk function, which is the sum of the current risk functions of each individual train situation.

$$\varepsilon(t) = \int_{t_o}^{t} R_Z(t)dt, \tag{4}$$

where t_0 - is the start time of the primary delay; $R_z(t)$ - is a generalized risk function that takes into account current risks operating on the site at time t.

The total delay time, taking into account the risks, probability of the event and the fill factor of the graph, is shown in Fig. 2.

As can be seen from the Fig. 2, when the boundary values of risk and fill factor, which correspond to the red zone of Table 1, are reached, the total delays increase exponentially. Therefore, it is necessary to apply all available operational measures to prevent the achievement of boundary values.

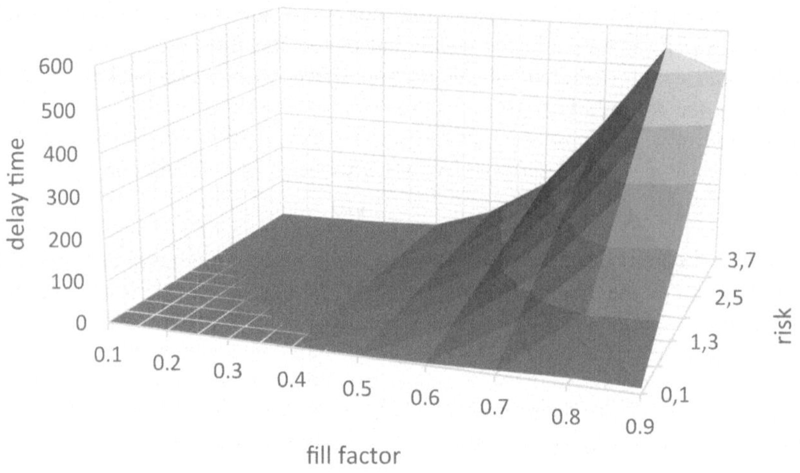

Fig. 2. The total delay time of trains on the section.

4 Conclusion

The proposed methodology makes it possible to estimate the total time of train delays depending on the coefficient of filling in the train movement schedule based on the risk value for the given coefficient of filling.

Assessment of the risk depending on the coefficient of filling in the train movement schedule can take values from 0.1 to 4.5. The risk value indicates the necessity of taking operational measures to influence the situation. A high value of risk is considered to be >2.5, which is indicative of taking immediate management measures. Otherwise, the train delay time in the event of conflict situations will increase with a geometric progression.

Since the value of the risk should be reassessed after the introduction of control measures, situations are possible in which the value of the risk does not change. This does not mean that management measures should not be implemented. In order to manage risk, it is necessary in any case to test the adoption and application of all possible management measures.

A risk management plan is necessary for the effective implementation of management measures. A risk management plan should contain mechanisms for monitoring and the proper functioning of management measures. Responsibility for management measures should be defined along with requirements for monitoring and evaluating the effectiveness of risk management measures.

The proposed methodology can be realized as a software product for the automated workplace of the train dispatcher in decision support systems.

References

1. Prokhorchenko, H., Scherbatska, A., Tkachuk, M.: Research effect of the reserve time reliability hight speed train schedule. In: 2017 USURT, pp. 205–213, Kharkov (2017)
2. Schittenhelm, B.: Planning with timetable supplements in railway timetables. Review **1** (2011). http://www.trafikdage.dk/papers_2011/63_BerndSchittenhelm.pdf. Accessed 21 Aug 2019
3. Kroon, L.G., Dekker, R., Vromans, M.J.: Cyclic railway timetabling: a stochastic optimization approach. In: 2007 Railway Optimization, pp. 41–66 (2007)
4. Cerreto, F., Nielsen, O.A., Harrod, S., Nielsen, B.F.: Causal analysis of railway running delays. Review **1** (2016). https://orbit.dtu.dk/files/125784518/790.pdf. Accessed 18 Aug 2019
5. Cerreto, F.: Micro-simulation based analysis of railway lines robustness. Review **6** (2015). fhttp://orbit.dtu.dk/fedora/objects/orbit:140777/datastreams/file_112408001/content. Accessed 15 Aug 2019
6. Instructions for determining station and inter-train intervals. Transport of Ukraine (2001)
7. Abramov, A.: Train Movement Schedule and Accessibility, 2nd edn. Russian State Open Technical University of Railway Engineering (2012)
8. Butko, T., Prokhorov, V., Chekhunov, D.: Forming of model of risk management at the switchyard station at operation railcar with dangerous goods on the basis of risk exposure. In: Proceedings of the International Scientific and Practical Conference Globalization of Scientific and Educational Space. Innovations of Transport, pp. 14–16 (2018)

Visual Analytics in Mobility, Transportation and Logistics

Kawa Nazemi[1(✉)], Dirk Burkhardt[1], Lukas Kaupp[1], Till Dannewald[2], Matthias Kowald[2], and Egils Ginters[3]

[1] Darmstadt University of Applied Sciences, Haardtring 100, 64295 Darmstadt, Germany
{kawa.nazemi,dirk.burkhardt,lukas.kaupp}@h-da.de
[2] Wiesbaden University of Applied Sciences, Kurt-Schumacher-Ring 18, 65197 Wiesbaden, Germany
{till.dannewald,matthias.kowald}@hs-rm.de
[3] Riga Technical University, Kalku Street 1, Riga 1658, Latvia
egils.ginters@rtu.lv

Abstract. Mobility, transportation and logistics are more and more influenced by a variety of indicators such as new technological developments, ecological and economic changes, political decisions and in particular humans' mobility behavior. These indicators will lead to massive changes in our daily live with regards to mobility, transportation and logistics. New technologies will lead to a different mobility behavior with new constraints. These changes in mobility behavior and logistics require analytical systems to forecast the required information and probably appearing changes. These systems have to consider different perspectives and employ multiple indicators. Visual Analytics provides both, the analytical approaches by including machine learning approaches and interactive visualizations to enable such analytical tasks. In this paper the main indicators for Visual Analytics in the domain of mobility transportation and logistics are discussed and followed by exemplary case studies to illustrate the advantages of such systems. The examples are aimed to demonstrate the benefits of Visual Analytics in mobility.

Keywords: Visual Analytics · Mobility behavior · Data analytics

1 Introduction

Mobility, transportation and logistics provide a variety of indicators for analytics systems. These indicators include issues for analyzing the infrastructure, technological development, mobility behavior and many more aspects. Visual Analytics couples machine learning methods to interactive visualizations and thus provides a good way to handle mass and heterogenous data, predict future scenarios or detect patterns. The analytical capabilities of Visual Analytics are due to the mixed usage of machine and human intelligence much greater than exclusive machine learning approaches, but still limited to the correct choice of data and methods.

E. Ginters et al. (Eds.): ICTE ToL 2019, LNITI, pp. 82–89, 2020.
https://doi.org/10.1007/978-3-030-39688-6_12

In this context, indicators for Visual Analytics are essential. The domain of mobility, logistics and transportation therefore is investigated to set up a first classification of useful and important indicators. These should be seen as a first attempt to identify indicators for Visual Analytics.

First the main idea of Visual Analytics is discussed with the most popular and widely used definitions and also described on the base of most influential model. Thereafter, classification of indicators are introduced and it is justified why these indicators are in particular important for mobility, transportation and logistics. Two different case studies that are using different indicators to illustrate the added value are involved in the paper. The main contribution of the article is the classification and introduction of the relevant indicators with related data examples.

2 Visual Analytics

Visual Analytics applications and approaches are increasingly used to analyze different data in a variety of analytical tasks. Even if the term is often used synonymously to information visualization, Visual Analytics gained clear definitions for a clear differentiation. An early and most influential definition of Visual Analytics was proposed by Thomas and Cook:

> *"Visual analytics is the science of analytical reasoning facilitated by interactive visual interfaces"* [1].

Their definition emphasizes the "overwhelming amounts of disparate, conflicting, and dynamic information" [1] in particular for security related analysis tasks. One of the main focuses of Visual Analytics is to "detect the expected and discover the unexpected' [1] from massive and ambiguous data. They outlined that the main areas of the interdisciplinary field of Visual Analytics are:

- Analytical reasoning techniques: for obtaining insights and support analytical tasks such as decision making;
- Visual representations and interaction techniques: for enabling users to explore and understand large amounts of data, and interact with them with their visual perception abilities;
- Data representations and transformations: to convert all types of data, even conflicting and dynamic, to support visualization and analysis;
- Production, presentation and dissemination: to provide a reporting ability for a broader audience and communicate the analysis results [1, 2].

This definition gained a series of revisions to precise the abstract formulation [3–7]. Keim et al. commented that the definition of such an interdisciplinary field is not easy and tried to precise it as follow:

> *"Visual analytics combines automated analysis techniques with interactive visualizations for an effective understanding, reasoning and decision making on the basis of very large and complex datasets"* [7].

Keim et al. stated more precisely the interdisciplinary nature of Visual Analytics by introducing and outlining the combined use of automated analysis techniques and interactive information visualizations. In addition, it emphasizes the challenge of data amount, thus this confines Visual Analytics to "very large" data-sets. The main characteristics of solving analytical tasks with interactive information visualizations still remain. Furthermore, they introduced a model that couples the stated techniques in an iterative. Figure 1 illustrates their model that targets on providing a tight coupling of visual and automatic analysis methods through human interaction to enable human to gain insights and knowledge [7].

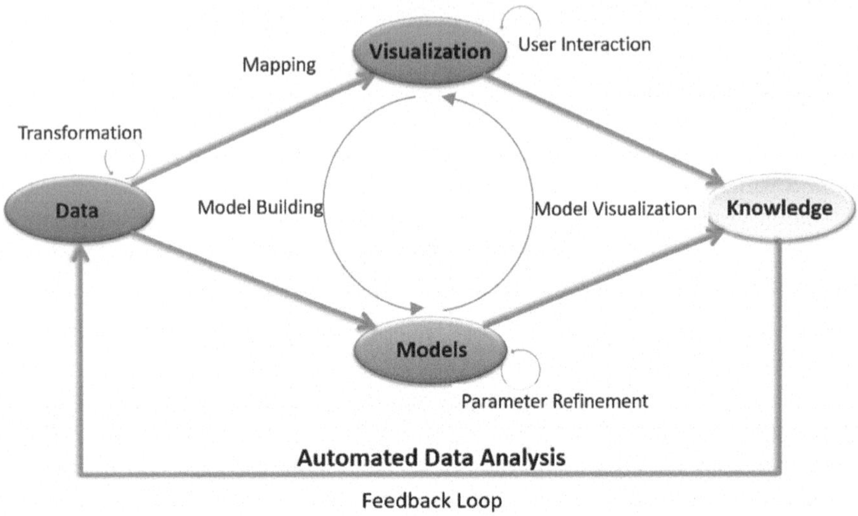

Fig. 1. The Visual Analytics model (adapted from [7]).

The Visual Analytics process models the different stages represented by oval forms and their transitions with arrows. The process starts with the data that may need to be preprocessed and transformed to an adequate way (indicated with the transformation arrow). After the transformation stage, a visual pipeline and model pipeline indicates the different approaches that are combined. The automatic analysis was proposed as techniques for data mining that are applied to generate models from the underlying data. These models can further be evaluated, refined, or specified by interacting with data [7]. Visualizations are used to interact with the models and manipulate and refine the parameters. Further the selection of alternating models can be visualized to evaluate the findings out of the generated model.

Nowadays the "model" is more related to approaches of machine learning and artificial intelligence respectively. Through this enhancement, Visual Analytics combines machine learning methods with interactive visualizations to reveal insights. The role of human and the possibilities to interact in the stages of analytical processing remains as they are proposed in the reference model for visualization [8]. The authors summarize that Visual Analytics is the science of analytical reasoning empowered by interactive visualizations and combines interactive visualizations with models and approaches of machine learning and artificial intelligence. They thus allow to solvi complex analytical tasks by uncovering hidden patterns in data by leveraging human perception of the visual space. The research on Visual Analytics is closely related to the topics of Data Science. Both areas seek to enhance the knowledge discovery process using machine learning, data mining and artificial intelligence methods. Visual Analytics, however, allows commonly a direct manipulation of the underlying models through graphical representations.

3 Indicators for Analytical Tasks in Mobility

The authors could outline that Visual Analytics combines methods of interactive visualization with machine learning to enable complex task solving based on huge amount of data. Employing the methods on Visual Analytics in the field of mobility and transport makes it necessary to identify and categorize indicators. These indicators have a strong relationship to data in Visual Analytics. Based on data and task a first classification of such indicators may be possible. The authors therefore identify indicators and the related data that lead to solving such tasks in mobility, transportation and logistics. These indicators that can be gathered from the variety of data should first be identified and categorized for the different analytical tasks in this domain. The authors identify following categories of data and related indicators for the domain of mobility, transportation and logistics:

- *Behavioral Indicators*: uses data that allow to gather mobility behavior;
- *Sensor Indicators*: uses data that allow to gather insights in production, usage and life-cycle in the domain of mobility, logistics and transportation;
- *Environmental Indicators*: uses data that allow to get insights of the impact of mobility to our environment: e.g. the built environment in terms of population density or diversity of usage, and accessibility;
- *Technology and Innovation Indicators*: uses data that allow to gather insights of the current trends in mobility and predict possible future scenarios;
- *Geographical Indicators*: uses data that allow to gather information about the geographical infrastructure of mobility, transportation and logistics.

Behavioral indicators enable Visual Analytics systems to analyze the mobility behavior in a general manner. This kind of data are gathered through mobile devices of drivers, navigation systems, stationary counters on street and a variety of further sources. The European Commission has already published a regulation for so called "eCall"-Systems that is mandatory for all cars manufactured later than April 2018 [9]. Sensor indicators enable Visual Analytics systems, in particular with machine learning approaches, to gather not only information about the manufacturing and predict maintenance but also to gather usage data. A huge number of internal sensors in manufacturing, cars, rails etc. provide sufficient information for Visual Analytics. Furthermore, external devices could be placed to gather these data. Environmental indicators enable Visual Analytics systems to give an insight of the impact of mobility, transportation and logistics to our environment. Therefore, different data gathering stations are available and also cars, trains etc. provide that information. Technology and Innovation indicators enable Visual Analytics systems to provide information about early technological signals [10], identify emerging trends [11], predicting upcoming trends etc. This leads to strategic decision making in industrial manufacturing and investing to more future related concepts and technologies. This area of Corporate Foresight plays more and more an increasing role in mobility, transportation and logistics. Geographical indicators are the most wide-spread data. These commonly occur as longitudes (east-west position) and latitudes (north-south position), and points of interest. These points of interest may be streets to uncover the overall infrastructure or analyze specific points-of-interest and enable solving analytical tasks.

4 Case Studies of Visual Analytics in Mobility

The authors introduce in this section with a Visual Analytics system that makes use of different indicators as described above. The main goal is to provide different examples for Visual Analytics in particular in the domain of mobility.

4.1 A Case Study on Analyzing Technological Trends in Mobility

The identification of technological trends in mobility enables decision makers from manufacturing, politics and strategic analysts to detect emerging technological trends and adapt their strategic directions [11]. For this purpose, patent data and scientific publications are appropriate. These data would allow to identify technologies at an early stage and react to the market developments appropriately. To illustrate the Visual Analytics system, scientific publications from different sources are used, e.g. DBLP, Springer, Eurographics, IEEE etc. Thus, the indicators for this case study are “Technological and Innovation indicators”.

In a first step, the DBLP data are used as baseline. These data do not provide any kind of information beside the authors’ names, titles, years and in most cases a DOI. Using that DOI, the data are enriched through web-mining methods [11, 12]. Further data from other sources are gathered and compared either through the DOI or through the combination of title and authors to eliminate duplicates. Thereafter, topics are generated through different approaches, e.g. LDA [13] and LSA [11]. With the enriched information a Visual Analytics system can be provided that models the underlying data in different ways, e.g. semantics model, geographical model and temporal model [11]. With this data the emerging technologies are identified [11] and provided on macro- and micro-level. On the micro-level an overview about all publications related to mobility are illustrated in Fig. 2a. Further the technological developments in mobility are illustrated in Fig. 2b. With the semantic data model and an appropriate visualization, the most influential author and the co-author relationship can be revealed as illustrated in Fig. 2c. This interactive visual interface further provides the temporal spread of the author’s publications that may indicate, if he or she is still working on this topic. The geographical data model provides information about the amount of publication from each country. Thereby the saturation of the country color is used as indicator, whereas the exact amount of publication can be gathered through a mouseover interaction. Figure 2d illustrates the geographical spread of publications about mobility.

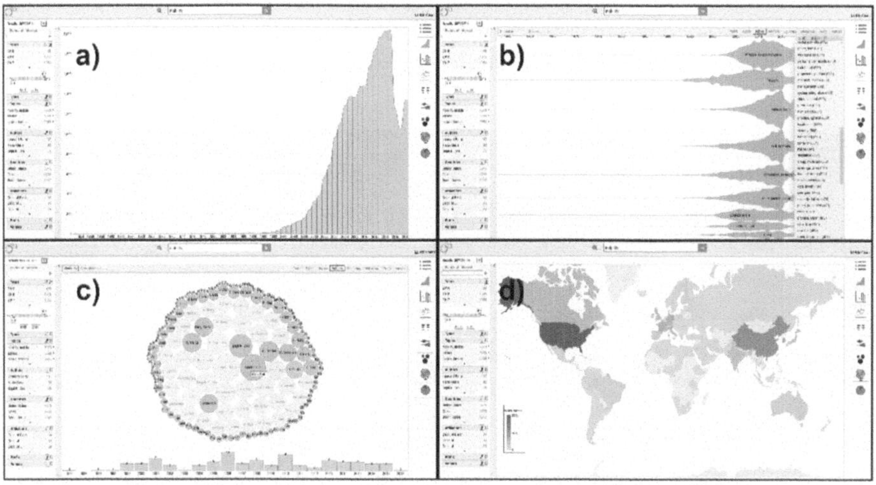

Fig. 2. A Visual Analytics system for trend analysis using technology and innovation indicators.

The above described case study just introduced a small number of analytical possibilities. A deeper insight can be found in [11, 12, 14].

4.2 A Case Study of Visual Analytics for Analyzing the Geographical Spread of Charging Stations for eMobility

Geographical indicators play for a variety of tasks in mobility, transportation and logistics an important role. The major aspect here is to detect paths or points in the infrastructure that are relevant to improve aspects of mobility, transportation and logistics. These could be streets, bridges, railway roads and many more. The authors use in this case study, with the same Visual Analytics System as described above, plug in charging stations worldwide. The goal is to provide an interactive visual system for charging station providers but also for political decision makers to analyze not only the charging behavior but also the needs for further charging stations.

The data used here are from different providers. The authors tried to use in particular first-hand data from the charging provider themselves. Redundancies were eliminated and the data were enriches using data from service providers. As the Visual Analytics system is not a dedicated system just for plug in charging station, the user starts with a search (which is also provided in the overview) for plug charging. Thereafter a world map with a heatmap-visualization is provided to see where the most charging stations are as illustrated in Fig. 3a. Thus, this is a work-in-progress system, the data may have lacks in particular in Asia. The system allows zooming or filtering through the faceted search on the left bar. Zooming into a country shows first the overall coverage, which seems to be pretty good for Germany as illustrated in Fig. 3b.

But it also shows that the north-east part of Germany has a lot of lacks. A further zoom reveals the lacks and illustrates clearly that beside greater cities like Berlin or Dresden the coverage is not that good as illustrated in Fig. 3c. The system makes use of a kind of semantic zoom and visualizes each charging station as a colored dot as illustrated in Fig. 3d. Green color dots are used for free charging stations, whereas red dots are use to indicate that these charging stations are fee-based.

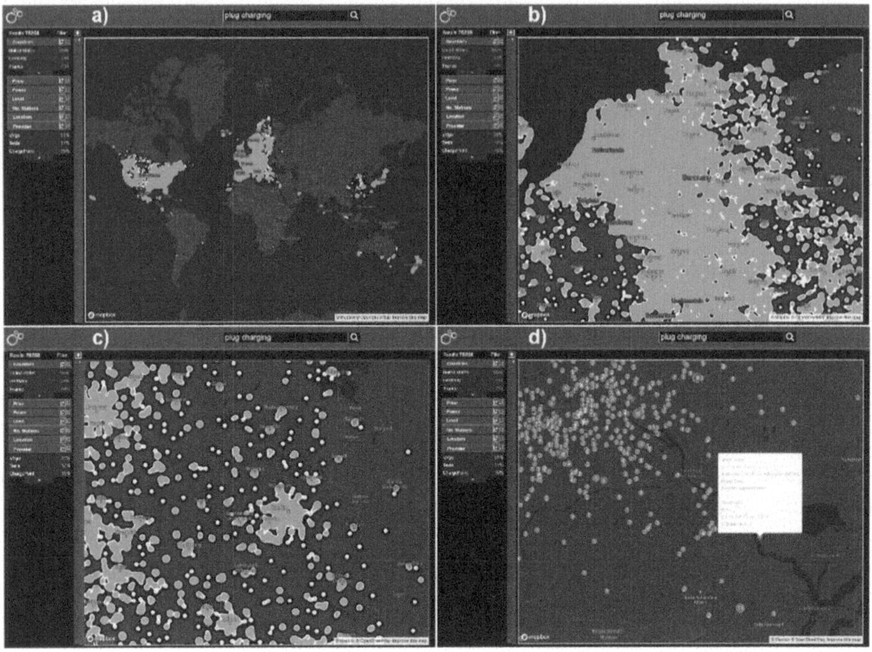

Fig. 3. A Visual Analytics System using geographical indicators.

5 Conclusion

Mobility, transportation and logistics are influenced by a variety of indicators such as new technological developments, ecological and economic changes, political decisions and in particular humans' mobility behavior. The authors therefore investigated the entire domain of mobility, logistics and transportation to set up a first classification of useful and important indicators. First the main idea of Visual Analytics with the most popular and widely used definitions and a model are introduced. Thereafter, the classification of indicators are provided and justified why these indicators are in particular important for mobility, transportation and logistics. The authors added two different case studies that are using different indicators to illustrate the added value.

Acknowledgements. This work was partially funded by the Hessen State Ministry for Higher Education, Research and the Arts within the program "Forschung für die Praxis" and was conducted within the research group on Human-Computer Interaction and Visual Analytics (https://vis.h-da.de). The authors would like to thank the students Svenja Lehmann and Walter Oster for their implementation contributions.

References

1. Thomas, J.J., Cook, K.A.: Illuminating the Path: The Research and Development Agenda for Visual Analytics. National Visualization and Analytics (2005)
2. Nazemi, K.: Adaptive Semantics Visualization. Studies in Computational Intelligence, vol. 646 (2016)
3. Keim D., Andrienko, G., et al.: Visual analytics: definition, process, and challenges. In: Information Visualization. LNCS, vol. 4950, pp. 154–175. Springer, Berlin (2008)
4. Keim, D.A., Mansmann, F., et al.: Visual analytics: scope and challenges. In: Visual Data Mining: Theory, Techniques and Tools for Visual Analytics. LNCS, vol. 4404, pp. 76–90. Springer, Berlin (2008)
5. Thomas, J.: Visual analytics a grand challenge in science – turning information overload into the opportunity of the decade. In: Proceedings of APVIS 2007 (2007)
6. Thomas, J., Kielman, J.: Challenges for visual analytics. Inf. Visual. J. **11**, 309–314 (2009)
7. Keim, D., Kohlhammer, J., Ellis, G., Mansmann, F.: Matering the Information Age Solving Problems with Visual Analytics. Eurographics Association (2010)
8. Card, S.K., Mackinlay, J.D., Shneiderman, B.: Readings in Information Visualization: Using Vision to Think, 1st edn. Morgan Kaufmann, Massachusetts (1999)
9. European Union: Regulation 2015/758. https://eur-lex.europa.eu/legal-content/EN/TXT/PDF/?uri=CELEX:32015R0758. Accessed 21 Oct 2019
10. Ansoffi, H.: Managing strategic surprise by response to weak signals. Calif. Manag. Rev. **18**(2), 21–33 (1975)
11. Nazemi, K., Burkhardt, D.: Visual analytics for analyzing technological trends from text. In: Proceedings of IV2019, pp. 191–200. IEEE (2019)
12. Nazemi, K., Retz, R., Burkhardt, D., et al.: Visual trend analysis with digital libraries. In: Proceedings of the 15th International Conference on Knowledge Technology and Data-driven Business, Graz, pp. 14:1–14:8. ACM (2015)
13. Blei, D.M., Ng, A.Y., Jordan, M.I.: Latent Dirichlet allocation. J. Mach. Learn. Res. **3**, 993–1022 (2003)
14. Nazemi, K., Burkhardt, D.: A visual analytics approach for analyzing technological trends in technology and innovation management. In: Advances in Visual Computing, pp. 283–294. Springer, Cham (2019)

Analysis of Method for Calculating the Throughput Capacity of Station Neck

Oksana Ischuka[1(✉)], Denis Lomotko[2], and Natalja Muracova[3]

[1] Transport Institute of Riga Technical University, Azenes Street 12-316, Riga 1048, Latvia
oksieca@inbox.lv

[2] Ukrainian State University of Railway Transport, Feuerbach Sq. 7, Kharkiv, Ukraine
den@kart.edu.ua

[3] ICFSD of Riga Technical University, 1 Kalku Street, Riga, Latvia
natalja.muracova@rtu.lv

Abstract. In this article the comparative analysis of the methods for calculating the throughput capacity of the station's neck is presented. Analysis of methods has revealed the main factors affecting the accuracy of the calculation. Besides, not all local conditions are taken into account in the analytical formulas. Nowadays, the issue of improving the calculation of the throughput capacity of the station's neck, which will take into account the variety of the local conditions at the technical station in each case, is actual. The article proposes to use the calculating method based on the discrete timetable theory.

Keywords: Railway Transport · Railway switch · Throughput capacity assessment

1 Introduction

At present, one of the main tasks in the railway enterprises is the correct determination of the throughput capacity of station necks. The value of the throughput capacity of the station necks directly depends on the time of maneuvers [1–3].

In study [4], it is said that the movement of group trains occurs according to the timetable. The its main feature is the importance of determining the departure time from the initial station and, accordingly, the arrival time of the car group exchange station, where the timetable should take into account the time of maneuvers at the station. On the railway, much attention is also paid to solving the task of creating car traffic with a minimum duration of stay at the stations of the transport system [5]. Mainly, in this study have described one of the main distinctive features of this approach is focusing on optimizing the time of car handling. So, on the railway, the determination of the car's stay time at the station is the main task.

This article proposes to consider three ways to determinate of the throughput of the station neck - the analytical [1–3] and graph-analytical method [3, 6], as well as simulation modeling [7]. The calculation of this throughput is considered based on the

E. Ginters et al. (Eds.): ICTE ToL 2019, LNITI, pp. 90–98, 2020.
https://doi.org/10.1007/978-3-030-39688-6_13

example of the odd station neck of the Daugavpils marshalling station with the parallel arrangement of the sorting and receiving-departure yards.

This article is proposed a new approach for determining the throughput capacity of the station neck using the theory of scheduling [8–11]. In this regard, the proposal on improving the method for determining the throughput of the station neck is of a huge interest. The solution to this problem can be further used to develop an optimization mathematical model of the railway network capacity control based on the Ford-Fulkerson method [12].

2 Problem Statement

It is assumed that the capacity of the neck depends on the structure of rail car flows, the number of travel routes in the neck and their mutual influence, the time it takes for the elements to move in one movement, such as station and distillation devices of automation, telemechanics and other reasons [1]. Considering the complexity of the directional development of station necks, only the most loaded railway switches groups are used for calculation [2], which substantially simplifies the solution of the problem. It was revealed that many existing stations have an insufficient useful length of the receiving and departing tracks for processing full trains, so it is necessary to make additional shunting movements with the occupation of station necks. All of these additional conditional factors are not taken into consideration in analytical formulas. Thus, it is proposed that all these diverse local conditions at technical stations should be considered when calculating the throughput of the station neck.

3 Description of the Object of Research

In this article, to calculate the throughput of the neck, as an example, the path development of the odd station neck of the sorting station with the parallel arrangement of the sorting and receiving parks is considered (see Fig. 1).

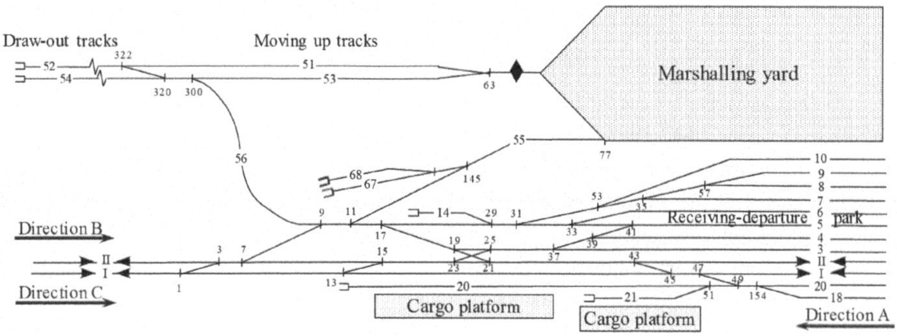

Fig. 1. Track development of the odd station neck of the marshalling yard with a parallel arrangement of marshalling and receiving-departure parks.

The neck is divided into railway switch elements, the number of which must be not less than the maximum possible number of train and shunting movements simultaneously performed in the neck. The neck is divided into the following switch elements: 1st element (1), 2nd element (13), 3rd element (45, 47); 4th element (3.7); 5th element (15, 21, 23); 6th element (43); 7th element (19, 25, 37, 39); 8th element (41); 9th element (9, 11, 17); 10th element (29, 31, 33, 35, 53, 57); 11th element (300, 320); 12th element (322); 13th element (63, 77); 14th element (145).

In accordance with the scheme of the station in the neck, no more than 4 (four) parallel routes are possible:

- Reception or departure of a train on the I main track from direction C;
- Receiving a train from II main track 3–5 receiving and departing tracks from direction B;
- Pulling the train from 6–10 pick-up routes to 54draw-outtracks;
- Thrust of the train from 52 sorting track to 51 moving up track for dissolution of the train.

According to the results of processing statistical data from certain directions, 31 freight trains arrive in the receiving-departure yard: 12 transit trains without processing, 17 transit trains with processing, 2 combined trains.

4 Comparative Analysis of Methods for Calculating the Capacity of the Neck

Currently, to determine the throughput of the station neck, there are various methods: analytical, graph-analytical and simulation.

In calculating the throughput of the station neck, the analytical method is the most common. When designing stations, two main types of analytical calculations can be used: direct calculation [1]; using the neck throughput capacity coefficient [2]. For the same initial data, the calculation results in both ways coincide.

The calculation of throughput using the coefficient K of using the capacity of the neck is determined by the formula:

$$K = \frac{\sum t_{chang}^{neck}(1 + \beta_{neck})}{1440 \propto_{neck} - \left(\sum t_{perm}^{neck} + t_{tech}^{neck}\right)} \quad (1)$$

where $\sum t_{chang}^{neck}$ – total time occupation of the busiest neck's element by train and shunting movements (changing operations) per train, min; β_{neck} – coefficient taking fully into account the failures of electric interlocking devices ($\beta_{neck} = 0,01$); $\propto_{neck}$ – coefficient taking into account the influence of possible interruptions in the use of railway switches of the calculated element due to the presence of hostile movements along the remaining elements of the neck; $\sum t_{perm}^{neck}$ – time of occupation of the calculated element of the neck by constant operations; t_{tech}^{neck} – the occupational time of the design element of the neck with constant operations (routine maintenance, planned repairs, snow removal, etc.), min.

The coefficient considering the effect of possible interruptions in the use of the loaded neck element is determined by the formula:

$$\propto_{neck} = 0,944 - 0,0103 \cdot \omega \quad (2)$$

where ω – a value characterizing the complexity of the work of the proposed neck, which is calculated by the formula:

$$\omega = \frac{M_0 - M_1}{Э_0 - 1} \quad (3)$$

where M_0 – total number of routes in the neck considered; M_1 – the number of routes with the occupation of the calculated element of the neck; $Э_0$ – the greatest possible number of simultaneous movements in the neck considered.

For station routes associated with operations that depend on the size of train traffic, the throughput is determined by the formula.

$$n_i = \frac{n_i'}{K} \quad (4)$$

where n_i' – the number of trains of a certain category arriving from the corresponding direction.

For calculating the performance indicators of the Daugavpils station [13] the Table 1 was compiled.

Table 1. Calculation of the most loaded element of the station neck.

The number of switch element	Group of switches included in this element	Neck element occupation time		Throughput capacity, K
		$\sum t_{chang}^{neck}$	$\sum t_{perm}^{neck}$	
1	1	0	5	0,000
2	13	0	0	0,000
3	45, 47	40	5	0,036
4	3, 7	76	65	0,070
5	15, 21, 23	94	49	0,085
6	43	60	42	0,054
7	19, 25, 37, 39	140	37	0,126
8	41	61	0	0,054
9	9, 11, 17	344	122	0,327
10	29, 31, 33, 35, 53, 57	195	84	0,183
11	300, 320	309	72	0,289
12	322	183	0	0,165
13	63, 77	137	32	0,126
14	145	52	12	0,147

Table 1 reflects the results of calculating the time it took for the elements to take on constant and changing operations, as well as the throughput capacity coefficient of the neck. By comparing the values of K for the used elements of the station neck, the busiest calculated element was found. Such an element is element No. 9, which contains switches 9, 11, 17. Element No. 9 has a load of 0.327. Using the calculated data, the capacity of the neck of the station was determined, which is reflected in Table 2.

Table 2. Throughput capacity calculation of the neck of the station.

Freight train category	Neck throughput for direction		
	From "A"	From "B"	From "C"
Transit without processing	3	33	0
Transit with processing	3	48	0
Partial	0	3	3
Total	93		

Thus, the throughput neck capacity, which is 93 trains per day, was obtained by using analytical method. The next method is to calculate the throughput by the graph-analytical method, which gives a visual representation of the implementation of the neck utilization, allows, to a greater extent than the analytical one, to consider the interaction of the switch elements, as well as the features of the station circuit. This method is recommended to be used with a high level of uneven arrival of trains at the station, the presence of periods of intense train movement with minimal intervals and a complex neck structure with parallel routes [3]. This method gives a more accurate result, especially with complex necks, but is more time consuming.

Using the same initial data, the throughput was also calculated by the graph-analytical method. For verification, the period of intensive work of the station lasting 4–6 h [3] is usually taken. Based on the analytical method, the throughput is 93 trains per day, which means 4 h – 93: 24 • 4 = 15 trains. Figure 2 shows the daily schedule of the marshalling yard for the first 4 h, so 15 freight trains entered there.

As a result of drawing a chart, the load of the calculated station elements is determined by adding the occupation time on a time scale. This time is shown in minutes to the right at the end of the chart in a specially selected column. The calculated amount of time is divided by the value of the calculation period and the percentage of load of this element is determined. From the chart calculation it follows that the busiest element is element 9, which includes switches 9, 11, 17 were loaded for 175 min within time frame of 250 min. Thus, you can find the loading of element 9: $T_{load}^{9el.} = 175 \cdot 100\%/250 = 70\%$. By analyzing this chart calculation, it is concluded that the directional development of the station neck is sufficient for a given size of movement and the most loaded element is used at 70%.

The next method is simulation modeling, the advantage of which is the ability to display real objects that are not amenable to strict formalization. Currently, specialists have created systems for the automated construction of simulation models, as well as methods for optimizing calculations [3]. The disadvantage of using this method is associated with the high complexity of building the model.

One of the simulation models is the ISTRA transport simulation system, which is focused on research and optimization of technological processes, and with its help it is possible to determine the capacity of the neck.

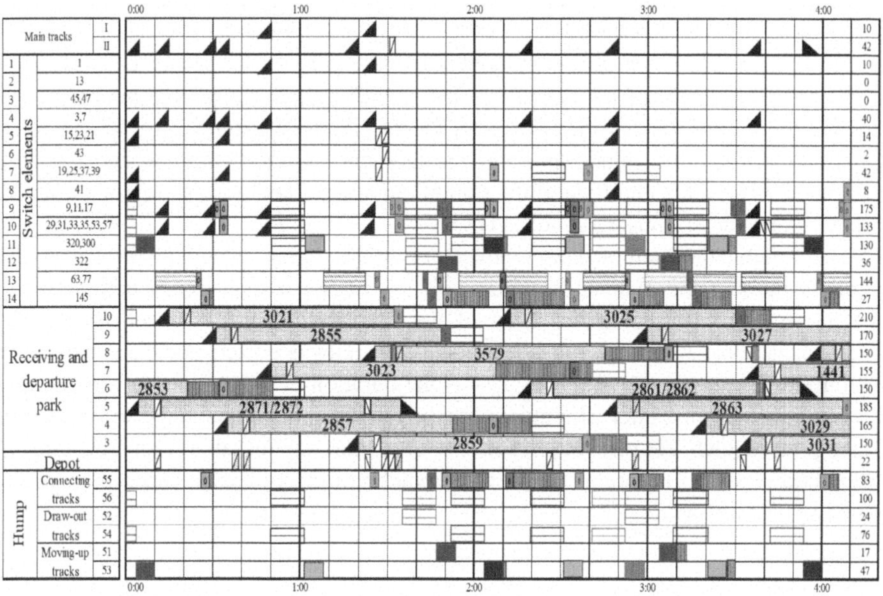

Fig. 2. Chart calculation of the track development and loading of the odd neck of the sorting station.

When making calculations, it should be considered that in the simulation system, as in a real station, the throughput and processing capacity can be affected by unevenness – the arrival of trains and their decomposition, operations, etc. As a rule, dispatch control and its implementation in the model "smoothes out" the negative impact of unevenness on the operation of the plant and the resulting indicators, but does not completely eliminate it. Therefore, the result of one calculation may incorrectly display the processing ability of the station of significant uneven arrival of trains [7].

5 Recommendations

The article proposes to calculate the throughput of the neck using the scheduling theory, which is a section of discrete mathematics that deals with ordering problems [8–11].

At the sorting station, all trains and shunting movements are not carried out simultaneously, with the exception of some movements, therefore, according to the method of setting the input information to determine the capacity of the neck, a dynamic problem (on-line) will be considered. For the type of task – a one-stage task with deadlines, the schedule is built in real time, i.e. before starting to solve the problem, the values of all parameters are not known. The schedule is built in parts as new information arrives. Thus, it is necessary to solve the dynamic problem with the simultaneous receipt of requirements.

Formulation of the problem. It is given: $m = \{m_1, \ldots, m_i, \ldots, m_m\}$ – the set of rail switch elements along which it is necessary to make station movements (cars); $j = \{j_1, \ldots, j_j, \ldots, j_n\}$ – a set of station movements along the switch elements (work); r_i – the moment of receipt of the demand in the service system (this value represents the moment the station begins to move from the receiving-departure yard; d_i – planned (deadline) period (this value represents the moment by which the ***i*** work should be completed, that is, the end time of the station movement, set by some reasons external to the system under consideration) [11].

The possible duration of station movements at the station, which are the sum of the requirements and the planned time: $a_i = d_i + r_i$. Station movements *I* consists of g_i operations. For each operation, a set of the following values is specified: m_{i1}, t_{i1}; $m_{i2}, t_{i2}; \ldots; m_{ig_i}, t_{ig_i}$, where m_{ij} – is the number of the railway switch element at which the station movement j, $1 \leq m_{ij} \leq m$, t_{ij} – is the duration of the station movement, i.e., the length of the time interval required for the movement corresponding to movement j along the switch element m_{ij}. Then the total duration of all station movements i (duration of work) according to the corresponding railway switch elements is determined by the formula:

$$t_{ij} = \sum\nolimits_{j=1}^{g_i} t_{ij} \tag{5}$$

It is required to determine the total expectation of the beginning of the station movement of rolling stock, given the optimal schedule for processing freight trains at the marshalling yard. It is denoted by W_{ij} – the waiting time for the beginning of the station movement ij, i.e. time interval between the end of the $(j-1)$ and the beginning of the j station movement. Then the total waiting time for all station movements is determined by the formula:

$$W_i = \sum\nolimits_{j=1}^{g_i} W_{ij} \tag{6}$$

The result of scheduling will be the definition of a set of numbers W_{ij}.

In most cases, the performed station movements have priority over each other, and this should be reflected in the criteria for evaluating the schedule, for each movement the so-called weight is set – the value u_i. Thus, as a criterion for evaluating the schedule, the criterion of timely execution of work (streamlining in accordance with the planned period) is applied. For example, priority will be determined: if there is a final group of wagons in the train; receiving a freight train is more important than shunting movements; uncoupling and hitching a train locomotive is more important than moving a shunting locomotive; if according to the forecast two trains of their formation will be ready to depart simultaneously, the first one is formed that will go earlier along the line of the schedule, and the second will be formed later, etc. These conditions will determine the priority of one station movement over another for timely execution of technological operations. It is assumed that for each work, the quantities u_i that characterize its relative priority are given. These quantities will be used in the criteria for evaluating the schedule. Then the total waiting time for all station movements can be set using one of the following formulas:

$$\bar{W}_u = \sum\nolimits_{i=1}^{n} u_i W_i \tag{7}$$

Therefore, considering the formula No. 1 to determine the use of the capacity of the neck, the following values can be put:

- Total time occupation of the busiest neck element by train and shunting movements (changing operations) per train $\sum t^{\text{neck}}_{\text{chang}}$ per expression $\sum\nolimits_{j=1}^{g_i} t^{\text{chang}}_{\text{ij}}$;
- Occupation of the calculated element of the neck with constant operations $\sum t^{\text{neck}}_{\text{perm}}$ on the expression $\sum\nolimits_{j=1}^{g_i} t^{\text{perm}}_{\text{ij}}$;
- Coefficient that considers the influence of possible interruptions in the use of railway switches of the calculated element due to the presence of hostile movements along the remaining elements of the neck 1440 ▪$\propto_{\text{neck}}$ on the expression $1440 - \sum\nolimits_{i=1}^{n} u_i W_i$.

Thus, the formula for determining the coefficient of neck throughput capacity is improved:

$$K = \frac{\sum\nolimits_{j=1}^{g_i} t_{ij}^{chang}(1 + \beta_{neck})}{1440 - \sum\nolimits_{i=1}^{n} u_i W_i - (\sum\nolimits_{j=1}^{g_i} t_{ij}^{perm} + t_{tech}^{neck})} \tag{8}$$

6 Conclusion

Finally it can be noted that when comparing the methods the analytical method has a point like estimate of throughput, but graph-analytical method has a cumbersome construction and the dependence of the quality of the assessment on the qualifications of the researcher who performs the calculation. Otherwise simulation gives the most accurate result by eliminating the disadvantages inherent in the above methods, but also has a drawback – this is the complexity of building the model and the dependence of the result on a combination of source data.

Studies have shown that while doing calculations using the analytical method, the capacity of the neck is affected by factors – a coefficient that considers the effect of possible interruptions in the use of the loaded neck element neck, and a value characterizing the complexity of the neck operation. Then, making a graph-analytical method of checking the throughput of the neck, it is clear that it is not used for 100%, but only for 70%. When calculating by the graph-analytical method, it can be summarized that in this case when hostile routes appear, there are unproductive delays on the receiving and sending routes. They have a direct effect on the throughput capacity of the neck; however, these delays are not reflected in the analytical formula. As a result, it is proposed to determine the throughput of the neck using a discrete scheduling theory. The article presents an improved dynamic model with the simultaneous receipt of requirements, which reflects unproductive delays of station movement by train on receiving and departure tracks. As a result, this new approach will make it possible to more accurately and adequately assess the throughput capacity of the station neck.

Acknowledgements. The gratitude is expressed to the colleagues from Ukrainian State University of Railway Transport who provided information on the methods for calculating the throughput capacity of station neck, as well as for the valuable advice and comments on the article.

References

1. Efimenko, Yu.I., Loginov, S.I., Sukhodoev, V.S.: Railway stations and junctions: manual for students of secondary vocational education institutions, pp. 295–296. Publishing Center Academy, Moscow (2006)
2. Russian Railways OJSC: Instructions for calculating the throughput capacity of railways, pp. 70–80. Russian Railways, Moscow (2010)
3. Apatsev, V.I.: Railway stations and junctions: textbook, pp. 304–314. FSBEI Educational and Methodological Center for Education in Railway Transport, Moscow (2014)
4. Prokhorchenko, A., Parkhomenko, L., Kyman, A., Matsiuk, V., Stepanova, J.: Improvement of the technology of accelerated passage of low-capacity car traffic on the basis of scheduling of grouped trains of operational purpose. Procedia Comput. Sci. **149**, 86–94 (2019)
5. Panchenko, S., Ohar, O., Shelekhan, G., Skrebutene, E.: Optimization of transport system operation using ranking method. Procedia Comput. Sci. **149**, 110–117 (2019)
6. Ischuka, O., Lomotko, D., Freimane J., Sansyzbajeva, Z.: Main directions of improvement of the method for calculation of idle time of cars at technical stations. In: 2018 ICTE in Transportation and Logistics, vol. 149, pp. 475–482 (2019)
7. Kashcheeva, N.V.: Modeling of Transport Systems: Guidelines, p. 16. UrGUPS, Yekaterinburg (2017)
8. Malakooti, B.: Operations and Production Systems with Multiple Objectives. Wiley, Hoboken (2013)
9. Levin, V.I.: Structural-logical Methods in the Theory of Schedules: Monograph, p. 124. Publishing House Penza State Technological Academy, Penza (2006)
10. Butko, T.V.: Scientific approaches to formalizing the process of automated scheduling of freight trains. Collection of scientific articles of Ukr. state Acad. railway tr., no. 131, pp. 31–37. Ukrdazt, Kharkov (2012)
11. Conway, R.V., Maxwell V.L., Miller L.V .: Schedule Theory. Science, Moscow (1975)
12. Dolgopolov, P., Konstantinov, D., Rybalchenko, L., Muhitovs, R.: Optimization of train routes based on neuro-fuzzy modeling and genetic algorithms. Procedia Comput. Sci. **149**, 11–18 (2019)
13. State Stock Company Latvian railway, LDZ Cargo Ltd: Joint Technological Process of Daugavpils Station and Daugavpils Freight Terminal, Riga, pp. 213 (2011)

Improvement of the Procedure for Determining the Duration of a Passenger Trip on the Railways of Ukraine

Valerii Kuleshov[1], Ganna Shapoval[1], Maksym Kutsenko[1(✉)], and Jelena Stepanova[2]

[1] Ukrainian State University of Railway Transport, Feyerbakh sqr. 7, Kharkiv 61050, Ukraine
valerijkuleshov2015@gmail.com, ann.shapoval@ukr.net, maksimus84@meta.ua

[2] Riga Technical University, Azenes Street 12, Riga 1048, Latvia
jelena.stepanova@vdzti.gov.lv

Abstract. The development of multimodal passenger traffic requires a high level of consistency from its participants at different stages of the process. This will ensure the most efficient functioning of the passenger complex as a complete technological system. For this purpose, it is proposed to use modern experience in the creation of transport hubs on the basis of existing passenger stations. For their effective functioning, it is necessary to improve the system of provision of information services on the state of the passenger train. To be able to inform passengers in a timely manner about changes in the schedule of passenger trains or other types of transport, it is proposed to use the developed procedure for determining the duration of the passenger trip. It allows you to take into account the number of stops of passenger trains on the route and, accordingly, the time spent on accelerating and slowing the train when heading it on the section. Carriers of various forms of ownership will be able to use the results obtained for calculating the use of railway infrastructure.

Keywords: Multimodal transportation · Transport system · Passenger complexes

1 Introduction

Multimodal passenger transportation is a form of interaction between transport operators of different types of transport with consumers of transport services. The development of multimodal transportation in Ukraine creates the preconditions for the development of projects of interaction of passenger railway companies with urban carriers, which will help to increase the attractiveness of railways for passengers. Recently, transport hubs (TH) have become widespread in Europe [1–3]. The purpose of their creation is to increase the comfort of passengers when using several types of transport.

For multimodal transportation the technology of simultaneous sale of tickets for the main type of transport (railway or air) and for additional (bus, ferry) is used.

E. Ginters et al. (Eds.): ICTE ToL 2019, LNITI, pp. 99–103, 2020.
https://doi.org/10.1007/978-3-030-39688-6_14

Foreign transport hubs offer a wide range of additional services for passengers:

- Purchase of tickets for all directions;
- Places for car parking;
- Car rental services and insurance policies;
- Special places for bicycles;
- Devices for charging mobile phones;
- The ability to pay for utilities, mobile communications and the Internet;
- Landline phone for free taxi booking;
- Book crossing libraries in waiting rooms;
- Issuance and delivery of online purchases.

Passengers may be offered the option of arranging a multimodal trip using suburban or regional trains to the airport with a minimum waiting time and reasonable cost.

The effective functioning of the transport hubs requires forecasting the time of arrival of trains at the station and interconnecting it with the timetable of other types of transport. The purpose of the study is to improve the procedure for defines the duration of trip by a passenger. The results can be used by railway passenger companies and carriers of various forms of ownership for calculating the use of railway infrastructure.

2 Problem Statement

The passenger station is a main subsection in the area of railway traffic management. In its activities, the passenger station is guided by state and network regulations. The choice of the location of the passenger station and transport hub is defined by the conditions for organizing of passenger trains [2].

Increasing the speed of passenger trains makes it possible to be competing railways with road transport in the distance of 250–500 km, and in the distance of 500–1000 km with the aviation [3]. But Ukraine's rail transport does not fully meet the requirements for the duration of passenger trips. The maximum speed of passenger trains is 120 km/h. The exception is the Kiev-Kharkov and Kiev-Dnepr routes, which achieve a maximum speed of 160 km/h. The maximum speed of passenger trains in the main directions is only 55–65 km/h.

It is important for passengers to provide timely information on any changes that occur along its route. To resolve this issue, the an Automated system for the transmission of voice information and control of information boards (AS TVICIB) was introduced at the regional branch of Southern Railway of the JSC «Ukrzaliznytsia» . The purpose of creation of the AS TVICIB is to provide information services to passengers and railway workers. The development of multimodal passenger transportation requires the further introduction of the AS TVICIB and the possibility of its use in the calculation of passenger transportation operators of different forms of ownership for the use of railway infrastructure. The use of constituents of AS TVICIBs installed at passenger stations as the main means of informing passengers about train schedules requires the establishment of appropriate regulations for the maintenance and operation of this type of information systems [4].

Questions of increasing the efficiency of passenger complexes, improving the technology of organizing passenger transportation, rational using of railway infrastructure of passenger stations were paid much attention in the works of domestic and foreign scientists: Butko [5], Gruntov, Danko, Ilovaysky, Mironenko, Nehrey, Ohar [2], Prokhorchenko [5], Primachenko [6], Shterson, Pothoff and others.

The quality of the transport hubs operation depends directly on the reliability of the railway infrastructure, which should provide opportunities for the formation of technological routes during the operations of the acceptance and departure of trains. In turn, to ensure the ability of a passenger to trip with the use of several modes of transport, it is necessary to strictly adhere to the established timetable for all participants of multimodal transportation.

To evaluate the quality of work of the passenger complex as a technological system, such indicators as the occupancy rate of the passenger car, the technical speed of movement, the average daily run of wagons and locomotive are used. The use of these indicators does not allow to be taking into account the possible change of rolling stock operator-owners and the evolution of automated control systems for passenger stations. Wagon turnover may be used to evaluate the quality of the transport process. It allows you to evaluate the robustness of the transport process for a particular subsystem or system as a whole.

The need for calculations for carriers of different ownership (rolling stock owners) to determine the payment for the use of railway infrastructure requires predicting the time of arrival of a passenger train at a passenger station or at a transport hub. To do this, a procedure for determining the length of a passenger trip or the length of passage of a passenger train in a direction is proposed. Given the development of high-speed traffic and multimodal transportation, the following formula can be used to determine the duration traffic of a passenger high-speed train on a particular section of direction.

3 The Method of Determining

The need for calculations for carriers of different ownership (rolling stock owners) to determine the payment for the use of railway infrastructure requires predicting the time of arrival of a passenger train at a passenger station or at a transport hub. To do this, a procedure for determining the length of a passenger trip or the length of passage of a passenger train in a direction is proposed. Given the development of high-speed traffic and multimodal transportation, the following formula can be used to determine the duration traffic of a passenger high-speed train on a particular section of direction:

$$F(T_{ps_i}) = f\left[\frac{L_i}{\mu_1 v_{x_i}} + (K_{ps}\mu_2 + 1)(t_s^{ps} + t_{ov}^{ps} + t_{sd}^{ps})\right] \to \min \tag{1}$$

where L_i - the length of the section of movement of the passenger high-speed train, km; K_{ps} - number of stations with stops of passenger high-speed trains (without excluding starting and ending stations); t_s^{ps}, t_{ov}^{ps}, t_d^{ps} - duration of stops, overclocking, deceleration of passenger high-speed trains, hours; μ_1 - coefficient taking into account speed increase of passenger high-speed trains; μ_2 - coefficient taking into account the reduction of the number stops of passenger high-speed trains; v_{x_i} - average speed at the section of movement of the passenger high-speed train, km/h; with (1) restrictions:

- Permissible speed of the passenger train:

$$60 \leq v_{x_i} \leq 160 \tag{2}$$

- Number of passenger train stops:

$$0 \leq K_{ps} \leq 3 \tag{3}$$

Coefficients as μ_1 and μ_2 allow to take into account the effect of increasing the speed of movement of the passenger train and reducing the number of stops on the way of its direction. This is quite relevant for high-speed passenger transportation.

For the conditions of the Kharkov-Kiev high-speed train (Intercity) train, the calculation of duration of the passage of the specified section was conducted. The calculations took into account:

- The train runs daily;
- Length of the section of movement of the passenger high-speed train $L_i = 586,47$ km;
- Number of high-speed train stops along the route $K_{ps} = 2$;
- Duration of stops $t_s^{ps} = 0{,}03$ h;
- The overclocking duration $t_{ov}^{ps} = 0{,}02$ h;
- Duration of deceleration of passenger high-speed train $t_d^{ps} = 0{,}02$ h;
- Coefficient taking into account the increase in the speed of passenger trains $\mu_1 = 1{,}33$;
- Coefficient taking into account the reduction of the number of stops of passenger high-speed trains $\mu_2 = 0{,}76$ (indicators obtained expertly);
- Average speed at the section of movement of the passenger high-speed train $v_{x_i} = 96{,}14$ km/h:

$$T_{ps_{719}} = \frac{586{,}47}{1{,}33 \cdot 96{,}14} + (2 \cdot 0{,}76 + 1)(0{,}03 + 0{,}02 + 0{,}02) = 4{,}77\,\text{h}.$$

According to the train schedule at the specified section, the passenger travel time is 5,3 h. At the same time, the estimated amount of time the railway infrastructure use differs from the timetable. This has a certain impact on the amount of payment to operator of passenger transportation.

4 Conclusion

The development of multimodal transportation stimulates all participants in the transportation process to maximize adherence to the timetable. For the passengers is relevant the information about changes that would occur in the train schedule. The use of the Automated system for the transmission of voice information and control of information boards (AS TVICIB) gives an opportunity to inform passengers who are in the transport hub about the time of arrival of trains and possible changes in timetable. The automated control system at the transport hub can be built on the functional basis of existing automated systems in order to provide accurate, timely and quality passenger service.

The proposed approach for trip calculation allows to predict the time of arrival of passenger trains to the passenger station, which is an integral part of the transport hub. The procedure developed takes into account the additional effect from increasing the speed of the passenger train and reducing the number of stops on the route. The results obtained can be used by railway passenger companies and carriers when calculating the use of railway infrastructure. The proposed approach can be implemented as a software product in the decision support systems of the dispatch staff. This will reduce the risks in managing the transport systems that may arise from multimodal passenger transportation using different modes of transport.

References

1. Dalton, G.: Perspective structures of management on the European railways. Eur. Rev. **200**, 62–75 (2001)
2. Ohar, O., Drebot, M., et al.: A mathematical model for determining the rational location of a passenger station for service of high-speed trains at a large railway junction. In: USURT 2016, Kharkov, pp. 23–30 (2016)
3. Vickerman, R.: High-speed rail and regional development: the case of intermediate stations. J. Transp. Geogr. **42**, 157–165 (2015). http://www.sciencedirect.com/science/article/pii/S0966692314001227-cor1
4. Kuleshov, V., Shapoval, G., Kuleshov, A., Gromov, O., Lysenko, M.: Improvement of satellite navigation systems in the conditions of development of passenger complex at high-speed transportation. In: USURT 2017, Kharkov, pp. 96–106 (2017)
5. Butko, T., et al.: Modeling of distribution of passenger traffic on trains on the basis of swarm intelligence. East.-Eur. J. Enterp. Technol. 154–161 (2010)
6. Primachenko, G.: Development of proposals for increasing the profitability of passenger rail transportation in Ukraine. East.-Eur. J. Enterp. Technol. 33–39 (2015)

The Use of Programmable Logic Controllers in Railway Signaling Systems

Mieczyslaw Kornaszewski(✉)

Kazimierz Pulaski University of Technology and Humanities in Radom, Malczewskiego 29, 26-600 Radom, Poland
m.kornaszewski@uthrad.pl

Abstract. The railway computer signaling systems should be realized using devices fulfilling specific requirements: high level of reliability, verification to enter safety conditions (errors detecting and the proper reaction), modularity, possibility to link to the network, possibility to connect many railway external devices (train sensors, signaling lights, track occupancy control circuits), possibility to link the external railway infrastructure to different control devices, possibility of work in the severe exploitation conditions (temperature, dampness, vibration). Such requirements can be ensured by the use of Programmable Logic Controllers (PLC). Computer devices based on PLC controllers are small sized, and have great possibilities of controlling, testing, diagnostic, and self-diagnostic. This paper presents computerized railway signaling systems used in Poland, which operate based on modern PLC controllers. The railway signaling systems' software consider different requirements and fit to local topographic conditions, according to the imposed high level of safety.

Keywords: Programmable logic controllers · Railway signaling system · Safety

1 Introduction

In the late 20th century, the modern railway signaling systems based on microprocessor technology and microprocessor controllers have been implemented in Poland. These systems possess sophisticated self-control mechanisms, events recording, technical diagnostic allowing remote capturing information concerning events and faults [1, 4, 12].

The paper presents the configuration of safe computer railway systems based on scientific research conducted on modern, safe solutions of level crossing systems. The level crossing should be understood as a place of roads intersection for cars and railways lines at one level, whereas the railway level crossing system is intended to secure the traffic in this place. These systems belong to safe systems and they have the forth classification level CENELEC. The rules of used software by them are strictly defined by the norm CENELEC EN 50 126, EN 50 128 and EN 50 129. The Safety Integrity Levels (SIL) included in the norms of the European Committee of Normalization in CENELEC is important criterion.

E. Ginters et al. (Eds.): ICTE ToL 2019, LNITI, pp. 104–111, 2020.
https://doi.org/10.1007/978-3-030-39688-6_15

According to CENELEC recommendations the computer controlling system should be designed to provide all phases of the system's life: assumptions, designing, starting and the exploitation. Every modification of the system refers to the equipment and software. The negative result of analysis after every stage or detected errors during testing, starting and the exploitation cause the return for establishments [1, 4, 14].

These rules are presented in the following norms: EN 50 126 – taking back to all railway devices and it is establishing requirements from the scope of the reliability, the availability, keeping abilities and the safety of action, determined with abridgement of RAMS (Reliability, Availability, Maintainability and Safety) as well as the independence and the quality; EN 50 128 – applying to the model of creating providing steering and protecting software for train systems; EN 50 129 – a process of designing the software and the equipment for individual systems which can function as part of the comprehensive system of steering and protecting is defined. SIL-4 is the highest safety level and determines the damage intensity (the probability of damage occurrence in a unit time) for a single system element 10E-11 [5, 6, 9, 14].

Currently in Poland the scientific research is carried out on the implementation of innovative technologies in railway transport. The research deal with signaling lights with LED diodes, photovoltaic cells supplying railway signaling devices, open radio communication standard for railway signaling systems and dedicated railway signaling circuits implemented in Field-Programmable Logic Devices (FPLD) [8].

The aim of the paper is to provide a technical and functional analysis of PLCs, which are most frequently used in computer based railway signaling systems in Poland. The audience of the results presented in the paper are professionals interested in safety of railway transport and, in particular, specialists in the field of railway signaling.

2 Characteristics of Control Methods in Railway Signaling Systems

In contemporary times the speed of the data processing and clear presentation of controlled processes on monitors are being realized by computers used in controlling systems. The first solution was replacing controlling relay desktop in adjustable computer with the presentation of the motor situation depicted on the monitor. Fulfilling more developed controlling techniques required replacing relay-automation fully computer. As a rule, computer railway signaling systems should be executed as fail-safe, what means that a single damage (of equipment or software) or interference cannot cause a dangerous situation, assuming negligibly small probability of double (multiple) damage occurrence. Additionally, single errors are assumed to be detected in a relatively short time and the damage detected by an appropriate system's response to the fact.

For computer railway signaling systems it is necessary to consider issues of safety and reliability on levels: of technical devices, creating the infrastructure of railway signaling system and of system software.

To meet the requirements of safety the system has to consist of two computers linked together in a suitable structure, which at least performs appropriate data processing and mutual check-up possible, etc. There are also other systems based just on one unit. For obtaining required conditions of safety the other computer is utilized as the hot reserve. Appropriate software operations are executed for the safety of single-channel systems. They are relying on encoding and converting data through two programs in one unit which are testing each other. The best situation is when applications are written by different groups of programmers.

Multi-channel systems, mostly two-channel or three-channel, are named “2 out of 2” and “2 out of 3” respectively, which safety is ensured by redundancy of hardware and software. In solutions with two computers the results are compared, and according to safe work condition of system “2 out of 2” only full compatibility effects of all calculations are obtained on output of active channels. In the “2 out of 3” system the negative result of compare makes switch to activate the third computer, which result of converting is taking under consideration alike previously on two computers [3, 4, 7].

3 PLC Applied in Polish Railway Signaling Systems

3.1 The Advantages of PLC Application

PLC controllers are industrial computers, which under the control of real time operational software realize executing application algorithms on the basis of obtained data of controlling processes and device [2, 11].

The main feature of utilizing PLC controllers to the processes control is the placement of operational logic of device in the PLC memory of free access, which are the part of controller’s memory. The operation of the controller involves monitoring of digital and analog inputs, making decisions based on the main control program and controlling outputs. Controlling circuit, program memory and supplier constitute the central unit of the controller. Besides program memory, controllers possess also data program and intermediate results memory. Presented systems constitute the basis structure of controller. In most application communication devices serving as communicators to other controllers or operator’s stations supplement this basis structure. Other commonly used extension of the structure is initial input signal processing units [5, 6, 8].

3.2 PLC Usage in Automated Railway Crossing Level Systems

The considerations regarding the PLC controllers’ usage in Polish railways could be presented in the best way based on the technical solutions used in automated railway crossing level systems. There is presented geographical layout and work organization of the typical level crossing system (SPA-5 system manufactured by Bombardier Transportation Poland) at Fig. 1, where: I, II, III, XI, XII, XIII – railway sensors, S1, S2, S3, S4 – roads signaling lights, N1, N2, N3, N4 – railway barriers, 1 W, 2 W, 1 N, 2 N – warning targets on level crossing, Dz1, Dz2 – acoustic signaling, ERR-10/11 – container cabinet with devices for controlling the level crossing, EZG-17 – diagnostic device, ERP-6/7 – remote control device, PC – Computer Control Center [4, 6].

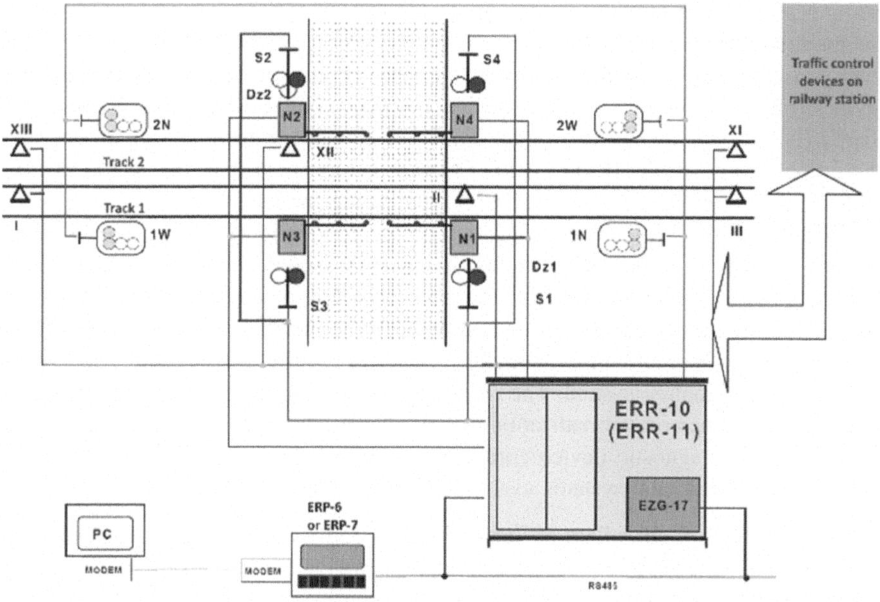

Fig. 1. Configuration of devices in automatic railway level crossing system on the example of SPA-5 type solution.

Switching on the warning of the vehicle drivers is carried out at incoming the train. At the certain distance from roads crossing the switching on railway sensors I (XIII – for driving of the train on an incorrect railway track) and III (XI) are sed. Incoming the train to the level crossing system is dependent on direction of the movement and is recorded by one of these devices. Signal of incoming of the train to roads crossing is being recorded by sensors I (XIII) and III (XI) and is being sent to controlling devices, which cause switching on lights (S1, S2, S3, S4) and acoustic signals (Dz1, Dz2) and closing the railway barriers (N1, N2, N3, N4). On the basis of input signals from railway sensors, controlling devices recognize the direction of the train movement. Switching off signalization and opening the railway barriers occur at the moment when the moving train finishes the interaction on II and XII placed in the vicinity of the level crossing. Moving away train should not cause the warning, interacting on the last sensor III and XIII (I/XI). If the one direction movement is planned then devices III and XIII is redundant [4, 6].

In the paper the analysis of three most commonly used types of PLCs in the Polish railway industry is presented. All of the analyzed controllers meet requirements of highest safety integrity level (SIL-4) and are typically implemented in construction of railway level crossing systems or other railway signaling systems.

Modular Processors of Scheidt & Bachmann GmbH. In the microprocessor automatic level crossing system of BUES 2000 the control of the devices and correct operation are performed in three levels: diagnostic, supervising and executing. Diagnostic level consists of diagnostic module and diagnostic hub [7, 10]. Elements of management level supervising all processes relevant to protection functionality of level crossing are: central modular processors, signaling light, barrier modular processors, line modular processors, service keyboard and main program memory [5, 6]. Executing level consists of: I/O modules (railway line sensors, connecting to railway signaling devices, pulpits to control outdoor devices), signal boards, LSTR boards and CAN boards 10/200 (railway barrier drivers). The basic element of the modular processor is a microprocessor controller (INTEL CPU 80188 EB microcontroller) that works with six controllers Control Area Network (CAN) type during the data transfer.

Controlling of the level crossing system is performed by two channels. Each channel consists of central unit, lights and barrier module, diagnostic module and railway module. Each module has the own doubled processors, which processes independent fragments of program in real time. An application of appropriate software of Scheidt & Bachmann GmbH modular processors ensures keeping high safety level of the railway signaling system. The device is versatile, it means, it accepts errors not having the influence on the realization of basic functions. Thanks to an application of components of diagnostic device complying with PC standard, it is possible to use software and operational systems available on the market [3, 9, 10, 13].

MINICONTROL Controllers of Series 2003. MINICONTROL controllers of series 2003 are used in the SPA-5 type level crossing system. Each controller checks the state of toll gate and warning signal lights. The goal of controllers is generation of safe controlling signals of level crossing system and output devices based on incoming controlling signals of railway sensors. The information about the state of level crossing system is transmitted in cycling manner to the remote control unit. The software contains function testing of all elements of level crossing system. Single controller consists of electronic modules (see Fig. 2) encased in the plastic case. Modules (electronic boards) are placed in the case in suitable places (according to required configuration) [5, 6, 12].

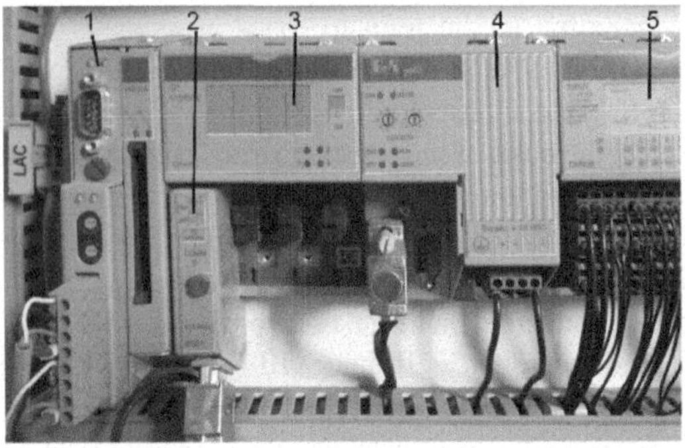

1. Transmission module CAN (with ELS type wheel sensors),
2. Transmission module RS-485,
3. Central unit (for configurations with EOC type train sensors or configurations with ELS type wheel sensors),
4. 24V DC power supply module,
5. Input\Output module.

Fig. 2. Construction of MINICONTROL 2003 series PLC controller used in SPA-5 level crossing signaling (channel A).

In Fig. 2 the view of the double PLC controller has been presented, which configuration corresponds to standard level crossing system with redundant number of railway sensors. Types, number and arrangement of digital inputs and outputs modules connected to the central processing unit CPU of PLC controllers are selected depending on the number and configuration of external devices occurring on the railway level crossing. Controlling system of SPA-5 consists of two identical controlling channels.

Each channel consists of PLC controller, interfaces allowing receiving of input and output signals amplifying these signals to the level required by output devices [4, 6, 11, 12].

GE Fanuc Controllers of Series 90-30. The control of level crossing system of RASP-4Ft type (produced by KOMBUD Radom in Poland) is realized on PLC controller GE Fanuc Automation of series 90-30. Controllers have central unit IC693CPU350. Their operation is based on information exchange and synchronization of the operation through two channels of data bus (see Fig. 3).

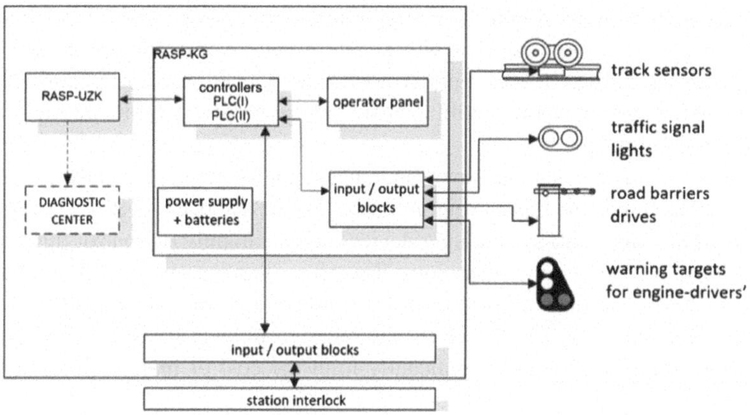

Fig. 3. Simplified operating scheme of RASP-4Ft railway signaling.

In the RASP-4Ft level crossing system, the ordinary EOC type rail sensors have been replaced with the systems of axle counters (track occupancy control) based on the Frauscher RSR-180 type wheel sensors [3, 9]. The GE Fanuc series 90-30 PLCs have a typical modular design, but are now becoming less used. An application of controllers of series 90-30 allows both executing the program and the realization other tasks such as: diagnostic, Input/Output (I/O) scanning and alarms processing. System program also contains procedures to communicate with the programmer. These procedures allows the user to read and erase controlling programs, read the status information and ensure the steering of the controller. Occurrence of damages or circumstances having the influence on the operation of the system are indicated by errors visualization [7, 14].

4 Conclusion

The PLC controllers are used in the modern railway signaling systems, especially in the automatic level crossing systems where the high level of safety and reliability is required. Advantages of controllers such as the reliability, flexibility of the application and ability to build complex automatic systems causes the new and existing devices to be fitted with controlling circuits. Software of railway signaling systems allows taking into account different user's requirements and strict matching the operation of system to local conditions. Higher safety of the automatic level crossing systems is the result of application of modern technologies based on two channels control, diversity of applications in both channels, possibilities of immediate faults detection in devices and applications (self-testing), and also on possibility of monitoring of the system operation by both registrations of events and failures. The newest computer railway signaling systems require application software adjusted to configuration of railway signaling devices in the field. An application of PLC controllers allows for increasing controlling functions.

Future scientific research on the use of programmable logic controllers in modern railway signaling systems should be carried out towards greater stability of individual railway traffic control devices. This will ensure greater safety for people and freight moving along the railway tracks. Furthermore, the application of programmable logic controllers in railway transport should increase safety of railway traffic through limiting mistakes made by people.

Research is also carried out on a limited extent for dedicated railway signaling circuits implemented in Field-Programmable Logic Devices (FPLD). Algorithms implemented in many railway signaling devices consist mainly of control functions and simple storage and processing operations. Therefore, they can be readily implemented in hardware. Such approach allows for fast implementation of algorithms [8]. Research on the implementation of innovative technologies to railway transport in Poland will increase train traffic safety and significantly limit the cost of maintaining the railway infrastructure.

References

1. Carlson, A.H., Frincke, D., Laude, M.J.: Railway security issues: a survey of developing railway technology. In: Proceedings of International Conference on Computer, Communication and Control Technologies, vol. 1, pp. 1–6. International Institute of Informatics & Systemics, Orlando (2003)
2. Georgieva, M., Nenov, I.: Implementation of microcontrollers in railway automatic traffic control systems. In: Proceedings of 28th International Spring Seminar on Electronics Technology. International Spring Seminar on Electronics Technology ISSE, pp. 60–63. IEEE, New York (2005)
3. Kornaszewski, M., Lukasik, Z.: Analysis of setups of the control of rail microprocessor systems. Adv. Electr. Electron. Eng. **1–2**(7), 223–226 (2008)
4. Kornaszewski, M., Pniewski, R.: Impact of new informatics solutions using in railway transport on its safety. In: Proceedings of 22nd International Scientific Conference Transport Means 2018, vol. 2, pp. 996–1001. Kaunas University of Technology (2018)
5. Kornaszewski, M.: Digest of modern PLC controllers applied in modern level crossing systems. In: Computer Systems Aided Science and Engineering Work in Transport, Mechanics and Electrical Engineering, pp. 153–160. Kasimir Pulaski Technical University, Radom (2008)

6. Kornaszewski, M.: Microprocessor technology and programmable logic controllers in new generation railway traffic control and management systems. Int. Sci. J. Arch. Transp. Syst. Telematics **11**(2), 18–23 (2018)
7. Kornaszewski, M.: Safe computer solutions applied in new generation railway traffic control systems. In: Computer Systems Aided Science and Engineering Work in Transport, Mechanics and Electrical Engineering, vol. 122, pp. 279–284 (2008)
8. Kornaszewski, M., Bojarczak, P., Pniewski, R.: Introduction of world innovative technologies to railway transport in Poland. In: Globalization and Its Socio-Economic Consequences, vol. 2, pp. 962–969. University of Zilina (2016)
9. Lu, D., Cai, B., Wang, J., Liu, J., Toro, FG.: Safety margin estimation using risk assessment method for satellite-based train localization unit. In: Proceedings of the ASME Joint Rail Conference, AMER SOC Mechanical Engineers, New York (2016)
10. Nowakowski, W., Ciszewski, T., Lukasik, Z.: The concept of railway traffic control systems remote diagnostic. In: Mikulski, J. (ed.) Communications in Computer and Information Science, Smart Solutions in Today's Transport, vol. 715, pp. 471–481. Springer (2017)
11. Olderog, E.R.: Correct real-time software for programmable logic controllers. In: Olderog, E.R., Steffen, B. (eds.) Correct System Design: Recent Insights and Advances. Lecture Notes in Computer Science, vol. 1710, pp. 342–362. Springer, Berlin (1999)
12. Pede, G., Agostini, A.: Innovative technologies for public transport. In: Brebbia, C.A. (ed.) Urban Transport XIII: Urban Transport and the Environment in the 21st Century, vol. 96, pp. 449–458. WIT Press, Southampton (2007)
13. Zhang, Z.G., Li, X.F., Gan, Y.L.: Railway crossing intelligent safety alarming system design based on PLC. In: Proceedings on 26th Chinese Control and Decision Conference 2014 (CCDC), pp. 5045–5048. IEEE, New York (2014)
14. Zhou, R.Y., Liu, A., Peng, W., Shen, Z.: Research on safety control system for enterprise railway crossing based on programmable logic controller. In: Progress in Safety Science and Technology Series, vol. 7, pp. 351–356. Science Press, Beijing (2008)

Application of Logical Diagnosis to Identify Faults in Computerized Railway Traffic Control Systems

Waldemar Nowakowski(✉), Tomasz Ciszewski, and Zbigniew Lukasik

Faculty of Transport, Electrical Engineering and Computer Science, Kazimierz Pulaski University of Technology and Humanities, Malczewskiego 29, 26-600 Radom, Poland
{w.nowakowski, t.ciszewski, z.lukasik}@uthrad.pl

Abstract. Modern railway traffic control systems commonly use information and communication technologies. It allows for a significant enhancement in the functionality of these systems. One of the trends in the development of railway traffic control systems is equipping them with automatic diagnostics. This makes it possible not only to improve the reliability of these systems, but also effectively raises the railway traffic safety. The authors discussed the problem of using logical diagnosis, which on the base of the technical knowledge about the diagnosed systems, allows to detect faults in railway traffic control systems. In order to take uncertainty of diagnostic inference into account in the presented methodology Bayesian inference was used. As an example of the diagnosed system, the interface of the level crossing protection system and interlocking system were used.

Keywords: Railway traffic control systems · Logical diagnostics · Bayesian inference

1 Introduction

Railway traffic control is an element of the railway transport system that significantly affects the safety and efficiency of the process of moving people and goods [3, 4, 13, 14]. Despite the continuous development of railway traffic control systems, like other technical systems, can be damaged [8, 17]. Therefore, the maintenance and renewal processes of these systems are of particular importance [6]. The actions, carried out by the railway infrastructure manager, that allow maintaining the equipment in full efficiency is the diagnostic process, which also influences ensuring the determined level of railway traffic safety [2, 9, 11]. In general, the diagnosis process is based on examination and diagnostic inference. Diagnostic examination means the observation of diagnostic signals related to a given object, containing information about its condition. The test results, usually in the form of a set of values of recorded signals or the course of specific functional dependencies, constitute the basis for diagnostic inference, the aim of which is to determine the state of the object. A review of current scientific research and

E. Ginters et al. (Eds.): ICTE ToL 2019, LNITI, pp. 112–119, 2020.
https://doi.org/10.1007/978-3-030-39688-6_16

proposed technical solutions regarding the diagnostics of railway traffic control systems shows that the existing methods are insufficient. Therefore, the goal of the conducted research was an attempt to implement logical diagnostics in the evaluation of the technical condition of selected railway traffic control systems. The authors of the paper have noticed that this type of analysis is associated with the risk of insufficient access to diagnostic data. One of the best known methods of processing uncertain knowledge is probabilistic inference, also known as Bayesian inference [1, 5, 7, 16]. It is the method of processing uncertain knowledge based on the direct use of the probability theory, in which individual statements are assigned the probability of their occurrence. The results of the research can be useful for engineers and designers in the process of development a new diagnostic systems dedicated to railway automation. The proposed methods should also improve the operation of the Maintenance and Diagnostics Centers in the future.

2 Logical Diagnostic on the Example of the Level Crossing Protection System Interface

Logical diagnostics is based on the analysis of symptoms and failures with the use of technical knowledge regarding the diagnosed system [5, 15]. In the general case, assuming the set of n symptoms $s_1, s_2,\ldots, s_n$ and set of m failures $f_1, f_2,\ldots, f_m$, can be defined the Boolean function of symptoms:

$$V[s_1,\ldots,s_n] \tag{1}$$

and the Boolean function of failures:

$$F[f_1,\ldots,f_m] \tag{2}$$

as well as the Boolean function of symptoms and failures:

$$E[s_1,\ldots,s_n,f_1,\ldots,f_m] \tag{3}$$

The methodology of logical diagnosis will be presented on the example analysis of failures that can occur in the case of cooperation between the Level Crossing Protection System (LCPS) and interlocking system using a digital interface [10, 12]. It is intended to make the indications of station exit and entry semaphores dependent on the state of the LCPS located in the area of the railway station, as well as for automatic activation of the warning at the LCPS while train entrance, departure or passing through the station without stopping. The current state of LCPS system is transmitted to the computers of interlocking system using digital transmission, while "activate the warning" command is transmitted by them to LCPS. Communication is carried out independently in two channels (A and B). The communication protocol and the topology of the connections between the systems meet the safety requirements at SIL-4 (see Fig. 1).

The following failures can be distinguished for discussed interface:

- f_1 - failure of interface or error on "LCPS – interlocking system" link;
- f_2 - failure of single channel of interlocking system interface;
- f_3 - failure of interlocking system;

and the associated symptoms:

- s_1 - activation of the warning by the interlocking devices, without a prior command;
- s_2 - switching on a warning for a single channel;
- s_3 - no response from the data exchange modules;
- s_4 - activation of the warning by wheel sensors, without a prior activation from interlocking devices;
- s_5 - simultaneous activation of the warning by the interlocking devices for all directions.

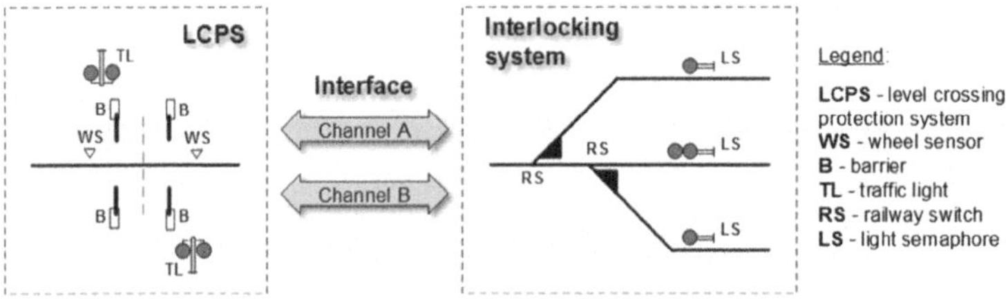

Fig. 1. Digital interface between the LCPS and the interlocking system.

The methodology of logical diagnosis requires in the next step to create a logical base that will take into account all possible combinations of symptoms and failures, included the all possible interface states. For n symptoms and m failures 2^{n+m} states are obtained, which may significantly hinder the use of this method for a large number of analyzed failures or and symptoms (in this case it is $2^{5+3} = 256$ states). At the beginning it is necessary to create a logical base (see Tables 1 and 2) for failures, containing 2^m elements (C_0, C_1,..., C_7) and separately for symptoms, containing 2^n elements (C^0, C^1,..., C^{31}).

Table 1. Logical base of interface failures.

	C_0	C_1	C_2	C_3	C_4	C_5	C_6	C_7
f_1	0	1	0	1	0	1	0	1
f_2	0	0	1	1	0	0	1	1
f_3	0	0	0	0	1	1	1	1

The following convention of symbols has been adopted in built logical bases: C_i - set of failures corresponding to the ith column in the failures logical base, and C^k - set of symptoms corresponding to the kth column in the symptoms logical base.

Table 2. Logical base of failures symptoms.

	C^0	C^1	C^2	C^3	C^4	C^5	C^6	C^7		C^{28}	C^{29}	C^{30}	C^{31}
s_1	0	1	0	1	0	1	1	1		0	1	0	1
s_2	0	0	1	1	0	0	1	1	…	0	0	1	1
s_3	0	0	0	0	1	1	0	1		1	1	1	1
s_4	0	0	0	0	0	0	0	0		1	1	1	1
s_5	0	0	0	0	0	0	0	0		1	1	1	1

The adopted symbols allow for their easy interpretation: i in binary notation means a set of failures, and k in binary notation means a set of symptoms. For example, C_3 stands for a set of failures $f_1 \cdot f_2 \cdot \overline{f_3}$, a C^4 means a set of symptoms $\overline{s_1} \cdot \overline{s_2} \cdot s_3 \cdot \overline{s_4} \cdot \overline{s_5}$. Based on the logical base for symptoms and the logical base for failures, it is possible to create a common logical base for failures and their symptoms (see Table 3).

Table 3. Logical base of interface failures and their symptoms.

	C_0^0	C_0^1	C_0^2	C_0^3		C_0^{28}	C_0^{29}	C_0^{30}	C_0^{31}	C_1^0	C_1^1	C_1^2	C_1^3		C_1^{28}	C_1^{29}	C_1^{30}	C_1^{31}
s_1	0	1	0	1		0	1	0	1	0	1	0	1		0	1	0	1
s_2	0	0	1	1		0	0	1	1	0	0	1	1		0	0	1	1
s_3	0	0	0	0		1	1	1	1	0	0	0	0		1	1	1	1
s_4	0	0	0	0	…	1	1	1	1	0	0	0	0	…	1	1	1	1
s_5	0	0	0	0		1	1	1	1	0	0	0	0		1	1	1	1
f_1	0	0	0	0		0	0	0	0	1	1	1	1		1	1	1	1
f_2	0	0	0	0		0	0	0	0	0	0	0	0		0	0	0	0
f_3	0	0	0	0		0	0	0	0	0	0	0	0		0	0	0	0

	C_6^0	C_6^1	C_6^2	C_6^3		C_6^{28}	C_6^{29}	C_6^{30}	C_6^{31}	C_7^0	C_7^1	C_7^2	C_7^3		C_7^{28}	C_7^{29}	C_7^{30}	C_7^{31}
s_1	0	1	0	1		0	1	0	1	0	1	0	1		0	1	0	1
s_2	0	0	1	1		0	0	1	1	0	0	1	1		0	0	1	1
s_3	0	0	0	0		1	1	1	1	0	0	0	0		1	1	1	1
s_4	0	0	0	0	…	1	1	1	1	0	0	0	0	…	1	1	1	1
s_5	0	0	0	0		1	1	1	1	0	0	0	0		1	1	1	1
f_1	0	0	0	0		0	0	0	0	1	1	1	1		1	1	1	1
f_2	1	1	1	1		1	1	1	1	1	1	1	1		1	1	1	1
f_3	1	1	1	1		1	1	1	1	1	1	1	1		1	1	1	1

Table 4. Reduced logical base of interface failures and their symptoms.

	C_1^1	C_1^2	C_1^3	C_1^4	C_1^5	C_1^6	C_1^7	C_2^2	C_4^8	C_4^{16}	C_4^{24}
s_1	1	0	1	0	1	0	1	0	0	0	0
s_2	0	1	1	0	0	1	1	1	0	0	0
s_3	0	0	0	1	1	1	1	0	0	0	0
s_4	0	0	0	0	0	0	0	0	1	0	1
s_5	0	0	0	0	0	0	0	0	0	1	1
f_1	1	1	1	1	1	1	1	0	0	0	0
f_2	0	0	0	0	0	0	0	1	0	0	0
f_3	0	0	0	0	0	0	0	0	1	1	1

The columns of this table form products $C^k \cdot C_i$ and have been marked with the symbol C_i^k. Unfortunately, the size of the logical base increases significantly along with the increase in the number of analyzed elements. Therefore, it is necessary to reduce the logical base by taking into account the technical knowledge of the analyzed interface. For this purpose, the function *E* is used, whose task is to select in the set of all elements of the logical base, only those that are possible. As a result of these actions, the reduced logical base is obtained (see Table 4).

Note that the occurrence of failures signatures: $C^1 = s_1 \cdot \overline{s_2} \cdot \overline{s_3} \cdot \overline{s_4} \cdot \overline{s_5}$, $C^3 = s_1 \cdot s_2 \cdot \overline{s_3} \cdot \overline{s_4} \cdot \overline{s_5}$, $C^4 = \overline{s_1} \cdot \overline{s_2} \cdot s_3 \cdot \overline{s_4} \cdot \overline{s_5}$, $C^5 = s_1 \cdot \overline{s_2} \cdot s_3 \cdot \overline{s_4} \cdot \overline{s_5}$, $C^6 = \overline{s_1} \cdot s_2 \cdot s_3 \cdot \overline{s_4} \cdot \overline{s_5}$, $C^7 = s_1 \cdot s_2 \cdot s_3 \cdot \overline{s_4} \cdot \overline{s_5}$ leads to failure f_1, while failure signatures: $C^8 = \overline{s_1} \cdot \overline{s_2} \cdot \overline{s_3} \cdot s_4 \cdot \overline{s_5}$, $C^{16} = \overline{s_1} \cdot \overline{s_2} \cdot \overline{s_3} \cdot \overline{s_4} \cdot s_5$, $C^{24} = \overline{s_1} \cdot \overline{s_2} \cdot \overline{s_3} \cdot s_4 \cdot s_5$ leads to failure f_3. Unfortunately for the signature $C^2 = \overline{s_1} \cdot s_2 \cdot \overline{s_3} \cdot \overline{s_4} \cdot \overline{s_5}$ function of failures has the form $F = C_1 + C_2$. This means that when observing signature C^2, it cannot clearly be determined which failure has occurred. In this case, however, using the Bayesian rule, the question of which of the possible failure is most likely to occur can be answered.

3 Bayesian Rule Application in Diagnostics

In the process of operating technical objects cause and effect chains can be found. If the cause of the event is denoted with A_k and its effect with *B*, then the conditional probability of the occurrence of event A_k under the condition of the occurrence of event *B*, when $P(B) > 0$ is defined by Bayes formula [5]:

$$P(A_k|B) = \frac{P(A_k)P(B|A_k)}{\sum_{i=1}^{n} P(A_i)P(B|A_i)} \tag{4}$$

where: $k = 1,2,\ldots, n$.

Probability $P(A_k)$ is called *a priori probability*, whereas probability $P(B|A_k)$ is called likelihood, because it represents the degree of belief in occurrence *B* if the event A_k is true. Since the values of these probabilities are usually unknown, their estimators are used. Finally, using the Bayesian formula, the probability $P(A_k|B)$ is calculated, i.e. the probability of the A_k event occurring provided that the event *B* has occurred. Probability $P(A_k|B)$ is called *a posteriori probability*.

The observation of the system and the analysis of the recorded states allow using the Bayesian rule to make the most reliable diagnosis for the symptoms occurring in the system. For example, the conditional probabilities $P(f_k|s_j)$ of a f_k failure at the observed symptom of s_j, can be determined using the formula:

$$P(f_k|s_j) = \frac{P(f_k)P(s_j|f_k)}{\sum_{i=1}^{K} P(f_i)P(s_i|f_i)} \tag{5}$$

where: *K* - number of possible failures.

Determination of the occurrence of failure in the system occurs when the condition of symptoms is consistent with the signature of the given failure, then:

$$P(f_k|s_1, s_2, \ldots, s_J) = \frac{P(f_k)P(s_1, s_2, \ldots, s_J|f_k)}{\sum_{i=1}^{K} P(f_i)P(s_1, s_2, \ldots, s_J|f_i)} \quad (6)$$

where: K - number of possible failures, J - number of failure signature elements

Assuming the independence of symptoms, formula (6) can be expressed in the following way:

$$P(f_k|s_1, s_2, \ldots, s_J) = \frac{P(f_k)P(s_1|f_k)P(s_2|f_k)\cdots P(s_J|f_k)}{\sum_{i=1}^{K} P(f_i)P(s_1|f_i)P(s_2|f_i)\cdots P(s_J|f_i)} \quad (7)$$

If the nth symptom does not have to occur, according to the failure signature, then formula (7) is inserted into following formula:

$$P(\bar{s}_n|f_k) = 1 - P(s_n|f_k) \quad (8)$$

In the logical diagnosis the possibility of occurrence different combinations of symptoms (failures and signatures) is taken into account. Then it becomes important to answer the question which failure is most likely for a given signature. While preserving symbols adopted in logical diagnostics, where i^{th} set of failures is marked as C_i, and k^{th} set of symptoms as C^k, based on the Bayesian rule (4) it is obtained:

$$P(C_i|C^k) = \frac{P(C_i)P(C^k|C_i)}{\sum_{j=0}^{2^m-1} P(C_j)P(C^k|C_j)} \quad (9)$$

where: $P(C_i)$ - probability of occurrence of the failures combination C_i, $P(C^k|C_i)$ - probability of observation of the symptoms combination C^k associated with the combination of failures C_i, m - number of failures.

In general case, for function of failure F and function of failure signatures V the Bayesian formula becomes [5]:

$$P(F|V) = \frac{\sum_{k \in V} \sum_{i \in F} P(C_i)P(C^k|C_i)}{\sum_{k \in V} \sum_{j=0}^{2^m-1} P(C_j)P(C^k|C_j)} \quad (10)$$

For case of interface failure considered in the previous chapter, when there is a lack of a clear indication of the type of failure, which results from the function of failure

$F = f_1 + f_2$ for the signature $V = \overline{s_1} \cdot s_2 \cdot \overline{s_3} \cdot \overline{s_4} \cdot \overline{s_5}$, it can be estimated using the following equations:

$$P(f_1|V) = \frac{P(f_1)P(\overline{s_1}|f_1)P(s_2|f_1)P(\overline{s_3}|f_1)P(\overline{s_4}|f_1)P(\overline{s_5}|f_1)}{\sum_{i=1}^{3} P(f_i)P(\overline{s_1}|f_i)P(s_2|f_i)P(\overline{s_3}|f_i)P(\overline{s_4}|f_i)P(\overline{s_5}|f_i)} \quad (11)$$

$$P(f_2|V) = \frac{P(f_2)P(\overline{s_1}|f_2)P(s_2|f_2)P(\overline{s_3}|f_2)P(\overline{s_4}|f_2)P(\overline{s_5}|f_2)}{\sum_{i=1}^{3} P(f_i)P(\overline{s_1}|f_i)P(s_2|f_i)P(\overline{s_3}|f_i)P(\overline{s_4}|f_i)P(\overline{s_5}|f_i)} \quad (12)$$

The calculated probabilities allow to indicate which of the possible failures is the most probable. For example, on the basis of the maintenance data, acquired by the authors, regarding the described interface, a total of 271 different failures were observed in the analyzed period, including 262 failures f_1, 8 failures f_2 and 1 failure f_3. For the functions of failures $F = f_1 + f_2$ and signature V, the probabilities of the individual failures have been calculated using Eqs. (11) and (12) and the values of: $P(f_1|V) = 0.9762$ and $P(f_2|V) = 0.0238$, respectively, have been obtained. On the basis of the above, it can be concluded that the most probable cause of failure f, with observed symptoms $\overline{s_1} \cdot s_2 \cdot \overline{s_3} \cdot \overline{s_4} \cdot \overline{s_5}$, is failure f_1. Therefore, the technician should first check the occurrence of f_1 failure.

4 Conclusion

The railway traffic control systems are used to guarantee the efficient and safe railway operation. These systems, like most of complex technical systems, consist of multiple components, modules and devices. Technical progress made in the area of railway traffic control systems causes that they are built as electronic systems, with common use of information and communication technologies. This enables the development of IT systems supporting the diagnostic process, which not only increases the reliability of railway traffic control systems, but also improves the safety of railway traffic. Currently, manufacturers in the railway automation sector offer a wide spectrum of diagnostic systems which are suitable for their solutions and devices. Therefore, scientific research is conducted and activities related to the implementation of new diagnostic methods dedicated to railway traffic control systems are undertaken. Nowadays, due to the significant impact on the safety of high-speed railways, the greatest attention is paid to the diagnosis of switches. Attention is also paid to the necessity of standardization of the interfaces for obtaining diagnostic data. The paper discusses the possibility of applying logical diagnosis and Bayesian reasoning in the diagnostic process of such systems. The authors were unable to find the researches of other authors' concerning the proposed methods in the field of the railway automation diagnostics. The presented example of the diagnostics of the digital interface linking the LCPS system and the interlocking system confirmed the practical possibility of implementing the proposed diagnostic methods. Therefore, positive research results encourage the authors to include other types of railway traffic control systems in the planned research. As an example, the digital interface of the link between the level crossing protection system and the interlocking system was used. Positive effects of the conducted research induce the authors to extend the scope of the studies to other types of railway traffic control systems.

References

1. Alzbutas, R., Iešmantas, T.: Application of Bayesian methods for age-dependent reliability analysis. Qual. Reliab. Eng. Int. **30**(1), 121–132 (2014)
2. Durmus, M.S., Ustoglu, İ., Tsarev, R.Y., Schwarz, M.: Modular fault diagnosis in fixed-block railway signaling systems. IFAC-PapersOnLine **49**(3), 459–464 (2016)
3. Felice, F., Petrillo, A.: Methodological approach for performing human reliability and error analysis in railway transportation system. Int. J. Eng. Technol. **3**(5), 341–353 (2011)
4. Flammini, F.: Railway Safety, Reliability, and Security: Technologies and Systems Engineering. IGI Global, Hershey (2012)
5. Grabski, F., Jaźwiński, J.: Metody bayesowskie w niezawodności i diagnostyce. WKŁ, Warszawa (2001)
6. Kornaszewski, M., Bojarczak, P., Pniewski, R.: Introduction of world innovative technologies to railway transport in Poland. In: Kliestik, T. (ed.) Globalization and Its Socio-Economic Consequences, vol. 3, pp. 1036–1043 (2016)
7. Langseth, H., Portinale, L.: Bayesian networks in reliability. Reliab. Eng. Syst. Saf. **92**(1), 92–108 (2007)
8. Lewiński, A., Perzyński, T.: The reliability and safety of railway control systems based on new information technologies. In: Mikulski, J. (ed.) Communications in Computer and Information Science, vol. 104, pp. 427–433. Springer, Heidelberg (2010)
9. Łukasik, Z., Nowakowski, W., Ciszewski, T., Freimane, J.: A fault diagnostic methodology for railway automatics systems. Procedia Comput. Sci. **149**, 159–166 (2019)
10. Márquez, F.P.G., Pedregal, D.J., Roberts, C.: New methods for the condition monitoring of level crossings. Int. J. Syst. Sci. **46**(5), 878–884 (2015)
11. Nowakowski, W., Ciszewski, T., Łukasik, Z.: The concept of railway traffic control systems remote diagnostic. In: Mikulski, J. (ed.) Communications in Computer and Information Science, Smart Solutions in Today's Transport, vol. 715, pp. 471–481. Springer, Cham (2017)
12. Nowakowski, W., Ciszewski, T., Młyńczak, J., Łukasik, Z.: Failure evaluation of the level crossing protection system based on fault tree analysis. In: Macioszek, E., Sierpiński, G. (eds.) Recent Advances in Traffic Engineering for Transport Networks and Systems. Lecture Notes in Networks and Systems, vol. 21, pp. 107–115. Springer, Cham (2018)
13. Perpinya, X.: Reliability and Safety in Railway. InTech, Rijeka (2012)
14. Restel, F.J.: The Markov reliability and safety model of the railway transportation. In: Nowakowski, et al. (eds.) Safety and Reliability: Methodology and Applications. Taylor & Francis Group, London (2014)
15. Stawarz, P.: Bayesowskie modele w diagnostyce. Politechnika Wrocławska, Wrocław (2006)
16. Straub, D., Kiureghian, A.: Bayesian network enhanced with structural reliability methods: methodology. J. Eng. Mech. **136**(10), 1248–1258 (2010)
17. Tang, L.: Reliability assessments of railway signaling systems: a comparison and evaluation of approaches. Norwegian University of Science and Technology, Trondheim (2015)

Ensuring Railroad's Digital Automation Systems Resistance to Dangerous States

Sergii Panchenko, Anatolij Bojnik, Viktor Kustov(✉), and Vasyl Sotnyk

Ukrainian State University of Railway Transport, Feierbakh Sq. 7, Kharkiv 61050, Ukraine
panchenko074@ukr.net, at@kart.edu.ua, kvf@satep.com.ua, sotnyk@gmail.com

Abstract. Allowable functional safety levels of railroad's digital automation systems set by EN50129 standard, cannot ensure required safety of a train traffic, especially for high speed. This article proposes methods to define functional safety rules and to confirm their correctness at the initial set up pre-check and while operation phase. Proposed methods to verify hardware and software architecture, which includes special methods to estimate probability of critical failure plus functional safety test methods.

Keywords: Safety Integrity Level · Hazardous failure · Railway control systems

1 Introduction

Primary area of focus for railroad automation systems evolution is to upgrade them from old technologies based on electromagnetic relays to digital technologies, based on microprocessors. Since hazardous failure probability for electronic components is hundreds thousand times more than for Class-I electromagnetic relays, it must find a way to ensure functional safety (FS) of them. Way to solve this problem is to develop functional safety requirements for electronic components used.

2 Analysis and Problem Statement

Specialists in the field of development, standardization, certification and operation of electronic systems related to safety should correctly justify the acceptable safety levels for the developed systems. Errors arising from this, can lead to significant damage, therefore, the correct rationing of functional safety is of great importance.

International and European standards for systems, which may cause safety issues, defines allowed hazardous failure probability per hour or level of critical failure per safety function, which are named Safety Integrity Level (SIL) [1–4]. These levels are calculated as hazardous failure probability for expected operating time and then divided by this time [1]. Such requirements are acceptable for control systems with low count of safety features, but probability of hazardous failure, obviously spikes with growing

E. Ginters et al. (Eds.): ICTE ToL 2019, LNITI, pp. 120–128, 2020.
https://doi.org/10.1007/978-3-030-39688-6_17

in a count of safety functions in system and every such failure may endanger a whole system. For example, for railway control systems such a microprocessor centralization of railroad arrows and signals (MPC), it has around 10 safety functions per railroad arrow. So, for the whole MPC system which manages 100 railroad arrows, European safety standards (SIL1–SIL4) will be thousands of times less restrictive. This leads to the situation when functional safety requirements become less restrictive for complex systems with big number of safety features.

In Table 1 it listed theoretical allowed hazardous failure levels for MPC system with 1000 safety features and hazardous failure probabilities of system for exponential distribution of probability failures.

Table 1. Safety Integrity Level by standards and theoretical levels of functional safety for MPC of large railway station (100 railway arrows).

SIL by EN and IEC standards (probability of hazardous failure per safety feature)	Allowed theoretical hazardous failure probability level for SCS (per 1000 safety features)	Theoretical hazardous failure probability for MPC (per 1000 safety features) 10 years	Theoretical hazardous failure probability for MPC (per 1000 safety features) 20 years
SIL4 (10^{-9}–10^{-8})	SIL4 (10^{-6}–10^{-5})	0,0839–0,5835	0,1607–0,8267
SIL3 (10^{-8}–10^{-7})	SIL3 (10^{-5}–10^{-4})	0,5835–0,9998	0,8267–0,9999
SIL2 (10^{-7}–10^{-6})	SIL2 (10^{-4}–10^{-3})	0,9998–1	0,9999–1
SIL1 (10^{-6}–10^{-5})	SIL1 (10^{-3}–10^{-2})	1	1

Obviously, such train traffic safety requirements are not acceptable and must be reviewed by EN and IEC standards. It want to highlight that more strict safety rules will lead to development complexity and higher cost of MPC, but less strict safety requirements will lead to incidents and disasters. The goal of this article is to develop methods to define new safety rules and provide recommendations to review requirements of most important standards EN and IEC, for safety-critical electronic systems. Railway system MPC will be used as an example.

3 Methods of Standardization of the Safety Function

3.1 SIL Based on Railway Relay of the 1st Reliability Class

Considering that railroad relay systems provide acceptable train safety and relays of the 1st reliability class successfully fulfill hazardous safety functions, the requirements to permissible probability of dangerous failure per hour for safety levels SIL4 can be determined by the operational (actual) rate of hazardous failure of Class-I relay:

$$PFH_d \leq \lambda_{d.Class-Irelay}. \tag{1}$$

3.2 SIL Based on the Permissible Number of Hazardous Failures

Permissible probability of dangerous failure per hour for safety levels SIL1–SIL4 is determined by the acceptable number of hazardous failures during the period of operation:

$$PFH_d = \frac{n_d(t)}{tN_{sf}}, \tag{2}$$

where $n_d(t)$ – count of hazardous failure happened; t – system service lifetime; N_{sf} – safety features count used while systems operation (should be statistically enough for an acceptable results confidence level).

3.3 SIL Based on the Elimination of All Hazardous Failures

Permissible probability of dangerous failure per hour for safety levels SIL4 for a certain period of system operation (less than 1 hazardous failure occurrence) will be determined by the following expression:

$$PFH_d \leq .\frac{1}{tN_{sf}}. \tag{3}$$

3.4 SIL Based on the Permissible Hazardous Systems Failure Rate

Some standards define functional safety (FS) for a system in a whole or per some conditional meter, for example, per small or large railway station, per automatic traffic control system, per 1 km of railway, per 1 railway crossing, etc. [5–7]. Given that the number of critical functions N_{sf} in such systems will be significantly different, it is proposed to determine the permissible hazardous failure rate per safety function as follows:

$$PFH_d \leq \frac{PFH_{d.total}}{N_{sf}}, \tag{4}$$

where $PFH_{d.total}$ – accepted hazardous failure intensity by a system in a whole by abstract meter.

3.5 SIL According to the Risk Substitution Method

The permissible level of dangerous failures of the safety function is proposed to be determined on the basis of statistics on the number of dangerous failures operated safety systems acceptable to society:

$$PFH_d = \frac{n_{d.total}(t)K}{100tN_{sf}}, \tag{5}$$

where $n_{d.total}$ – number of hazardous failures of a large number of systems used for a certain period of operation t; K – the percentage of safety features critical failures caused by the components design to the total number of system's critical failures, according to their operation statistic; N_{sf} – amount of safety features used (the size of the statistical sample which must provide a high level of confidence).

4 Calculation of the SIL Based by Proposed Methods

4.1 Calculation of the SIL Based on Railway Relay of the 1st Reliability Class

According to operation data of railroads in former USSR, for over 5 years (43800 h) 15 million Class-I electromagnetic relays was recorded 9 hazardous failures [8], so their operational hazardous failure rate has to be determined as:

$$\lambda_{Class-Irelay} = 0{,}14 \times 10^{-10} \text{ per hour.}$$

Thus, the acceptable hazardous failure rate for the strictest SIL4 level per one safety feature of safety-related systems will be equal to:

$$PFH_d \leq 0{,}14 \times 10^{-10} \text{ per hour.}$$

4.2 Calculation of the SIL Based on the Elimination of All Hazardous Failures

To define FS requirements, it is recommended to take as an example Ukrainian railroads (14th world-largest), which operate about 45 thousand railroad arrows (respectively 450 thousand safety functions for arrows and signals control systems) and which is similar to Europeans railroads. In order to exclude the possibility of critical failures in microprocessor control systems for arrows and signals over a 20-year operation period (175200 h), the acceptable critical failure rate of the system for SIL4, according to expression (3) should be equal to:

$$PFH_d \leq 0{,}127 \times 10^{-10} \text{ per hour.}$$

4.3 Calculation of the SIL Based on the Permissible Hazardous Systems Failure Rate

In the standard [5], the FS level for railway automation systems defined as acceptable critical failure rate of 10^{-7} per hour. per train station with up to 22 railway arrows s and 10^{-9} per hour for stations with more than 22 railway arrows.

Then, for small railroad stations (up to 8 railway arrows, up to 80 safety functions), according to expression (4), the acceptable value will be calculated as: $PFH_d \leq 0{,}125 \times 10^{-8}$ per hour.

For large railroad stations (80 railroad arrows, 800 safety functions), the acceptable value will be calculated as: $PFH_d \leq 0{,}125 \times 10^{-11}$ per hour.

An analysis of the data obtained, raises concern related to decreasing safety requirements (a thousand times) for small (often intermediate) stations, at which train speed is significantly higher than at large stations and failure consequences will be significantly severe. Decreasing of railway switches amount from 22 to 23 leads to the same safety requirement decrease. This means that the standard analyzed [7] is not well fit to define safety requirements for railroad control systems.

In the standard [6], the FS standard for automatic railway crossing signaling is the permissible dangerous failure rate of 10^{-8} per hour railway crossing. Then for most system (up to 20 responsible functions) the permissible hazardous failure rate will be 20 times less and amount to $0{,}5 \times 10^{-9}$ per hour.

In the standard [7], the FS standard for railway automatic blocking systems at railway is the permissible failure rate of 10^{-9} per hour per one kilometer of the railway length. While for most systems (from 10 to 20 responsible functions per 1 km), the permissible intensity of the dangerous failures is from $0{,}5 \times 10^{-9}$ to 10^{-9} per hour.

According to the specified standards [5–7], the permissible hazardous failure rate for each safety function is a significant range of values from $0{,}125 \times 10^{-8}$ to $0{,}125 \times 10^{-11}$ per hour. Moreover, the consequences of accidents and catastrophes caused by dangerous failures of the indicated railway systems can be the same, which confirms the need to normalize not the safety indicators of the systems as a whole, but the safety indicators of individual responsible functions, depending on the possible consequences of their dangerous failures.

4.4 Calculation of the SIL Standard According to the Risk Substitution Method

According to the statistic for over 5 years of operation (43800 h) of ten thousand relay-based railway switches and signals control systems of the former USSR (2nd world-largest railroad, 220 thousand arrows, respectively 2.2 million safety features), 100 hazardous failures was observed [8] and only 8% was caused by operating results. In this case, the acceptable safety feature hazardous failure rate according to formula (5) for level SIL 3 is $PFH_d \leq 0{,}83 \times 10^{-10}$ per hour. For SIL 4, with x5 more restrictive safety requirements, acceptable safety feature hazardous failure rate for railroad station's automated control system will be $PFH_d \leq 0{,}166 \times 10^{-10}$ per hour.

Value calculated sets functional safety level of modern control systems to the same level as traditional relay-based systems, the safety of which was considered acceptable for many years (primarily safety improvement was almost impossible due to cost, e.g. redundancy of many relays required). In the same time even significant amount of hazardous failures (100 for 5 years) was considered acceptable, taking into account railroad size. In semiconductor-based systems even more strict safety rules are used, defined by Eqs. (1–3), since nowadays this will require significantly less overhead than before and cost of incidents remains the same at almost unacceptable level, especially for high-speed trains. For example, cost of high-speed train is about $25–30 million and in case of incident it mostly cannot be repaired. Consequences for passengers, infrastructure and railroad (due to fear of passengers for their safety) in general will be much more severe.

An analysis of the calculation results according to 4.1–4.4 shows that the most acceptable and promising way to ensure the FS is the method of defining the safety requirements by probability of safety function critical failure per certain period of operation. For the strictest SIL 4, rounded, it will be equal to $0{,}13 \times 10^{-10}$ per hour. In standard [9] for this safety level similar value are equal to $0{,}14 \times 10^{-10}$ per hour. When the standard [9] was developed in 2002, this value was calculated by the risk replacement method for systems used in former USSR, which was $0{,}7 \times 10^{-10}$ per hour for SIL 3 and $0{,}14 \times 10^{-10}$ per hour for SIL 4. The average of acceptable critical failure rate for 4.1., 4.2 и 4.4 also will be $0{,}14 \times 10^{-10}$ per hour.

Therefore, it is necessary to revise the safety requirements defined by the European standards CENELEC EN 50129 and the international standards IEC 61508. It is also necessary to exclude requirement to eliminate hazardous failure of the redundant channel before the appearance of a hazardous failure in another redundant channel, because if this is guaranteed, there will never be hazardous failure in redundant systems and there is no need to set requirements for them. It would be better to regulate the probability of resolving hazardous failure in one redundant channel until a hazardous failure will appear in another redundant channel or regulate acceptable diagnosis time and recovery from hazardous failure in one redundant channel.

For advanced safety requirements, it is necessary to determine requirements for the redundant channels based on their type, i/o logic and system type by failures possible [10].

5 Calculation of the Probability of Danger Failure

Minimum permissible value of probability of dangerous failure per hour of every redundant channel by structure «2» from «3» and «2» from «2» with safety logic solver should be defined, for exponential failure rate distribution, according to [11–13], by next equations:

$$PFH_{d.1.2v3} = \sqrt{\frac{PFH_d}{6(MTBM_d + MTBM_c)}}; \tag{6}$$

$$PFH_{d.1.2v2} = \sqrt{\frac{PFH_d}{2(MTBM_d + MTBM_c)}}, \quad (7)$$

where *PFH* – probability of dangerous failure per hour acceptable intensity of system critical failure (SIL); $MTBM_д$ – maximum acceptable mean time between maintenance of system; $MTBM_c$ – mean time between maintenance corrective of each channel.

The calculation according to formulas (6)–(7) is performed without the need to determine the intensity of dangerous failures of elements, therefore, in the absence of multiple dangerous failures and dangerous failures of logic solvers, it has maximum reliability with a reasonable law of the distribution of dangerous failures.

In case when present multiple hazardous failure of each channel redundancy and logic solvers, Eqs. (6) и (7) will be transformed to the next form:

$$PFH_{d.1.2v3} = \sqrt{\frac{PFH_d - \lambda_{d.m} - n_{ls}\lambda_{ls1} - n_{ls}\lambda_{ls.m}}{6(MTBM_d + MTBM_c)}}; \quad (8)$$

$$PFH_{d.1.2v2} = \sqrt{\frac{PFH_d - \lambda_{d.m} - n_{ls}\lambda_{ls1} - n_{ls}\lambda_{ls.m}}{2(MTBM_d + MTBM_c)}}, \quad (9)$$

where $\lambda_{d.m}$ – multiple hazardous failure rate of redundant channel; n_{ls} – logic solver amount; λ_{ls1} – rate of single hazardous failure of logic solver; $\lambda_{ls.m}$ – rate of multiple hazardous failure of logic solver.

For the different of channel redundancy in double structure, for maximumpermissible fixed values $MTBM_d, MTBM_c, \lambda_{d.1}, \lambda_{d.2} = const$, minimal probability of dangerous failure per hour is defined as:

$$PFH_{d.1мin} = \frac{PFH_d - \lambda_{d.m} - n_{ls}\lambda_{ls1} - n_{ls}\lambda_{ls.m}}{2\lambda_{d.2}(MTBM_d + MTBM_c)}; \quad (10)$$

$$PFH_{d.2мin} = \frac{PFH_d - \lambda_{d.m} - n_{ls}\lambda_{ls1} - n_{ls}\lambda_{ls.m}}{2\lambda_{d.1}(MTBM_d + MTBM_c)}, \quad (11)$$

where $\lambda_{d.1}$; $\lambda_{d.2}$ - hazardous failure rate of the 1st, 2nd redundant channel.

The main disadvantage of calculations using formulas (8–11) is the need to obtain reliable data on the rates of single and multiple dangerous failures of the redundant channels and logic solvers.

6 Calculation of the Permissible Mean Time Between Dangerous Failure

Minimally permissible mean time to dangerous failure $MTTF_d$ of structure «2» from «3», «2» from «2» with safety logic solvers, according to researches [10–13] and exponential distribution law for dangerous failures of redundant channel, proposed to define by the next equations:

$$MTTF_{d.1.2v3} \geq \sqrt{\frac{6(MTBM_d + MTBM_c)}{PFH_d}}; \tag{12}$$

$$MTTF_{d.1.2v2} \geq \sqrt{\frac{2(MTBM_d + MTBM_c)}{PFH_d}}. \tag{13}$$

With multiple dangerous failures of redundant channels and logic solvers, Eqs. (12) и (13) will be transformed to:

$$MTTF_{d.1.2v3} = \sqrt{\frac{6(MTBM_d + MTBM_c)}{PFH_d - \lambda_{d.m} - n_{ls}\lambda_{ls1} - n_{ls}\lambda_{ls.m}}}; \tag{14}$$

$$MTTF_{d.1.2v2} = \sqrt{\frac{2(MTBM_d + MTBM_c)}{PFH_d - \lambda_{d.m} - n_{ls}\lambda_{ls1} - n_{ls}\lambda_{ls.m}}}. \tag{15}$$

For the different redundant channels in double structure, at maximum acceptable fixed $MTBM_d, MTBM_c$, $\lambda_{d.1}, \lambda_{d.2} = const$, minimal acceptable mean time to dangerous failure of each channel equals to:

$$MTTF_{d.1\min} = \frac{2\lambda_{\mathbf{d.2}}(MTBM_d + MTBM_c)}{PFH_d - \lambda_{d.m} - n_{ls}\lambda_{ls1} - n_{ls}\lambda_{ls.m}}; \tag{16}$$

$$MTTF_{d.2min} = \frac{2\lambda_{\mathbf{d.1}}(MTBM_d + MTBM_c)}{PFH_d - \lambda_{d.m} - n_{ls}\lambda_{ls1} - n_{ls}\lambda_{ls.m}}. \tag{17}$$

If during operation of the systems the operating time to a dangerous failure of any backup channel is less than permissible according to formulas (12–17), a repeated examination of the functional safety of the entire system is carried out and the reason for the violation of the functional safety norms is determined.

7 Conclusion

Article proposes methods to define acceptable safety requirements for railroad automation systems and individual backup channels. Such methods will help to define safety levels for standards governing the functional safety. These methods can also be applied to other critical electronic systems, where safety features may lead to significant material damage or threat to human life and unacceptable environmental impact. Methods and recommendations developed should help to improve European and international safety standards like EN 50129, IEC 61508 and similar.

The results obtained, in comparison with others, allow not subjectively, but objectively approach the appointment of FS norms. Moreover, the levels of SIL have a sufficiently reliable justification.

The direction of further research is the use of the obtained research results for proposals for correcting the levels of SIL and other sections of the basic standards FS.

The research results can be used for electronic signaling systems beyond the railroad cases, to ensure their safety in various fields where safety-critical systems are used, e.g. nuclear plants, aviation, space, chemical industry, medicine, military systems etc.

References

1. IEC 61508-1. Functional safety of electrical/electronic/programmable electronic safety-related systems - Part 1: General requirements (2010)
2. CENELEC - EN 50129. Railway applications. Communication, signaling and processing systems. Safety related electronic systems for signaling (2018)
3. BS EN 50129. Railway applications. Communication, signaling and processing systems. Safety related electronic systems for signaling (2018)
4. DIN EN 50129. 2019–06. VDE 0831-129. Bahnanwendungen -Telekommunikationstechnik, Signaltechnik und Datenverarbeitungssysteme. Sicherheitsbezogene elektronische Systeme für Signaltechnik. Deutsche Fassung EN 50129:2018 + AC (2019)
5. GOST P 54897-2012. Railroad automation and telemechanic for stations. Safety requirements and control methods (2012)
6. GOST P 54898-2012. Railroad automation and telemechanic for crossings. Safety requirements and control methods (2012)
7. GOST P 54900-2012. Railroad automation and telemechanic for railways. Safety requirements and control methods (2012)
8. Certification and safety proving for railroad automation systems. In: Sapozhnikov, V.I. (eds.) Transport, 288 p. (1997)
9. DSTU 4178-2003. Train traffic control systems. Functional safety and reliability. Requirements and testing methods, 32 p. National Consumer Standard Organization of Ukraine, Kyiv (2003)
10. Kustov, V.F.: Nuances of functional safety estimate for semiconductor railroad automation systems while development and operation. Transport safety. In: VIII International Scientific-Practical Conference Materials, pp. 193–95. Gomel (2017)
11. Kustov, V.F.: Mathematical models of functional safety of microprocessor systems of railway automation: textbook for high schools, pp. 65–71. UkrDAZT, Kharkiv (2010)
12. Kustov, V.F.: Mathematical models of functional safety and resilience for refurbished technical systems in case of redundancy «2» from «3». In: Collection of Scientific Works of DonIZT, vol. 23, pp. 5–14. Donetsk (2014)
13. Kustov, V.F.: Features of ensuring the functional safety of microprocessor control and monitoring systems in railway transport. In: Railway Transport of Ukraine, vol. 1, pp. 22–30 (2012)

Mathematical Models of the System Integration and Structural Unification of Specialized Railway Computer Systems

Valentin Moiseenko, Volodymyr Butenko, Oleksandra Golovko, Oleksandr Kameniev, and Vitalii Gaievskyi(✉)

Ukrainian State University of Railway Transport, Kharkiv 61050, Ukraine
butenko@kart.edu.ua

Abstract. The article presents the method of the comparative evaluation of centralization systems based on the graph theory using the adjacency matrices and graph distance matrix of the system structure. For these matrices, the coefficient of structural redundancy, the coefficient of structural compactness, the number of the principal contours of the connected graph and the degree of the structure centralization of the control system are calculated. The obtained indicators make it possible to analyze the structures of the control systems and to draw conclusions regarding their quality properties, in particular safety. The unfolded explanation is given describing how the basic properties of the system showing its functional operability are presented in the graph structure. It also explains the connection between the structure elements and the hierarchy levels of the microprocessor centralization of the control system with the number and spatial location of knots, their distribution on correspondent groups according to levels. In accordance with the connection lines between subsections and elements, the number and location of arcs constructing a graph are determined. As a practical application of the offered method, the example of the comparison of two wide-spread systems of the microprocessor centralization, namely, "Impuls" and "Ebilock-950" systems, is given. It enables structurally to unify and on the system level to integrate railway specialized computer systems. The results of the research presented in the article are intended for specialists in the field of development and research of railway control systems.

Keywords: Graph distance matrices · Railway control system · Microprocessor-based controllers

1 Introduction

Practically all the existing domestic and foreign microprocessor control systems of electric train service fulfill equal functions, but yet all of them have individual peculiarities both at a software and hardware level. At present a technical and process staff

Research materials agreed by the team of authors, taking into account the distribution of copyright between each of them.

E. Ginters et al. (Eds.): ICTE ToL 2019, LNITI, pp. 129–136, 2020.
https://doi.org/10.1007/978-3-030-39688-6_18

doesn't pay attention to those peculiarities because an implementation and operation of those systems take place independently of one another, and there is little necessity of their integration. But new problems arise at the modern stage of development of the world transportation techniques. The active interaction between the system and facility complexes of different-purpose railway automation demands that they should be integrated into a uniform system, which, in its turn, demands that operation logic of different-purpose systems and facilities should be revised as the stage of mutual integration and unification. When developing the operation of railway systems, the practice of using the methods of other branches has made a good showing. For instance, for providing a small-capacity railway transport with adaptive control features the mathematic model [1] is optimized with the help of a coded genetic algorithm. For that, it is advisable to investigate the structures of some existing systems of modern microprocessor-based interlocking in order to show the interactions of structure elements and to evaluate their interference when designing and operating the management-information systems of railway automation. Such structures are most effectively analyzed when using the graph theory [2]. The graph theory has already been used by co-authors when solving different tasks, particularly, when using the maximal clique method in undirected graphs [3]. However, there is a problem of evaluating the state of microprocessor system facilities at the stage of minor statistical data.

2 Analysis and Definition of Research Problem

The model of facility state definition of the railway automation control system with limited statistical data is expanded in paper [4]. The model forecasts a technical state of railway automation facility under the limited statistical data. Paper [5] reveals the structure "weak points" and analyzes the system behavior. The use of such features is possible after building-up the system.

At the design phase the microprocessor-based interlocking structure is defined as a complicated one, which may be pre-estimated from different viewpoints. Just in paper [6] the centralization-degree classification of structures is presented, and kinematic structure properties are computed according to it. There is lack of common opinion on the system consisting of simpler structures, and some other non-kinematic structure characteristics are not considered. Paper [7] analysis has revealed that some investigation methods of systems are difficult for understanding or weren't be used for electric train operation control systems at the stations. So, for microprocessor-based interlocking system structure research the tabular-analytical method is advised to be used as it is simple and clear. The similar method is used in an adjacent area of carrying by air and is presented in papers [8–10]. In paper [8] special attention is paid to a classification of elementary structures. The paper also presents the evaluation parameters of activity of consolidated air traffic service centers based on sub-units work, i.e. flight service which, in its turn, carries out the basic work of the first-hand air traffic control. This approach is badly applied to microprocessor-based interlocking with another structure. Paper [9] presents communication problems among different sub-units of traffic control and evaluates communication effectiveness of operation posts which are of a great importance for safe flights. For microprocessor-based interlocking structure those methods are extreme and demand a revision for more effective application. Paper [10] presents an evaluation of a group work and cooperation principles with the help of graphs. To use microprocessor-based interlocking in evaluation area, it is necessary to reconsider the demonstration of those principles when interacting with different organization levels of the above structure in contrast to a specific character of air traffic control operation. The critical structure properties leading to drop of control system operation may be revealed on the basis of this method. The comparative structure analysis is used for the research.

3 Theoretical Background

Taken from a graph theory [2] the graph $G(X, U)$ is an aggregate of two sets: the X point set, named as vertices, and the U rib set. Every element $u_{ij} \in U$ is a couple of elements (x_i, x_j) ordered by X set, vertices x_i and x_j are named as endpoints or ends of rib u_{ij}. Named as the control system graph is the graph where the set of X vertices interprets the control system element set and U rib set – the set of links among them (2.1). It will be written as follows: $G = G(X, U)$, where X – control system element set; U – control system rib set.

For research, a graphical method (as a diagram) and a matrix method are used to represent a graph. The diagram graph vertices are represented as a circle with the umbers inside it. The graph structure description give an idea of the system structure. The routing and character of links are important.

The purpose of the article is to investigate system integration and structural unification of railway specialized computer systems by graph-analytical method.

4 The Railway Controllers Assessment Approach

The mentioned paper represents the structures of two microprocessor-based interlockings of switches and signals of "Impuls" and "Ebilock-950" that are in service of Ukrainian general use railways.

Let's determine the "Ebilock-950" system structure elements and associate them with control system "Ebilock-950" graph vertices (see Table 1) and develop "Ebilock-950" system structure graph presented in Fig. 1.

Every graph vertex which correspondы to a definite element of communications network hardware is designed by a circle with an object number. In this paper, understood as a matrix are the graph adjacencies with n vertices, a matrix with the element $a_{ij} = 1$, if vertex i is adjacent to j and $a_{ij} = 0$, otherwise (1):

$$A = \left\| a_{ij} \right\| nxn, \tag{1}$$

where n – the graph vertex; a_{ij} – graph matrix element.

Similarly, the graph of the "Impuls" system is constructed based on Table 2. In accordance with this table, similar rules are used to construct the system graph, identical to the principles given in Fig. 1.

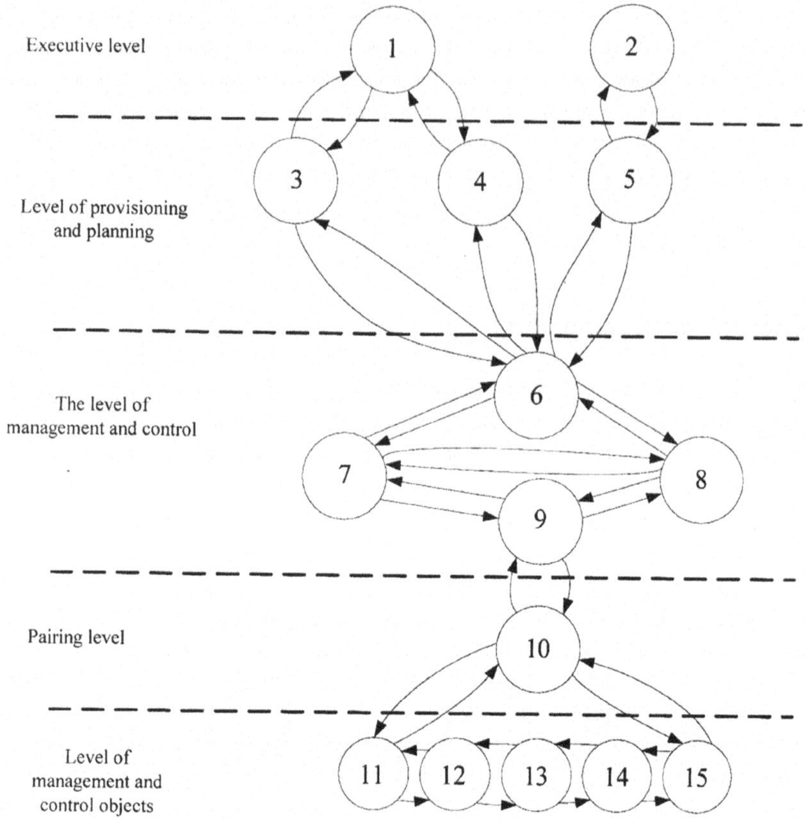

Fig. 1. "Ebilock-950" system structure graph.

Table 1. Correspondence of "Ebilock-950" system structure elements to graph vertices.

Hierarchy levels	Structure element	Graph vertex
Executor level	Operator	1
	Engineer	2
Provision and planning level	Operator workstation (main)	3
	Operator workstation (reserve)	4
	Engineer workstation	5
Net	Communication network	6
Control and check level	Central computer 1	7
	Central computer 2	8
Net	Communications network	9
Interface level	Concentrator	10
	Object controller	11, 12, 13, 14, 15

Table 2. Correspondence of "Impuls" structure elements to graph vertices.

Hierarchy levels	Structure element	Graph vertix
Executor level	Operator	1
	Engineer	2
	Tuner	3
Provision and planning level	Operator workstation (main)	4
	Operator workstation (reserve)	5
	Engineer workstation	6
	Distance dispatching workstation	7
Net	Communication switch board 1	8
	Communication switch board 2	9
Control and check level	Control calculator	10
	Control calculator	11
	Control calculator	12
Net	Communication module	13, 14, 15
Interface level	Microprocessor 1	16, 18, 20, 22, 24
	Microprocessor 2	17, 19, 21, 23, 25
	Wayside apparatus connection	26, 27, 28, 29, 30

As a result of representing the graphs of control system structures, it has becomes clear that they are oriented to that one by using a matrix method of presentation, and matrices may be got for their mathematical research. The system adjacency matrix of "Ebilock-950" and "Impuls" systems are specified.

In the built-up adjacency matrix, the sum of line elements is equal to a degree of outcome out of a corresponding vertex and the sum of column elements – to a degree of income. Formulas of degree of vertex outcomes:

$$p_i^+ = \sum_{j=1}^{n} a_{ij}, \tag{2}$$

and degrees of vertex incomes:

$$p_j^- = \sum_{k=1}^{n} a_{kj}, \tag{3}$$

where p_i^+ i p_j^- - outcome and income degrees of i and j vertices correspondingly; n – number of graph vertices; a_{ij} – adjacent matrix element A.

The degree p_i of vertex is determined as a total amount of arcs incidental to this vertex, i.e. income degree and outcome degree of vertices:

$$p_i = p_i^+ + p_i^-, \tag{4}$$

The income and outcome degrees of vertices for of computer system structure graph are presented in Tables 3 and 4. The graph interpretation method described in this matrix allows it to be used to verify the integrity of the graph in order to diagnose the system.

Table 3. The vertex income-outcome degrees of "Impuls" structure graph.

Vertices	1	2	3	4	5	6	7	8	9	10	11	12	13	14	15
p_i^+	2	1	1	3	3	3	3	7	7	5	5	5	11	11	11
p_i^-	2	1	1	3	3	3	3	7	7	5	5	5	11	11	11
p_i	4	2	2	6	6	6	6	14	14	10	10	10	22	22	22

The value p_i is the result of summing p_i^+ and p_i^-. The maximal p_i value in the income and outcome degree Table indicates the most loaded element in the control system structure.

Table 4. The vertex income-outcome degrees of "Impuls" structure graph (continuation).

Vertices	16	17	18	19	20	21	22	23	24	25	26	27	28	29	30
p_i^+	4	4	4	4	4	4	4	4	4	4	2	2	2	2	2
p_i^-	4	4	4	4	4	4	4	4	4	4	2	2	2	2	2
p_i	8	8	8	8	8	8	8	8	8	8	4	4	4	4	4

The vertex income and outcome degrees of "Impuls" system structure graphThe vertex income and outcome degrees of "Ebilock-950" structure graph are presented in Table 5.

Table 5. The vertex income-outcome degrees of "Ebiblock-950" structure graph.

Vertices	1	2	3	4	5	6	7	8	9	10	11	12	13	14	15
p_i^+	2	1	2	2	2	5	3	3	3	3	2	2	2	2	2
p_i^-	2	1	2	2	2	5	3	3	3	3	2	2	2	2	2
p_i	4	2	4	4	4	10	6	6	6	6	4	4	4	4	4

The analysis of built-up graphs allows the determining of the fact that in "Ebilock-950" system the most loaded structure vertices are three communication modules which duplicate each other. In "Impuls" system the most loaded vertices are three vertices that are unreserved, and when one of them is faulty it's impossible to control the wayside apparatus. The values of structure redundancy and major loops are used when performing a large scope of work at the station, and relatively a large scope of information is transferred.

5 Conclusion

The control system analysis has shown up that both the systems have structure redundancy, and the value is not close to 0 (minimal value of structure redundancy). Besides, this “Impuls” value is considerably higher than that of “Ebilock-950”. This is the evidence of high survivability of the railway traffic control system, which is provided with facilities reservation.

The control loops define the most stable control system functional substructures with feedback. A high value indicates an increase of survivability and connectivity of microprocessor-based interlocking structure.

In such a case, it’s important to ensure the survivability of the system for its uninterrupted functioning when system elements or connectivity are faulty. If that’s the case, it is advisable to use the system with the largest values of the mentioned indices.

The next characteristic is structure compactness. The control system analysis has shown that both the systems have high structure compactness and it is close to 1, but in “Ebilock-950” it is lower. When analyzing the control loops amount, it has been defined that a larger amount of loops is in “Impuls” system and a smaller one in “Ebilock-950”. The built-up mathematical model of analyzing the control system gives the possibility of comparing quantitative characteristics of the control system and determining a more optimal one. The above-mentioned advantages when using the graph are usability, intelligibility and visualization of results.

In contrast to the traditional scientific approaches mentioned above, which were described in the introduction, the use of a graphoanalytic method provides simplification of the mechanism of research and much greater clarity and clarity of the results obtained. It is possible to quantitatively compare the characteristics of control systems and determine the best. Further development of the scientific results considered in the article may be the development of a mathematical model for the analysis and evaluation of automated control systems in the form of some vector-function, which would match the graph of some vector whose values characterize the system being followed.

The obtained scientific results can be used for certification studies of safety critical systems, development and design of microprocessor-based control systems for railway transport.

References

1. Prokhorchenko, A., Parkhomenko, L., Kymana, A., Matsiuk, V., Stepanova, J.: Improvement of the technology of accelerated passage of low-capacity car traffic on the basis of scheduling of grouped trains of operational purpose. Procedia Comput. Sci. **149**, 86–94 (2019). ICTE in Transportation and Logistics 2018 (ICTE 2018)
2. Tatt, U.: Teoriya grafov, p. 424. Mir, Moscow (1988). (in Russian)
3. Listrovoy, S., Butenko, V., Bryksin, V., Golovko, O.: Development of method of definition maximum clique in a non-oriented graph. East. Eur. J. Enterp. Technol. **5**(4), 12–17 (2017)
4. Moiseenko, V., Kameniev, O., Butenko, V., Gaievskyi, V.: Determination model of the apparatus state for railway automatics with restrictive statistical data. Procedia Comput. Sci. **149**, 185–194 (2019). ICTE in Transportation and Logistics 2018 (ICTE 2018)

5. Moiseenko, V., Kameniev, O., Gaievskyi, V.: Predicting a technical condition of railway automation hardware under conditions of limited statistical data. East. Eur. J. Enterp. Technol. **3**(9), 26–35 (2017)
6. Konovalov, V., Melihov, A., Kovtun, A.: Analiz ierarhicheskih struktur sistem mikroprocessornyh centralizacij zheleznodorozhnoj avtomatiki. Kinematika. Radioelektronni i komp'yuterni sistemi, no. 3(7), pp. 89–95. NAU im. N.E. Zhukovskogo «HAI», Kharkiv (2004)
7. Bochkov, K., Kovriga, A., Harlap, S.: Mikroprocessornye sistemy avtomatiki na zheleznodorozhnom transporte. Ucheb. posob., p. 255. BelGUT, Gomel (2013). (in Russian)
8. Reva, O., Vdovichenko, V., Ustimenko, I.: Ocinka strukturnoyi efektivnosti sluzhbi ruhu (na prikladi krasnoyarskogo centra obslugovuvannya povitryanogo ruhu). Aviacionno-kosmicheskaya tehnika i tehnologiya, no. 10, pp. 176–186. http://nbuv.gov.ua/UJRN/. Accessed 19 Nov 2019
9. Reva, A., Ustimenko, I., Koltakov, V.: Effektivnost organizacii kommunikacij v sisteme aeronavigacionnogo obsluzhivaniya poletov (na primere Krasnoyarskogo centra obsluzhivaniya vozdushnogo dvizheniya). Aviacionno-kosmicheskaya tehnika i tehnologiya, no. 7 (104), pp. 215–226. NAU im. N.E. Zhukovskogo «HAI», Kharkiv (2013). (in Russian)
10. Reva, O., Nevinicin, A., Biryukov, Y., Savonina, N.: Metodi teoriyi grafiv u strukturno-funkcionalnomu analizi organizaciyi diyalnosti dispetcherskoyi zmini, yak nevelikoyi grupi aviacijnih operatoriv. Naukovij visnik Hersonskoyi derzhavnoyi morskoyi akademiyi, no. 1 (8), pp. 267–275. HDMA, Herson (2013)

The Use of Digital Technologies to Provide Condition Monitoring and Modes of Vehicle Operation

Igor Gritsuk[1], Artur Kagramanian[2], Igor Khudiakov[1(✉)], Anastasija Pinchuk[3], Roman Symonenko[4], and Oksana Hrytsuk[5]

[1] Kherson State Maritime Academy, Kherson 73009, Ukraine
gritsuk_iv@ukr.net, igor.khudiakov563@gmail.com
[2] Ukrainian State University of Railway Transport, Kharkiv 61001, Ukraine
kartal2@ukr.net
[3] Riga Technical University, Kalku Street 1, Riga 1048, Latvia
midzun@gmail.com
[4] SE "State Road Transport Research Institute", Kiev, Ukraine
rsym1975@gmail.com
[5] H.S. Skovoroda Kharkiv National Pedagogical University, Kharkiv 61000, Ukraine
oxana.gri@gmail.com

Abstract. An integral part of the processes of management and communication support of modern transport and logistics systems is the use of technologies for remote information monitoring of the parameters of the technical condition when operating the vehicle. The task is to take into account the peculiarities of remote control by means of the IT modes of operation and rest of the driver while ensuring a control of the technical condition. The purpose of the study is analysis the interaction of monitoring tools to ensure compliance with the modes of operation taking into account road, transport, climatic conditions, etc. The initial data for monitoring is information about technical condition, safety and performance. A specialized digital tachograph equipped with a GPS monitoring unit is envisaged for the operation of the monitoring system. The results of joint interaction of technical means are presented.

Keywords: Automotive transport · Technical condition identification · Systems interaction

1 Introduction

In accordance with modern transport strategies, transport is the main part in modern countries economics. The effective functioning of this industry is necessary for ensuring the defense capability, improving the population living standards.

Nowadays, the transport sector satisfies the national economy in majority needs of cases transportation. The level of safety, quality and efficiency of passenger and cargo transportation, energy efficiency, man-made load on the environment is not convenient due to the modern requirements [1].

E. Ginters et al. (Eds.): ICTE ToL 2019, LNITI, pp. 137–144, 2020.
https://doi.org/10.1007/978-3-030-39688-6_19

The current state of informational and communication development and technology of the vehicle allows to ensure the solution of information workflow problems thanks to the rapid development of information resources and means of communication between vehicles [2]. The basic task of information transport operation tasks assigned practical realization of the synergistic association of computing resources of all road users in a single information space Internet; from a separate vehicle to transport enterprise-level organizations.

Numerous researchers have been involved in the formation of vehicle monitoring information systems. There exist some monitoring systems, developed in Kharkiv National Automotive and Road University (KHNARU) and National Technical University (NTU) [3], allow connecting vehicle identification, continuous automatic parameters measurement, characterizing its condition, diagnosis, serviceability, constructional elements, recognition malfunction and prevention their development. To sum up, technical condition to ensure maintenance system functioning [4]. These systems are a complex of on-board and software devices. However, parameters are not enough for monitoring, trouble-free operation.

The information system with the installed tachograph and tracker is based on the monitoring one [5]. Its peculiarity is the feature of regimes of work and rest periods (RWRP) of the driver, environmental indicators of vehicle, violations speed mode. That is the reason why this problem is being urgent at the moment.

The objective of the research is to solve the current issue of digital information usage for monitoring the technical condition and modes by means of tachograph and tracker system. Let's consider the possibility of obtaining information about technical condition of vehicle, and let's regard analysis and systematic interaction software monitoring tools to remotely operating of modes based traffic, transport, climate, and so on.

2 Outline of the Research

2.1 Information Support for Control of the Technical Condition and Vehicle Operation Modes

Hardware remote monitoring system is focused on major known onboard and stationary tools [3–5] and vehicle's technical condition regimes and rest are based on onboard known device [7]. In addition, remote identification and monitoring models for driving and resting modes of the driver were utilized using ITS and the Structured Analysis and Design Technique (SADT) methodology [8].

The initial data for monitoring the tachograph and tracker parameters in the system of technical condition is information about the vehicle technical condition, the safety and performance, the modes of operation, rest of the driver, design and operating conditions. All data is received remotely.

To implement the above provisions, the authors have developed the Motor Vehicle Safety and Performance Management (MVSPM) [9]. The MVSPM model was based on the structure and software application of the Information Program Complex (IPS) "IdenMonDiaOperCon (Identification, Monitoring Technical Condition, Diagnosis, Operating Conditions of the Vehicle under ITS) "HADI-15"" [5], which was used to obtain information about parameters of the technical condition of the vehicle under operating conditions.

2.2 Formation of Monitoring System Parameters

Model domain M_{comm} presented as the next set of components and component information system, namely the technical parameters of the vehicle engine, vehicle parameters, working hours and rest periods M_{tg}, vehicle condition and trailer (optional), vehicle environmental performance of M_{tr} [3–7]. The formula is in general:

$$M_{comm} = M_{v} + M_{tg} + M_{tr,} \tag{1}$$

where $M_{v} = \langle O_{v}, V_{v\ in.}, V_{v\ out.}, F_{v}, H_{v}, p_{v}, R_{v},\rangle$, according to information, which is shown in work [3]. In that work, in detail, system information and interaction for the vehicle as the first 79 elements of information was described. Further these characteristics with interaction tracker and tachograph will not be considered.

$$M_{tg} = \langle O_{tg}, V_{tg\ in.}, V_{tg\ out.}, F_{tg}, H_{tg}, P_{tg}, R_{tg},\rangle, \tag{2}$$

$$M_{tr} = \langle O_{tr}, V_{tr\ in.}, V_{tr\ out.}, F_{tr}, H_{tr}, P_{tr}, R_{tr},\rangle \tag{3}$$

where $O_{tg} = \left\{ o_{tgm} \middle| m_{tg} = \overline{1, M_{tg}} \right\}$ set of tachograph and tracker respectively automation objects in the system of monitoring the technical condition of the vehicle, which can be represented as separate parts of the blocks of collection and transmission of information such as: the identification of the vehicle, the condition of the vehicle, the time of the vehicle working, the speed of the vehicle, the condition of the trailer (optional equipment). All these parameters are forming a description of a model database of the information system for monitoring tachograph parameters in the system of technical condition of the vehicle:

- $V_{tg\ in} = \left\{ v_{tgl} \middle| \in L_{tg\ in} \right\}$, $V_{tr\ in} = \left\{ v_{trl} \middle| \in L_{tr\ in} \right\}$ are the set of input information elements for tachograph and tracker respectively;
- $V_{tg\ out} = \left\{ v_{tgl} \middle| \in L_{tg\ out} \right\}$, $V_{tr\ out} = \left\{ v_{trl} \middle| \in L_{tr\ out} \right\}$ are the set of output information elements for tachograph and tracker respectively;
- $V_{tg} = V_{tg\ in} \cup V_{tg\ out}$, $V_{tr} = V_{tr\ in} \cup V_{tr\ out}$ are the complete set of information elements for tachograph and tracker respectively;
- $F_{tg} = \left\{ f_{tgi} \middle| i_{tg} = \overline{1, I_{tg}} \right\}$, $F_{tr} = \left\{ f_{tri} \middle| i_{tr} = \overline{1, I_{tr}} \right\}$ are set functions use (automation functions) performed by the monitoring system parameters of tachograph and tracker respectively for tachograph and tracker respectively;
- $H_{tg} = \left\{ h_{tgj} \middle| j_{tg} = \overline{1, J_{tg}} \right\}$ $H_{tr} = \left\{ h_{trj} \middle| j_{tr} = \overline{1, J_{tr}} \right\}$ are a set of data processing tasks for the tachograph parameter monitoring system for tachograph and tracker respectively;
- $P_{tg} = \left\{ p_{tgk} \middle| k_{tg} = \overline{1, K_{tg}} \right\}$, $P_{tr} = \left\{ p_{trk} \middle| k_{tr} = \overline{1, K_{tr}} \right\}$ multiple users (number and composition of staff) that provide work with the tachograph and tracker parameters monitoring system;

- $R_{tg} = \{r_{tgy} | y_{tg} = \overline{1, Y_{tg}}\}$, $R_{tr} = \{r_{try} | y_{tr} = \overline{1, Y_{tr}}\}$ are the set of relationships between M_{tg} components of the tachograph and tracker parameters monitoring system (2).

If it is necessary to have compatible databases and integrate information systems, depending on the requirements of a specific variant of the task to create a domain, all functions, in terms of use, data processing tasks, information elements can be both reduced and enlarged in volume.

A general description of the tachograph and tracker parameter monitoring system for constituent sets is described in detail in the work [10, 11]. In the article only features of formation of monitoring system area are considered. Parameters of tachograph and tracker in system of technical condition of vehicle. The mathematical model should describe all the necessary information elements (input and output) for the respective automation object. A complete set of information elements $V_v = \{v_v | l_v = 1, L_v\}$ is shown at the Tables 1 and 2.

Table 1. Basic information elements of tachograph automation system.

Number	Designation	Name
80	v_{tg80}	Identification of the driver
81	v_{tg81}	ID cards and the country
82	v_{tg82}	Vehicle identification number, VIN, VRN
83	v_{tg83}	Country of registration and vehicle registration number (VRN)
84	v_{tg84}	Identification of the tachograph
85	v_{tg85}	Identification of the odometer
86	v_{tg86}	The range of engine rotation speed and duration of operating
87	v_{tg87}	Last control of the driver
88	v_{tg88}	The report about the day activities, information about the beginning and the end of the operation (time, location and odometer)
89	v_{tg89}	Activities with indicating the start time and end time
90	v_{tg90}	Date and time of last control of over speed Date and time of the first number of over speed
91	v_{tg91}	The five most serious over speed in the last 365 days Date, time and duration. Maximum and average speed
92	v_{tg92}	The most serious over speed for the last ten days Date, time and duration. Maximum and average speed
93	v_{tg93}	Status changes of the rear connectors D1/D2 and their duration

Preliminary information items from 1 to 79 are described in the work [5].

$$V_{tg} = \{v_{rg} | l_{tg} = 80, 93\} \tag{4}$$

$$P_{tg}(V_{tg}) = 14 \tag{5}$$

Table 2. Basic information elements of tracker automation system.

Number	Designation	Name
120	v_{tr120}	The load on the axle
121	v_{tr121}	The temperature in the body
122	v_{tr122}	Tire pressure
123	v_{tr123}	The temperature in the tires of the car
124	v_{tr124}	Control of the additional attachments. Allows determining the running time, idle time and analyzing the rationality of using additional equipment. The operation modes are controlled by the speed of the accessory engine
125	v_{tr125}	Control of position of the top attachment or working bodies of the special equipment. Allows to analyze the time of operation of special equipment
126	v_{tr126}	Control of lifting /lowering of a dump truck body. Allows to determine the number of flights and estimate the volume of cargo carried
127	v_{tr127}	Checking the opening of the tank manhole. Prevents fraud with fuel or other transported liquid, such as unauthorized selection, dilution, contamination
128	v_{tr128}	Mass emission of carbon monoxide
129	v_{tr129}	Mass emissions of hydrocarbons
130	v_{tr130}	Mass emissions of nitrogen oxides
131	v_{tr131}	Mass emissions of particulate matter

where P (V_{tg}) is the number of elements of the set V information system; l_{tg} is limiting function at the set of information systems that narrow definition of the function.

By the analogy:

$$V_{tr} = \{v_{tr} | l_{tr} = 120, 131\} \quad (6)$$

$$P_{tr}(V_{tr}) = 12 \quad (7)$$

The numbering starts with the 120-th element, because elements of 94 to 119 were left for the development and improvement of monitoring parameters tachograph automation system.

As a result of forming a description of the area for monitoring parameters with tachograph and tracker, information models were obtained, which ensure a constant unambiguous connection of the system objects with the information elements and with the objects of automation.

The next step was to form and analyze graphs of information structures of the MVSPM system model, which included the following interrelated operations: construction of sets of structural elements based on the system domain model; formation of a semantic adjacency matrix on a set of structural elements, construction of an oriented graph of its information structure; forming a semantic reach matrix on a set of structural elements; definition of information and group elements of structural sets; organizing groups of structural elements by hierarchy level, allocating and forming multiple keys and attributes in groups of data subsystems; construction of canonical models of database subsystems of the system [3, 5].

The relational model of the monitoring system based on the canonical structure of the database and the provisions [12], in accordance with the set of admissible values of the basic parameters of the tachograph and the tracker in the system of technical state of the vehicle, is shown in graph of canonical structure of technical condition parameters monitoring system (see Figs. 1 and 2).

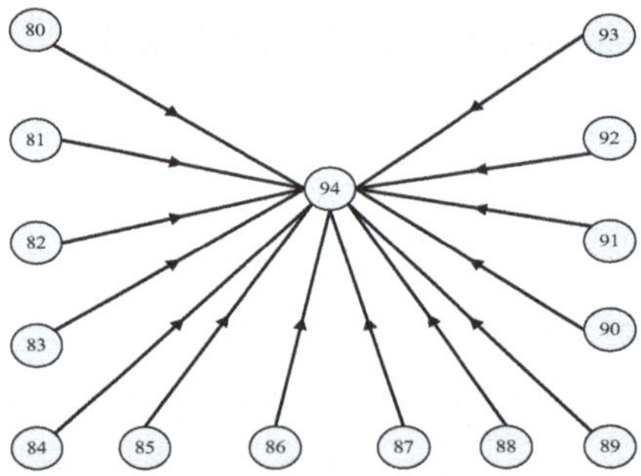

Fig. 1. Graph monitoring system with installed tachograph.

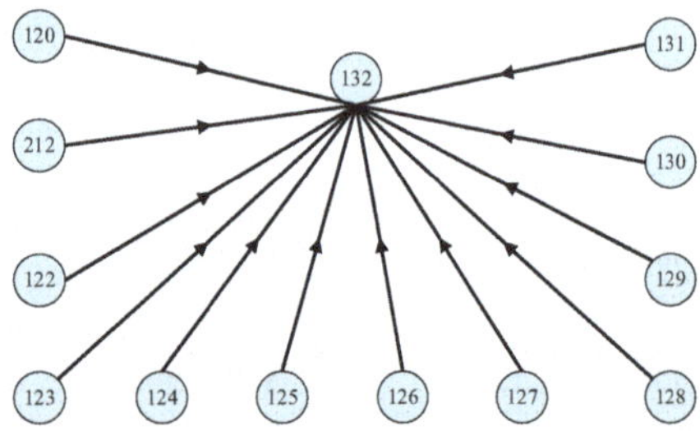

Fig. 2. Graph monitoring system with installed tracker.

Thus, the resulting information analysis is sufficient to create a relational-type database management system, including components of MVSPM.

System interaction of the main components of the monitoring process of the parameters of the vehicle condition to ensure and form the classification of operating conditions, namely, midrange ITS vehicle, IM ITS vehicle and BS ITS vehicle is proposed in the scheme (see Fig. 3), where:

- Midrange ITS is server part of IT;
- IM ITS is ITS vehicle Monitoring Infrastructure in Transport and Road Infrastructure;
- BS ITS is onboard part of ITS in the onboard information complex (OIC).

The parameters of the vehicle can be controlled by the information exchange between ITS elements of the vehicle with installed tachograph, tracker and highway infrastructure.

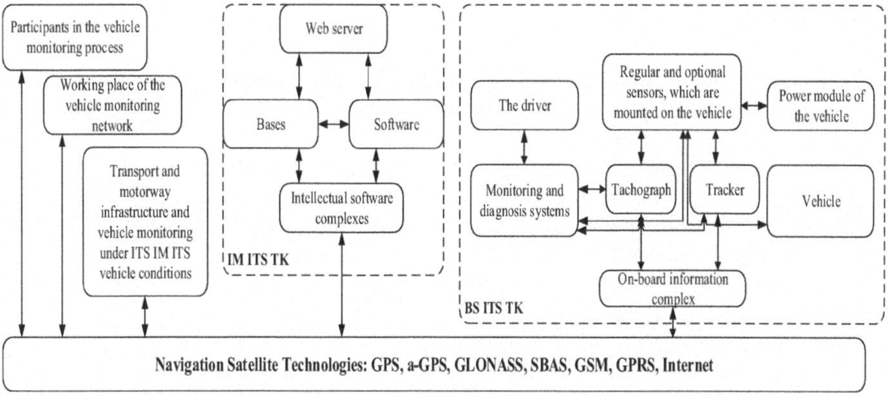

Fig. 3. Information exchange between ITS elements of a vehicle and between transport and highway infrastructure.

Thus, the information system of vehicle monitoring receives information not only about the status of the vehicle itself, but also about RWRP, environmental performance of the vehicle, violations of the speed regime, etc.

3 Conclusion

Thus, in modern transport and logistics systems the approach to the use the technologies of remote monitoring for the inspection the parameters of the technical condition during the operation of the vehicle is shown. This takes into account the features of remote control, using IT technologies, environmental performance of the vehicle, violations of the speed, etc. The analysis of the constant interaction of the monitoring means for observing the modes of operation taking into account the road, transport, and climatic conditions is carried out. The results of the interaction of the tachograph and the tracker in the monitoring information system are presented. As a result of the analysis, it can be concluded that this information is sufficient to create a relational type database for the control system, including the MVSPM components.

The development and use of the monitoring information system in the components of the MVSPM allows to control the modes of vehicles' operation more effectively, than the use of separate technical means, acting on their own algorithms separately from each other. In future research plans to test the possibility of using the MVSPM monitoring information system on special wheeled vehicles, water transport, as well as on stationary power plants, equipped with diesel generators.

References

1. Monitoring and Evaluation Strategy, Department for Transport. https://www.gov.uk/government/uploads/system/uploads/attachment_data/file/175300/monitoring-evaluation-strategy.pdf. Accessed 13 June 2017
2. Kuznetsov, E.: Management operation of cars. Transport, p. 224 (1982)
3. Gritsuk, I., Volkov, V., Mateichyk, V., Grytsuk, Y., et al.: Information model of V2I system of the vehicle technical condition remote monitoring and control in operation conditions. SAE Technical Paper 2018-01-0024 (2018). https://doi.org/10.4271/2018-01-0024
4. Govoruschenko, N.: Management operation of cars: Higher HQ, p. 312. Izdatel'stvo at Kharkov, Kharkov (1984)
5. Gritsuk, I., Gutarevych, Y., Mateichyk, V., Volkov, V.: Improving the processes of preheating and heating after the vehicular engine start by using heating system with phase-transitional thermal accumulator. SAE Technical Paper 2016 -01-0204 (2016). https://doi.org/10.4271/2016-01-0204
6. EUR-Lex. Access to European Union law. Directive 2002/15/EC of the European Parliament and of the Council of 11 March 2002 on the organisation of the working time of persons performing mobile road transport activities. Official J. L **080**, 0035–0039 (2002). https://eur-lex.europa.eu/legal-content/GA/TXT/?uri=CELEX:32002L0015. Accessed 30 Oct 2019
7. UNECE, European Agreement concerning the Work of Crews of Vehicles Engaged in International Road Transport (AETR), done at Geneva on 1 July 1970 (Consolidated text, version 2006, document ECE/TRANS/SC.1/2006/2). http://www.unece.org/trans/main/sc1/sc1aetr.html. Accessed 20 Oct 2019
8. Mark, D., Gowan, K: Methodology of structural analysis and designing SADT, Moscow, p. 191 (1993)
9. Gritsuk, I., Mateychyk, V., Simonenko, R., Khudiakov, I.: Features remote authentication modes of the driver information system monitoring vehicle. Monografia pod redakcjq naukowq KAZIMIERZA LEJDY, No.19, seria Transport System i Srodki Transportu Samochodowego, 12, 35-959, pp. 7–16. Oficyna Wydawnicza Politechniki Rzeszowskiej al. Powstancow Warszawy, Rzeszow (2019)
10. Khudiakov, I.: Models database information system monitoring the parameters of technical condition of vehicles. In: Scientific Notes, no.67, pp. 141–148. Lutsk National Technical University, Luck (2019)
11. Khudiakov, I., Simonenko, R., Manzheley, V., Chernenko, V.: Features of formation and analysis of information structures monitoring system parameters technical condition of vehicles in cooperation with the tachograph. In: Blatnický, M., Dizo, J., Gerlici, J., Hrytsuk, I. (eds.) Systems and Means of Transport. The Problems of Exploitation and Diagnostics, pp. 250–259. HDMA, Kherson (2019)
12. Atroshchenko, V., Shevtsov, P., Yatsynyn, R., Dyachenko, M., Pedko, Y.: Technical opportunities increase resources autonomous power machinery of power plants. Monograph, p. 192. Publishing House, Krasnodar (2010)

Motion Control of Smart Autonomous Mobile System Based on the Perception Model

Anatolii Kargin[1], Artem Panchenko[1], Oleksandr Ivaniuk[1(✉)], and Ruslans Muhitovs[2]

[1] Ukrainian State University of Railway Transport, Kharkiv 61001, Ukraine
{kargin, ivaniuk}@kart.edu.ua, panchenko074@ukr.net
[2] Institute of Transport, Riga Technical University, Azenes Street 12-316, Riga 1048, Latvia
ruslans.muhitovs@rtu.lv

Abstract. Modern systems, such as cooperative robotics, mobile computing systems, transportation systems, have to operate in highly dynamic and volatile environments. Smart and autonomous behavior becomes indispensable characteristics of such systems. The article deals with the task of managing movements along a given route with obstacles by a Smart Autonomous Machine (SAM). The possibilities of control are shown on the basis of a new perception model, supporting the such cognitive functions as calculating the meaning of the space-temporal data segment from sensors, tracking the context and switching attention. The knowledge bases are given in the form of rules with linguistic variables used in the perception model for situational motion control of the SAM. Protocols of computer experiments with a simulation of the perception model are given and discussed, allowing to check its adequacy and effectiveness. A description of field experiments with a mobile wheeled robot is given. The experiments were conducted in an artificially created space, including a set of distributed landmarks, which are used as route markers. The scenarios for different routes and possible obstacles are analyzed.

Keywords: Logistics · Smart Autonomous Machine · Perception model

1 Introduction

The Autonomous Systems (AS) concept is becoming more widespread when creating large-scale distributed in space objects. The AS includes a set of components (Smart Autonomous Machine, SAM) that implement a specific function in autonomous mode that requires a variety of information. Industrial ASs are assembled from automatic lines, CNC machines, transport systems, robots [1]. The railway [2] and automotive [3] industries in the long-term plans provide for the creation of AS based on smart trains and smart cars. The ASs that monitor distributed situations include mobile devices (robots, drones) for observation and receiving information. In smart city projects [3], as in all of the above, AS, along with other SAMs, use Mobile SAMs (MSAM) that move in space. Autonomy of MSAM is provided by the Intelligent Control System (ICS). One of the main tasks that ICS must solve is MSAM navigation.

E. Ginters et al. (Eds.): ICTE ToL 2019, LNITI, pp. 145–153, 2020.
https://doi.org/10.1007/978-3-030-39688-6_20

In mobile robotics, navigation is considered as an independent task, which is divided into three subtasks [4, 5]: Path Planning (PP), Localization (Lo), and building an environment map (Mapping, Ma). Particular attention in the navigation of robots is paid to the simultaneous solution of these three navigation problems. This area is known as SLAM (Simultaneous Localization and Mapping). [6, 7]. However, navigation models and methods are considered in isolation from the tasks of Motion Control (MC), Path Tracking, (PT), Target Searching (TS), Area Patrolling (AP) and others [8]. This approach limits the possibility of using ready-made solutions of both SLAM and MC, PT, TS when creating ICS for MSAM for the following reasons.

Primarily, the navigation models of global path planning [9] (building a sequence of control reactions that ensure the movement of MSAM to a predetermined position along a predefined route) cannot be used, since they assume the availability of complete information about the geometry of the environment, the use of traditional or modified program control methods for the implementation of the plan and, most importantly, the absence of disturbances is assumed. These shortcomings are deprived of the models of local path planning [10], when the environment is unknown in advance or interference may occur that violates the preliminary description of the environment. Local planning models can potentially be used in AS, since full knowledge of the environment is seldom available to MSAM and real environments are dynamic. Such models are related to the Sensor based Approach (SbA), since they are based on data from sensors about the state of the environment and are used for online decision-making based on this data. Two areas of SbA research can be distinguished for solving navigation problems: creation of heuristic algorithms for a specific application; building models of data processing from sensors borrowed from nature.

Heuristic algorithms for solving the problem of the local planning of the movements include Vector Field Histogram (VFH) and its further development – VFH+ , VFH*; Generalized Voronoi Graph (GVG) [11]. The capabilities of heuristic algorithms can be seen in the example of the Bug family of algorithms (Bug1, Bug2, Bug3, Dist Bug, Tangent Bug) [12, 13]. This algorithm allows the robot to reach the goal in the process of moving, if at any moment the information about the direction to the target and the distance to it is available to the robot, the robot can identify obstacles in the way of movement and perform their bypass. These conditions are supported by robot localizing in the environment method, for example, based on odometry in the presence of initial data on the initial coordinates of the robot and the target. However, for real applications, information about the final goal is not available to MSAM at all stages of movement, and in order to achieve the goal, the robot must follow the route through a sequence of defined subgoals. In a disordered environment, MSAM must be capable of situational planning. The model for the formation of control actions should be integrated with the model for situational planning. Obviously, the implementation of the indicated requirements for MSAM cannot be based on the mentioned heuristic algorithms.

Models of processing data from sensors borrowed from living nature are being intensively developed in the cognitive robotics [14] and embedded cognitive systems [15]. In [16], one of such Perception Models (PM) is presented, specialized for situational control of robot movement. The PM is integrated with a Rule-based system (RbPM). In RbPM, at the first stage of processing data from sensors, the PM interprets the spatio-temporal set of data about the environment in which the MSAM moves. The interpretation result is a structured fact, representing the situation at different levels of abstraction in the form of a fuzzy semantic description [17]. At the second stage, a fuzzy conclusion on the rules with Linguistic Variables (LV) forms the MSAM control actions. MSAM's knowledge about the environment, route, and methods for eliminating interference on the route are specified by fuzzy rules, which, in addition to domain LVs, include contextual and attention LVs. These special LVs support context tracking and attention switching, which allows for situational control of MSAM movements along a given route in an unknown environment.

The article deal with the implementation of RbPM for a class of tasks for which a context influence model covering several adjacent subgoals is required. Such approach will improve the quality of MSAM movement control along a given route in a wide class of interference.

2 Perceptual Model Implementation

The RbPM [16, 17] implementation for solving the problem of controlling the movement of MSAM along a given route under interference is based on the assumption that Landmark (LM) set $L = \{l_1, \ldots, l_m\}$ are distributed in the MSAM workspace. The i-th LM is characterized by a unique features set $Pr_i = \{pr_1, \ldots, pr_k\}$. On the set L, several routes can be given. The j-th MSAM travel route is defined as an ordered sequence of LMs $R_j = (l_{j1}, \ldots, l_{jp})$, where p is the length of the route. When moving MSAM along a route, interference may occur. Obstacles can be divided into two types: the obstacle occurs on the route between adjacent LMs l_{jn}, l_{jn+1} and is eliminated by "local detour" so that the LM l_{jn+1} is achieved; the obstacle makes it impossible to travel between adjacent LMs l_{jn}, l_{jn+1}. The LM l_{jn+1} can be reached through a different route passing through other LMs of the workspaces, it is possible that only those following l_{jn+1} LMs will be reached, for example, l_{jn+2} or l_{jn+3}.

RbPM is based on the following knowledge:

- Set of facts representing possible situations in the MSAM environment;
- Set of rules representing knowledge about possible routes, local map, MSAM workspaces, and control strategies;
- Set of MSAM strategies, algorithms and control reactions for typical situations.

2.1 Set of Facts

Data from sensors in PM is processed by the abstraction mechanism [17, 18] in order to obtain their description in verbal form as a multilevel fact at different levels of generalization. The fact, in addition to the verbal representation, has a Fuzzy Fact Characteristic (FFC), which shows how much the meaning of the verbal representation of the situation (prototype) corresponds to a specific data set from MSAM sensors. The FFC is a fuzzy L-R number with Gaussian L-R membership function. Based on the FFC, the certainty factor cf as a crisp characteristic is used [17, 18]. In the MSAM RbPM KB, for each LM there is a subset of facts $F_i = \{f_i^{Pr}, f_i^{Con}, f_i^{Env}, f_i^{Obs}, f_{il}^{Str}\}$, where f_i^{Pr} is a fact representing the unique attributes of LM $l_i \in L$; f_i^{Con} is a LM $l_i \in L$ context fact, reflecting in the environmental data processing, contextual links between LMs when MSAM moving along a route $R_j = (l_{j1}, \ldots, l_{jp})$; f_i^{Env} is a subset of facts describing different situations in the environment of the LM l_i; f_i^{Obs} is a subset of facts describing disturbances and obstacles; f_{il}^{Str} is a fact representing control strategy. Each of these facts at a high level of abstraction represents in RbPM KB a situation surrounded by MSAM. The concretization of the listed facts is given below when considering the experiments.

2.2 Route Knowledge

The route $R_j = \left(l_{j1}, \ldots, l_{jp}\right)$ is represented by set of p rules. The i-th rule associates the contextual fact of the i-th LM with m adjacent LMs. To represent the route, as well as all other knowledge in the RbPM KB, fuzzy rules in the form of Takagi-Sugeno were used.

$$\begin{aligned} &\boldsymbol{R_k^1}\ \boldsymbol{IF}\ event(f_i^{Pr})\ \boldsymbol{and}\ CF\_f_i^{Con} is\ high\ \boldsymbol{and}\ CF\_Rout\_j\ is\ high \\ &\quad \boldsymbol{THEN}\ cf\_f_{i+1}^{Con} = +1.0, cf\_f_{i+2}^{Con} = +0.5, cf\_f_i^{Con} = +0.25, \\ &\qquad cf\_f_{i-1}^{Con} = -1.0,\ cf\_f_i^{Mov\_Rout} = 1.0, i = 1, 2, \ldots, p. \end{aligned} \tag{1}$$

where $CF\_f_i^x$ is a LV that fuzzily represents the values of certainty factor cf of the fact f_i^x.

Rule (1) is written for $m = 2$. For contextual link length $m = 1$, in (1) only two contextual facts will be indicated in the ***THEN*** field $cf\_f_{i+1}^{Con} = +1.0$, $cf\_f_i^{Con} = -1.0$.

The RbPM uses an event model to process the rules: rule (1) is activated at the moment when the event $event\left(f_i^{Pr}\right)$ appears. It is the appearance of data from sensors about unique LM attributes.

2.3 Cartographic Knowledge of the Workspace

The LM l_i is directly connected by roads to a finite set of the workspace LMs $L_i = \{l_g\}_{g=1,q}$. The road adjacent to LM l_i and going from LM l_g has an identifier that reflects the name of its direction (north, east, south, west). Other direction signs may be used. The RbPM knowledge about this representation of a workspace map is described by a set of rules: each LM is described by $q \cdot (q-1)$ rules that link the direction of the road coming to LM l_i and the direction of the road coming from LM l_i and leading to LM l_g.

$$\begin{aligned} &\boldsymbol{R_k^2}\ \boldsymbol{IF}\ event(f_g^{Con})\ \boldsymbol{and}\ CF\_f_i^{Pr}\ is\ high\ \boldsymbol{and}\ CF\_f_{in}^{Road\_x}\ is\ high \\ &\quad \boldsymbol{THEN}\ cf\_f_{out}^{Road\_y} = +1.0. \end{aligned} \tag{2}$$

In rule (2), the target LM l_g is represented by the context fact f_g^{Con}, and the original LM is represented by three facts: f_i^{Pr}, $f_{in}^{Road\_x}$ and $f_{out}^{Road\_y}$. The last two facts describe the directions of the roads of entrance and exit accordingly. Rule (2) is activated by the event $event(f_g^{Con})$, generated by the appearance of the next target when MSAM is surrounded by LM l_i.

This rule activates the fact $f_i^{Road\_y}$, which contains a description of the road on which MSAM needs to continue driving. Rules (2) are not related to a specific route. They are activated by sensory MSAM location data when it is surrounded by LM l_i.

2.4 Knowledge About Conditions for Implementing Control Strategies

A strategy for controlling maneuver or movement along a route trajectory is implemented by controllers, for example, PID, neuro or fuzzy, which are based on some kind of direct or feedback control algorithm. The operation of the controllers is activated sequentially one after another as situations arise that are required to start the operation of the controller. For example, a turn was made (f_l^{Env}), a resolving sign (f^{Env_Go}) was detected, markup (f^{Env_z}) was located, and so on. Examples of rules for managing the implementation of the strategy are given below. The last rule in (3) switches the attention mechanism to another strategy when the situation is either unsafe (a hindrance has occurred) or the goal has been reached.

$$\begin{aligned}
&\boldsymbol{R_k^3}\ \boldsymbol{IF}\ event(f_{out}^{Road_y})\ \boldsymbol{and}\ CF_f^{Env_Go}\ is\ high\ \boldsymbol{and}\ CF_f_{in}^{Road_x}\ is\ high\\
&\quad \boldsymbol{THEN}\ cf_f_l^{Str} = +1.0;\\
&\boldsymbol{R_k^4}\ \boldsymbol{IF}\ event(f^{Env_Go})\ \boldsymbol{and}\ CF_f_l^{Env_z}\ is\ high\ \boldsymbol{THEN}\ cf_f_r^{Str} = +1.0;\\
&\boldsymbol{R_k^5}\ \boldsymbol{IF}\ event(f_l^{Env_z})\ \boldsymbol{and}\ CF_f_l^{Env_Go}\ is\ low\ \boldsymbol{THEN}\ cf_f_h^{Str} = +1.0;\\
&\boldsymbol{R_k^6}\ \boldsymbol{IF}\ event(f_{il}^{Obs})\ \boldsymbol{and}\ CF_U_{il}\ is\ high\ \boldsymbol{THEN}\ cf_U_{il} = -1.0,\ cf_Contr_{il} = -1.0.
\end{aligned} \tag{3}$$

2.5 Event-Driven Situation Management Mechanism

The situational management mechanism is based on an event model of data processing [16, 17]: at regular discrete time instants, based on data from MSAM sensor systems, the FFC is calculated for all RbPM facts that describe data from sensors. The conditions for events $event(f_i^{Pr}), event(CF\_f_{il}^{Obs})$, and so on are checked. When at least one event is identified, the mechanisms of context tracking and attention switching are activated. The context tracking mechanism at the first step makes a conclusion based on the fact bases $\{f_i^{Pr}\}_{i=1,m}$ and rules (1). The attention switching mechanism works with the rule base (3). In the next steps, the traditional fuzzy conclusion of Takagi-Sugeno works with the base of rules (2) and (3).

3 Experimental Study of the Model

To conduct experiments with PM, a model of a parametrized artificial workspace (PAW) was proposed and created. Variation of the PAW parameters allows to simulate different structures of the environment and situations in which the MSAM operates. This property provides the ability to conduct various experiments with the PM. The PM with the described properties is implemented on the basis of the polygon [19] (see Fig. 1a and b). To simplify the description of the simulation results in the article, the polygon is presented as orthogonal space. This simplification is not a limitation, since the PM uses the characteristics of the environment through facts and rules and is not directly dependent on the structure of the environment.

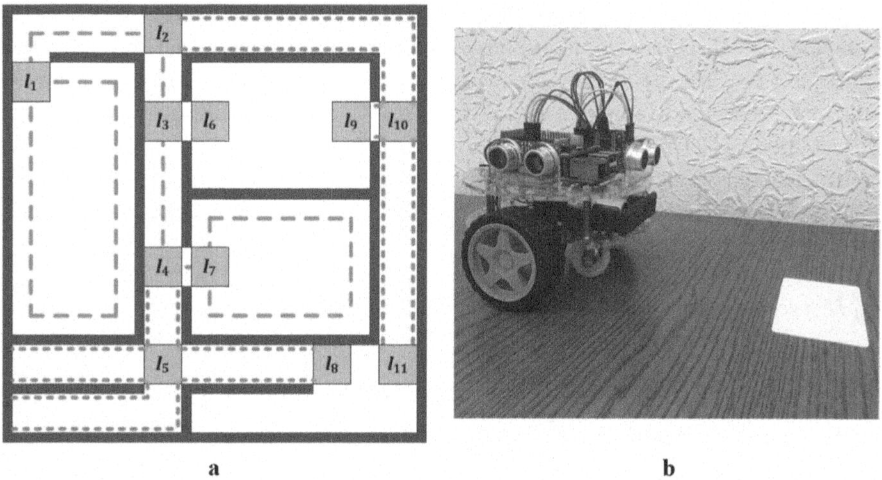

Fig. 1. a) Polygon for testing PM (the dashed line shows marking on the road, the dotted line shows bounding walls); b) MSAM used for experiments.

The MSAM workspace is characterized by two types of elements: sections of non-alternative movement (SNM) – straight sections, or sections that make turns; key points (KP) at which a decision is made on which way to continue moving. KPs are cross-roads of various types. The structure of the PAW is such that any two SNMs are conjugated via one KP. For the simulated the PAW, three types of the SNM are laid down in accordance with MSAM motion control method. First type – markings on the road, which is perceived by reflection sensors. Second type – bounding wall, which is perceived by ultrasonic distance sensors to the wall. Third type – unstructured space in which MSAM moves around the compass. The KPs of space are described by a vector $L = (p_N, p_E, p_S, p_W, t_N, t_E, t_S, t_W, d)$. The components p_N, p_E, p_S, p_W describe the configuration of the KP as an intersection (the presence of adjacent SNM to the KP) and are represented by the confidence factor *cf* of the presence of the adjacent SNM, respectively, from the north, east, south, and west. The components t_N, t_E, t_S, t_W characterize the type of SNM, the value of which is a factor of confidence in the presence of marking on the road, bounding walls, unstructured space, respectively. Component *d* is optional and reserved for the presentation of additional sensory data. An example of a specific value of the vector *L* for KP l_3 (see Fig. 1a) is given below $l_3 = (1.0, 1.0, 1.0, -1.0, 1.0, 0.0, 1.0, 0.0)$. By varying the *cf* values of the vector *L*, it is possible to simulate situations where the intersection structure is not strictly orthogonal, the SNM markup is fuzzy, and other deviations from the description of the situation prototype in the PM rules.

Radio-frequency identification (RFID) tags are used at the polygon as a hardware implementation of KPs of PAW. Tags have an internal memory and support the operation of writing and reading information using the appropriate modules. To construct the PAW, the Mifare 1k S50 series tags (1 kB of internal memory, operating frequency 13.56 MHz) and the MFRC-522 write/read modules were selected.

Each KP PAW is equipped with one RFID tag. The RFID tag stores information about the components of the vector *L*. These RFID tags are read by the MFRC-522 module installed on the MSAM. Based on this data, virtual MSAM sensors simulate the receipt of relevant data from environmental sensors and generate the FFC of set of facts F_i RbPM. Thus, in experiments, MSAM uses a hybrid sensor system. Identification of KP and their characteristics is based on data from virtual sensors (RFID tags). FFCs of part of set F_i are based on data from real sensors (physical object detection system (PODS) based on a set of ultrasonic sensors HC-SR04; marking detection system (MDS) based on a set of reflection sensors KY-033; north deviation detection system (NDDS) based on digital compass GY-9250 based on MPU-9250). A wheeled platform with two support and two driving wheels actuated by gear motors was used as a MSAM at the polygon (see Fig. 1b). The control is implemented on the basis of the Raspberry Pi 3B single-board microcomputer and WAVESHARE motor driver. The RbPM system software, as well as the MSAM movement control procedures for areas of non-alternative movement along the marking line, along the walls and azimuth movement, was developed in the Python 3 programming language.

Experiments with the RbPM were carried out using three MSAM routes as an example: $R_1 = (l_1, l_2, l_3, l_6, l_9, l_{10}, l_{11}), R_2 = (l_1, l_2, l_3, l_4, l_5), R_3 = (l_{10}, l_2, l_1)$. The routes are represented by 15 rules (1) for $m = 2$. The cartographic knowledge of 56 rules (2) represents the PAW shown in Fig. 1a. The sets of rules that represent knowledge about implementing a control strategy (3) include 12 rules type $\boldsymbol{R_k^3}$, 3 rules type $\boldsymbol{R_k^4}$, 3 rules type $\boldsymbol{R_k^5}$ and 2 rules type $\boldsymbol{R_k^6}$. The essence of the experiment was that as the MSAM moved along one of the routes, *cf* values were fixed in time for all the facts of the RbPM. The experiments were conducted for different values of *cf* of the facts $f_i^{Pr}, f_i^{Env}, f_i^{Obs}$, which simulated interference of a different nature. The quality of the model was evaluated by the relative number of experiments in which MSAM did not reach the route goal. In addition, the depth of context was varied.

4 Conclusion

The RbPM, which a Takagi-Sugeno fuzzy model integrated with a PM, can be implemented to solve the problems of navigating autonomous systems. Such perception model functions as abstraction from data from sensors, tracking context and switching attention support autonomous mode of control of the movements MSAM along a given route under hindrance using a local map. The RbPM improve the quality of MSAM movement along a given route for a wide class of obstacles. Firstly, in the case when, due to hindrance, several adjacent subgoals are unavailable for MSAM. Secondly, when the hindrance distorts the information required to solve the MSAM localization. And thirdly, when obstacles can interrupt the movement of MSAM along the route and situational re-planning of the route fragment is required. In all three cases, context and attention mechanisms allow MSAM to achieve the ultimate goal.

References

1. Mintchell, G.: Industry 4.0 Survey: Building the Digital Enterprise. https://themanufacturingconnection.com/2016/09/industry-4-0-survey-building-digital-enterprise. Accessed 25 Jan 2019
2. Rail Technical Strategy: Capability Delivery Plan, https://www.rssb.co.uk/rts/Documents/2017-01-27-rail-technical-strategy-capability-delivery-plan-brochure.pdf. Accessed 25 Jan 2019
3. Duarte, F., Ratti, C.: The impact of autonomous vehicles on cities: a review. J. Urban Technol. **25**, 3–18 (2018)
4. Stachniss, C.: Robotic Mapping and Exploration. Springer, Heidelberg (2009)
5. Rone, W., Ben-Tzvi, P.: Mapping, localization and motion planning in mobile multi-robotic systems. Robotica **31**, 1–23 (2012)
6. Bresson, G., Alsayed, Z., Yu, L., Glaser, S.: Simultaneous localization and mapping: a survey of current trends in autonomous driving. IEEE Trans. Intell. Veh. **2**, 194–220 (2017)
7. Fuentes-Pacheco, J., Ruiz-Ascencio, J., Rendón-Mancha, J.: Visual simultaneous localization and mapping: a survey. Artif. Intell. Rev. **43**, 55–81 (2012)
8. Hoy, M., Matveev, A., Savkin, A.: Algorithms for collision-free navigation of mobile robots in complex cluttered environments: a survey. Robotica **33**, 463–497 (2014)
9. Lv, T., Zhao, C., Bao, J.: A global path planning algorithm based on bidirectional SVGA. J. Robot. **2017**, 1–11 (2017)
10. Lu, Y., Xue, Z., Xia, G., Zhang, L.: A survey on vision-based UAV navigation. Geo-spat. Inf. Sci. **21**, 21–32 (2018)
11. Mac, T., Copot, C., Tran, D., De Keyser, R.: Heuristic approaches in robot path planning: a survey. Robot. Auton. Syst. **86**, 13–28 (2016)
12. Yufka, A., Parlaktuna, O.: Performance comparison of the BUG's algorithms for mobile robots. In: International Symposium on INnovations in Intelligent SysTems and Applications, INISTA 2009, pp. 416–421 (2009)
13. McGuire, K., de Croon, G., Tuyls, K.: A comparative study of bug algorithms for robot navigation. Robot. Autono. Syst. **121**, 103261 (2019)
14. Asada, M., Hosoda, K., Kuniyoshi, Y., Ishiguro, H., Inui, T., Yoshikawa, Y., Ogino, M., Yoshida, C.: Cognitive developmental robotics: a survey. IEEE Trans. Auton. Ment. Dev. **1**, 12–34 (2009)
15. Alippi, C.: Intelligence for Embedded Systems. Springer International Publishing, Cham (2014)
16. Kargin, A., Panchenko, S., Vasiljevs, A., Petrenko, T.: Implementation of cognitive perception functions in fuzzy situational control model. Procedia Comput. Sci. **149**, 231–238 (2019)

17. Kargin, A., Petrenko, T.: Internet of things smart rules engine. In: 2018 International Scientific-Practical Conference Problems of Infocommunications. Science and Technology (PIC S&T), pp. 639–644. IEEE (2019)
18. Kargin, A., Petrenko, T.: Abstraction and categorization in smart machines based on granular computations. Bulletin of the National Technical University "KhPI". A series of "Information and Modeling." 50, pp. 130–141 (2017)
19. Kargin, A., Ivaniuk, O., Galych, G., Panchenko, A.: Polygon for smart machine application. In: 2018 IEEE 9th International Conference on Dependable Systems, Services and Technologies (DESSERT). IEEE (2018)

Study of the Interaction of a Transport Vehicle with an Open Road

Kairatolla K. Abishev[1(✉)], Asylbek Zh. Kasenov[1], Karlygash B. Assylova[1], and Gali S. Gumarov[2]

[1] S. Toraighyrov Pavlodar State University, Pavlodar, Republic of Kazakhstan
kairatolla76@gmail.com
[2] M. Utemisov West Kazakhstan State University, Uralsk, Republic of Kazakhstan
ggs65@yandex.ru

Abstract. One of the ways to reduce the harmful effects of undercarriage of traction and transport vehicles on the ground and to improve their traction characteristics is the use of rubber-track propulsion. Experimental research have shown that the rubber-track propulsion provides a more even distribution of the specific pressures created by the traction-transport machine and reduces the peak loads on the ground compared to the metal track. When the caterpillar propulsion interacts with the ground, the force of gravity of the machine is transmitted to the ground through the propulsive support branch. The values of these reactions and the patterns of their distribution over the support surface reflect the physical essence of the processes of interaction between the propulsion unit and the ground. The article presents the results of the research of the distribution of normal pressure along the length of the support surface of the rubber track. The uniform distribution of pressure affects the permeability of the traction vehicle.

Keywords: Rubber-track propulsion · Interaction modeling · Traction-transport machine

1 Introduction

For operation in various conditions, transport vehicles of various types and designs with different technical characteristics have been created. The type and purpose of the machine determines the design of its running system.

One of the urgent tasks in the transport industry has always been the issue of increasing the efficiency of their use, a large role in the solution of which belongs to the improvement of running systems of machines.

Practice shows that with insufficiently high traction and coupling qualities of transport vehicles, a decrease in its productivity, an increase in fuel consumption and a deterioration in other operational properties of the machine are recorded. Currently, one of the ways to reduce the harmful effects of the running systems of towing vehicles on the ground and improve their towing qualities is the use of a rubber caterpillar [1]. Experimental studies [2, 3] showed that the rubber-tracked mover, in comparison with the metal caterpillar, provides a more uniform distribution of specific pressures created by

E. Ginters et al. (Eds.): ICTE ToL 2019, LNITI, pp. 154–163, 2020.
https://doi.org/10.1007/978-3-030-39688-6_21

the towing vehicle and reduces peak loads on the ground. Therefore, studies on the interaction of the running system of a transport vehicle with the road surface are relevant.

2 Improving the Traction Characteristics

When a track mover interacts with the ground, the gravity of the vehicle is transmitted to the ground through the bearing strand of the mover [4, 5]. From the ground the reaction forces effect on the mover. The values of these reactions and the laws of their distribution on the bearing surface reflect the physical essence of the processes of interaction between the mover and the ground. It is precisely these laws that determine the interaction parameters of the tracks with the bearing surface, as well as the performance indicators of the tracked vehicles [6].

Mathematical models that reveal the mechanism of interaction between a belt mover with the base and the nature of the distribution of specific pressures are developed for a metal belt mover and do not take into account such a feature of rubber tracks as their deformability.

To identify the patterns of pressure distribution, the scheme proposed by Professor V. V. Katsygin [7] is adopted, and a part of the rubber track mover, consisting of two lower track wheels and a rubber track is respected. The action of the tension spring of the directive wheel is replaced by the sprung first roller (see Fig. 1). It is assumed that the tracked vehicle moves uniformly along the horizontal surface of the track.

Let the load G_1 that equals to the gravity of the vehicle applied to the selected part of the mover effect the portion of the track AB. It is assumed that the caterpillar is a flexible belt and its speed is zero. The position of the tape depends on its tension and ground stress. In turn, ground stresses depend on the degree of precipitation of various sections of the belt [8, 9].

In Fig. 1 the coordinate axes is drawn so that the abscissa axis is horizontal (at the level of the path surface before the mover deforms ground), and the ordinate axis passes vertically down through the middle of the AB section.

The deflection of the rubber track between the rollers, and therefore the normal pressure on the rubber track between the rollers, is due to elastic deformations of the ground and the elements of the rubber track and are determined by linear laws:

- for ground:

$$p = kh_1, \tag{1}$$

where p – specific pressure, k – volumetric crumpling coefficient of ground, but h_1 – settlement of ground;

- for rubber track:

$$p = c_z h_2, \tag{2}$$

where c_z – normal stiffness of rubber track, but h_2 – normal deformation of elements of rubber track.

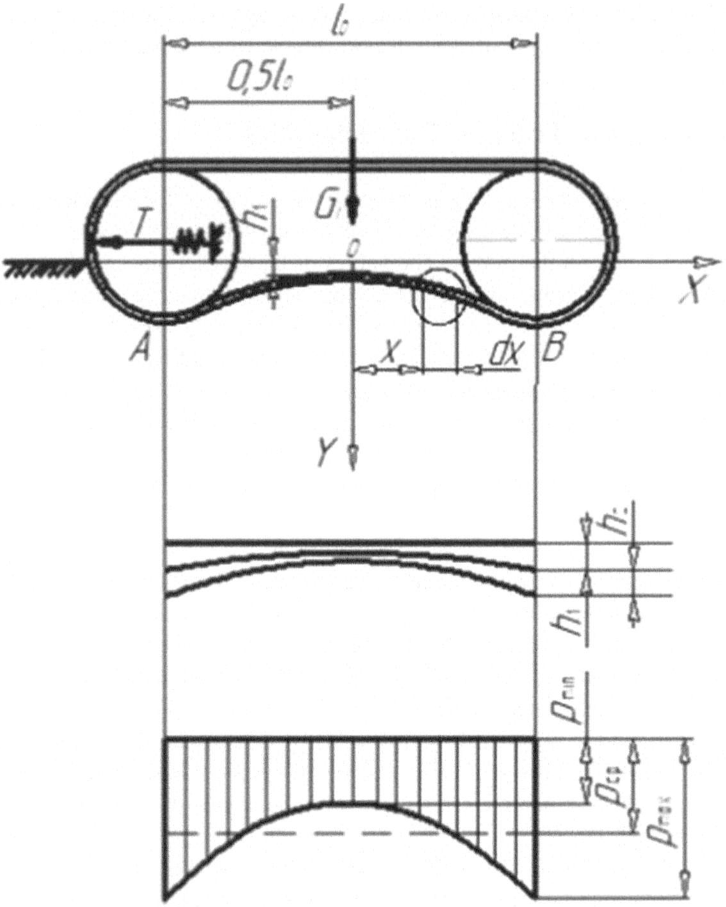

Fig. 1. Scheme to the conclusion of the law of pressure distribution along the length of the bearing strand of the rubber belt track.

In the joint solution of Eqs. (1) and (2) it is assumed that $y = h_1 + h_2$. As a result:

$$p = k_{st} y, \tag{3}$$

where k_{st} – specific stiffness coefficient, but y – joint deformation of ground elements of rubber track.

The specific stiffness coefficient depends on the volumetric crumpling coefficient of ground normal stiffness of elements of rubber track:

$$k_{st} = \frac{k c_z}{k + c_z}. \tag{4}$$

Using the obtained linear dependence (3) the law of distribution of normal pressure along the length of the bearing surface of the rubber track is determined.

Let us consider the equilibrium of an elementary strip spaced at a distance *x* from the origin of the coordinate system (the width *b* is equal to the width of the track, and the length is *dx*). This strip is in equilibrium under the action of the tension forces of the belt *F* and *dF*, as well as the forces of the ground dN_x:

$$dN_x = p_x b dx,$$

where p_x – pressure on *bdx*.

The indicated forces on the axes *OX* and *OY* (see Fig. 2) are shown. The sum of the projections of these forces on the axis is zero, i.e.

$$F\cos\theta + d(F\cos\theta) - F\cos\theta = 0.$$

or

$$d(F\cos\theta) = 0 \tag{5}$$

$$F\sin\theta + d(F\sin\theta) - F\sin\theta - p_x b dx = 0,$$

or

$$dF(\sin\theta) - p_x b dx = 0. \tag{6}$$

From Eq. (5) it follows that $F\cos\theta = \text{const}$, that is, the horizontal component of the belt tension is constant and equal to half the force of the tension spring $\frac{T}{2}$. The value $F = \frac{T}{2\cos\theta}$ in Eq. (6) is substituted. After reducing *dx*:

$$\frac{T}{2}\frac{d(tg\theta)}{dx} - p_x b = 0.$$

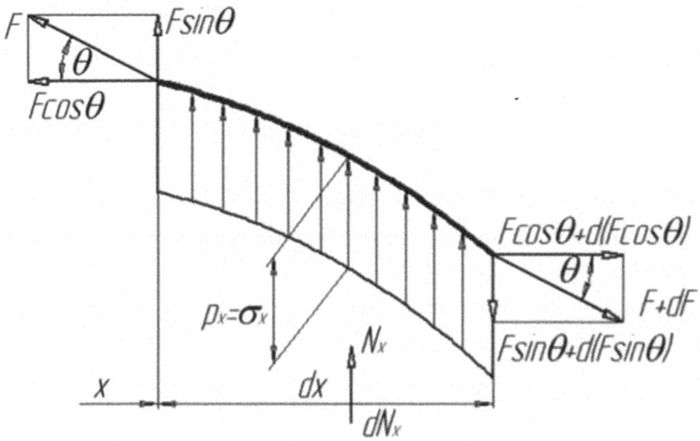

Fig. 2. Equilibrium of an elementary strip of a belt track.

The tangent of the angle of inclination of the belt is equal to the first derivative of precipitation y of x, $\frac{dy}{dx}$. Moreover, $p_x = k_{st}y$. Therefore:

$$\frac{T}{2}\frac{d^2y}{dx^2} - bk_{st}y = 0. \tag{7}$$

The resulting expression is a second-order homogeneous differential equation and determines the law of pressure distribution along the length of the bearing strand of the tracks. The solution in the form of $y = ce^{\lambda x}$ is provided.

Let us find particular solutions of the differential equation (7). After differentiating $y = f(x)$ twice, we get $\frac{d^2y}{dx^2} = ce^{\lambda x}\lambda^2$. Substituting the resulting expression in Eq. (7):

$$\frac{T}{2}ce^{\lambda x}\lambda^2 - k_{st}bce^{\lambda x} = 0,$$

therefore

$$ce^{\lambda x}\lambda^2 - \frac{2k_{st}b}{T}ce^{\lambda x} = 0$$

After the corresponding transformations:

$$ce^{\lambda x}\left(\lambda^2 - \frac{2k_{st}b}{T}\right) = 0,$$

wherefrom

$$ce^{\lambda x} = 0 \text{ and } \lambda^2 - \frac{2k_{st}b}{T} = 0.$$

From the last equality $\lambda = \pm\sqrt{\frac{2k_{st}b}{T}}$. Therefore, particular solutions of differential Eq. (7) are as follows

$$y_1 = c_1e^{x\sqrt{\frac{2k_{st}b}{T}}}; y_2 = c_2e^{-x\sqrt{\frac{2k_{st}b}{T}}}$$

In this case, a plus should always be put in front of the x coordinate, since with a negative value of the exponent, the settlement decreases with increasing x, which contradicts reality.

Let us find constants c_1 and c_2 under the following initial conditions $x = 0$, $y = h_1$. Then $h_1 = c_1e^0 = c_1$ and $h_1 = c_2e^0 = c_2$, i.e. $c_1 = c_2 = h_1$, but

$$y = h_1e^{x\sqrt{\frac{2k_{st}b}{T}}}.$$

Given that, the pressure on the ground is proportional to the settlement:

$$p_x = k_{st} h_1 e^{x\sqrt{\frac{2k_{st}b}{T}}}. \tag{8}$$

To estimate the values of pressures, it is necessary to determine the settlement h_i between bearing rollers. To do this, the elementary forces of the ground on the belt track between the points A and B (see Fig. 1) are summed and equate these forces to the force of gravity G_i of the same section:

$$G_i = \int_A^B dN_x = \int_A^B N(x)dx;$$

for the half-section:

$$\frac{G_i}{2} = \int_0^{l_0/2} bk_{st}h_1 e^{x\sqrt{\frac{2k_{st}b}{T}}}dx = h_1\sqrt{\frac{2k_{st}b}{T}}\left(e^{\frac{l_0}{2}\sqrt{\frac{2k_{st}b}{T}}} - 1\right)$$

Therefore: $h_1 = \dfrac{G_i}{2\sqrt{\frac{2k_{st}b}{T}}\left(e^{\frac{l_0}{2}\sqrt{\frac{2k_{st}b}{T}}} - 1\right)}$.

Substituting the value h_i in the formula (8), an equation characterizing the pressure distribution under the bearing strand of the track between the rollers is obtained:

$$p_x = \frac{G_i k_{st} e^{x\sqrt{\frac{2k_{st}b}{T}}}}{2\sqrt{\frac{2k_{st}b}{T}}\left(e^{\frac{l_0}{2}\sqrt{\frac{2k_{st}b}{T}}} - 1\right)}. \tag{9}$$

From Eq. (9) it follows that the pressure depends on gravity, the distance between the track rollers, the track width, tension, as well as the physical and mechanical properties of the ground.

To determine the non-uniformity of the pressure distribution along the length of the track bearing surface, the relative non-uniformity coefficient is used:

$$\xi_p = \frac{p_{max} - p_{min}}{p_{av}}, \tag{10}$$

where p_{max} – maximal pressure while $x = \frac{l_0}{2}$; p_{min} – minimal pressure while $x = 0$; p_{av} – mean pressure.

The mean pressure is determined by the formula

$$p_{av} = \frac{G_i}{bl_0},$$

where G_i – part of the mass of the vehicle falling on the considered section of the rubber track, b – rubber track width and l_0 – distance between rollers.

Substituting the value x in Eq. (9) the expression for determining the maximum and minimum pressure is obtained:

$$p_{max} = \frac{G_i k_{st} e^{\frac{l_0}{2}\sqrt{\frac{2k_{st}b}{T}}}}{2\sqrt{\frac{k_{st}bT}{2}}\left(e^{\frac{l_0}{2}\sqrt{\frac{2k_{st}b}{T}}} - 1\right)}, \tag{11}$$

$$p_{min} = \frac{k_{st} G_i}{2\sqrt{\frac{k_{st}bT}{2}}\left(e^{\frac{l_0}{2}\sqrt{\frac{2k_{st}b}{T}}} - 1\right)}. \tag{12}$$

Substituting the values p_{max}, p_{min} and p_{av} in formula (10):

$$\xi_p = \frac{l_0}{2}\sqrt{\frac{2k_{st}b}{T}}, \tag{13}$$

where T – tension force of the rubber track.

From this formula it follows that the pressure unevenness increases if the distance between the rollers and the width of the track are increased, and, conversely, decreases with increasing track tension [10].

Substituting the value k_{st} in the previous formulas:

$$\xi_p = \frac{l_0}{2}\sqrt{\frac{2kc_z b}{T(k + c_z)}}, \tag{14}$$

$$p_{max} = \frac{G_i k c_z e^{\xi_p}}{2(k + c_z)\sqrt{\frac{2kc_z bT}{2(k + c_z)}}\left(e^{\xi_p} - 1\right)}, \tag{15}$$

$$p_{min} = \frac{G_i k c_z}{2(k + c_z)\sqrt{\frac{2kc_z bT}{2(k + c_z)}}\left(e^{\xi_p} - 1\right)}. \tag{16}$$

These expressions characterize the law of distribution of normal pressure along the length of the bearing surface of a rubber track. The uniformity of the pressure distribution affects the cross-country capacity of the towing vehicle. The pressure distribution along the length of the bearing surface of the rubber track belt is of no small importance to the normal stiffness of the rubber track belt.

To determine the normal stiffness of a rubber track, an arbitrary track element ΔL limited to two sections 1–1 and 2–2 (see Fig. 3) is considered. Under the influence of the part of the mass of the vehicle G_i falling on the considered section, the rubber track is deformed by the value Δh.

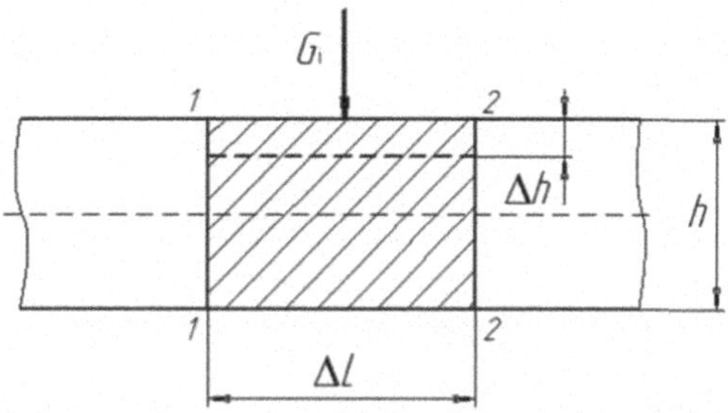

Fig. 3. Scheme for determining the normal stiffness of the track.

The normal deformation of the rubber track using Hooke's law is finded:

$$\Delta h = \frac{G_i h}{ES}, \tag{17}$$

where h – track thickness, but E – modulus of elasticity of the track material. For rubber E = 10 MPa. S – area of an arbitrary part of a track, $S = b \cdot \Delta L$.

The coefficient of normal stiffness of a rubber track is determined by the formula

$$C_z = \frac{G_i}{S_{\Pi} \cdot \Delta h} \tag{18}$$

where S_{Π} – cross sectional area of the track, $S_{\Pi} = h \cdot b$.

Figures 4 and 5 show the curves constructed by the formulas (14)–(16) for rubber track mover on the clay loam with $k = 0.5 \cdot 10^6$ N/m^3; $l_0 = 0.4$ m; $b = 0.35$ m; $T = 10$ kN; $G_i = 40$ kN.

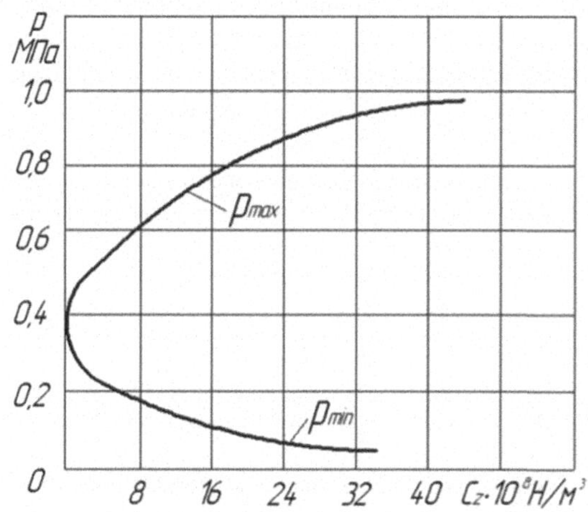

Fig. 4. The dependence of the specific pressures on the normal stiffness of the rubber tracks.

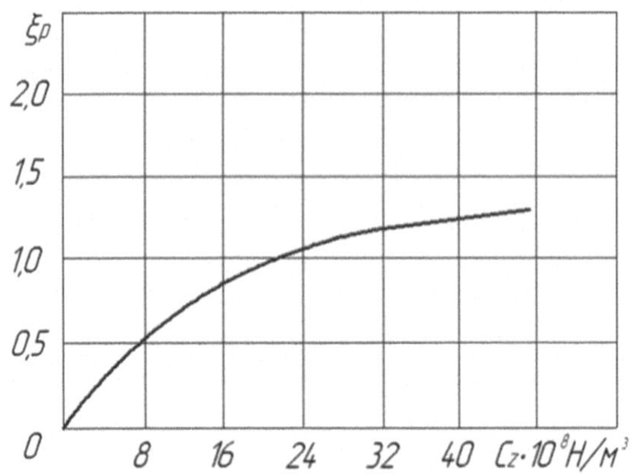

Fig. 5. Dependence of the coefficient of relative non-uniformity of the pressure distribution on the normal stiffness of the rubber track.

From the obtained dependences it follows that an increase in the normal stiffness of the elements of the rubber track causes an increase in the maximum specific pressure with an asymptotic approximation to its value for an absolutely rigid track.

The minimum specific pressure decreases, also asymptotically approaching its value for an absolutely rigid track. In accordance with this, there is an increase in the non-uniformity of the pressure distribution along the length of the rubber track.

3 Conclusion

As a result of the study, dependencies were obtained to determine the main structural parameters of the track running system, the nature of the distribution of normal pressure along the length of the supporting surface of the rubber track was determined. The obtained dependencies take into account such a feature of rubber tracks as their deformability. These studies allow to determine the optimal parameters of the running system of the transport machine at the design stage, which will ensure the preservation of the roadway and movement on concrete and asphalt surfaces.

Acknowledgements. The authors express their gratitude to Kasenov A.Zh. and Baltabekova A. N., who have articles in this journal [11, 12], for assistance in the preparation and editing of this article.

References

1. Rusanov, V.A.: Problems of soil re-compaction by movers and effective ways of its solution. Monograph, p. 368 (1998)
2. Skotnikov, V.A., Yantcov N.D.: Preservation of soil fertility under the influence of running systems. Mechanization and Electrification of Agriculture, vol. 8, pp. 43–44 (1982)

3. Turner, R.J.: Comparison of two and four track machines to rubber tire tractors in prairie soil conditions. Belt and Tire Traction in Agricultural Vehicles, SP-1291. SAE (1997)
4. Abishev, K.K., Bekenov, T.N.: Agricultural vehicle, has replaceable mover whose drive wheel is provided with cross-section pipe, where diameter of drive wheel is greater than diameter of guide wheel, and reinforced resin track matched with cross-section pipe. Patent KZ18188-A. Derwent Innovations Index No. 2019-512412 (2019)
5. Abishev, K.K., Bekenov, T.N., Mukanov, R.B.: Wheeled and tracked mover integrated agricultural transport facility, has power unit mounted on frame for moving frame corresponding to another frame, and cardan gear installed between power unit and transmission line. Patent KZ22165-A4. Derwent Innovations Index No. 2019-36810M (2019)
6. Abishev, K.K., Nurzhauov, A.N.: Some features of the kinematics and dynamics of the caterpillar engine. Bull. Semey Shakarim State Univ. **4**(56), 46–49 (2011)
7. Guskov, V.V.: Tractors. Theory. Learning guide, p. 384 (1977)
8. Abishev, K.K., Mukanov, R.B., Kasenov, A., Baltabekova, A.N.: An issue of intelligent road transport in Kazakhstan. Acta Polytech. **12**, 1–4 (2017)
9. Abishev, K.K.: Mathematical model of interaction of a rubber-tracked mover with a support base. Bull. S. Toraighyrov Pavlodar State Univ. **2**, 9–16 (2011). Series Energy
10. Abishev, K.K.: On the nature of the pressure distribution along the length of the support surface of the rubber track. Sci. J. "Search" **3**, 289–291 (2011). Series of Natural and Technical Sciences
11. Mukanov, R., Kasenov, A., Itybayeva, G., Musina, Z., Strautmanis, G.: Modeling of the cutting head for treating holes in the railway. Procedia Comput. Sci. **149**, 355–359 (2019)
12. Buss, D., Abishev, K., Baltabekova, A.: Driver's reliability and its effect on road traffic safety. Procedia Comput. Sci. **149**, 463–466 (2019)

Assessment of Compaction of the Road Base from Repeated Exposure to Frequently Repeated Loads of Self-propelled Road Roller

Zhandos Tassybekov[1], Tassybek Bekenov[1](✉), and Zhanibek Nussupbek[2]

[1] L.N. Gumilyov Eurasian National University, Munaitpasov Street 5, Nur-Sultan, Kazakhstan
zhandos.org@mail.ru, tas-bek@mail.ru

[2] S. Seifullin Kazakh AgroTechnical University, Zhenis Street 62, Nur-Sultan, Kazakhstan
nussupbek@gmail.com

Abstract. The analysis of the sources shows that there are no analytical dependences that describe the process of soil compaction with sufficient accuracy, taking into account the repeated effects of the loads of self-propelled road rollers. The absence of such laws complicates the study of the process of multiple compaction of soil by road rollers. Since the intensity of the compaction is significantly affected by the contact time of the road roller with the ground, the time factor is of significant importance, since the movement of the road roller is a consistent loading of the soil by the first, second, third, etc. passes. Therefore, the equations on the basis of which the calculations of multiple compaction of the soil from the impact of repeated loads by road rollers to stabilize its compaction are proposed. At the same time, the number of passes to complete compaction was calculated depending on the physical and mechanical properties of the soil and from the parameters of the road roller (load, imprint area, speed, respectively, the time of contact of the wheel with the ground, etc.).

Keywords: Automotive transport · Self-propelled vehicles · Load influence assessment

1 Introduction

All over the world it is customary to assess the economic development of the country on the state of the automobile road network. Poor quality road surface greatly increases the cost of transportation. Therefore, the importance of automobile roads in the modern world is very high, as the better the condition of automobile roads, the more efficiently used road transport. Modern road operating conditions, characterized by increasing vehicle speeds, increasing axial loads and the introduction of new road structures, impose high demands on the strength and stability of the roadbed.

Since the most important indicator of the transport and operational condition of the road is its strength, the road must be, first of all, resistant to loads from cars. Since car loads are dynamic, the effects of such loads are dangerous for both roadway and

E. Ginters et al. (Eds.): ICTE ToL 2019, LNITI, pp. 164–170, 2020.
https://doi.org/10.1007/978-3-030-39688-6_22

roadbed. Very large traffic loads or a significant weakening of the strength of the soil base, will eventually lead to the complete destruction of the road surface. It is known that the main cause of intensive wear and destruction of the road is the under-compaction of its base, it's happening due to the imperfection of the road equipment used, inability to regulate its force pressure on the compacted medium. Under-compaction of the base by only a few percent reduces the strength of the coating and durability by tens of percent. Compaction of road-building materials is both a part of the technological process of construction of road, and the main operation to ensure their strength and stability. The purpose of the seal is to obtain a dense and durable structure of soil and other road materials that can further resist external influences.

The purpose of this work is a brief review and analysis of existing research in the field of compaction of soil base, in justification of features and increased compactibility of the substrate and in the equation to assess the compactibility of the road base and its estimate from repeated exposure often-repeated loads self-propelled road roller. The second section presents an overview and analysis of existing works in the field under study. The third section presents a study of the compactibility of the foundation of the road from repeated exposure to loads self-propelled road roller, which includes a rationale for the features and increased compact ability of the foundation and equations to assess the compactibility of the road base and its estimation from repeated exposure to loads self-propelled road roller. The fourth section presents the conclusion of the work.

2 Review and Analysis of Existing Researches

It is known that the increase in dynamic loads of automobiles contributes to the intensive accumulation of residual deformations in the roadbed, which often occurs even on relatively new roads.

The analysis of the main causes of compaction of the roadbed is carried out in the works [1–3]. The analysis showed that the main factors influencing the degree of soil compaction include the type of compacted soil, its moisture content, the thickness of the compacted layer, the specific load transferred to the soil and others. In the process of compaction, the roadbed must be compacted to the regulatory requirements for different types of soil. The analysis of the existing works shows that the formula existing in the literature for determining the required number of passes on one track is not quite accurate, due to the fact that the number of passes is taken in accordance with the standards for compaction. The process of compaction is perceived in it as a process of increasing density, and not as a process of deformation. When the necessary coefficient corresponding to the norm is reached, it is entered into the calculation. While not taking into account such important properties of the soil as the bearing capacity of the soil and the soil deformation module. Often, the density of the roadbed can increase under the influence of various factors, including without compressive loads. Therefore, the state of soil bases for the purposes of design and testing of rollers is advisable to evaluate the methods that allow to determine both their strength properties and deformation properties.

It is known that the establishment of the values of the roller force of the road roller acting on the compacted material alone is not yet sufficient for a complete understanding its sealing efficiency. To do this, it is necessary to know the contact pressures and deformations (precipitation) under the roller, which creates a roller on the surface of the compacted material, and compare them with the strength and permissible deformation characteristics of the material itself. At present, there are also no analytical dependencies that would directly link the seal characteristic in the form of the running load and the compaction index in the form of the seal coefficient.

The analysis of existing works shows that the stress state of the earth bed has not been widely studied, and little attention has been paid to studies of multiple compaction of the bases. In addition, there is no method for calculating the multiple compaction of the base, which makes it impossible to reasonably recommend a certain number of passes of road rollers to stabilize the compaction of a particular type of soil. All this confirms the relevance of the development of the estimated assessment of compaction methods of the soil base from repeated exposure to loads of road rollers.

3 The Compaction of the Road Base from Repeated Exposure

3.1 Justification of the Equation for the Assessment of Compaction

To work as a features the seal is offered the specific load, which represents the ratio of normal load to the contact area of drum road roller with base. As a measure of compactibility base offers a stable value of its precipitation from the effects of repeated load rollers from road rollers, with their passages on the same track. Based on the proposed characteristics of compaction and compaction index of the soil the design scheme of multiple compaction of the soil from the impact of repeated roller loads is justified [4, 5]. Well known at short-term effects on the soil loads the length of time the soil in a state of stress is shorter than the time required for complete flow wave from the load current value. The properties of soil that affect its compactibility under the influence of the load taken also depend on time of action of the load on the ground. From the analysis, it is clear that the compactibility road roller of soil should determine how the properties of soils at a given moment, changes the soil properties as repeated passages of road rollers and design features of the road roller.

The authors propose a model of compaction of road foundations based on the mechanics of soil deformability, based on the design scheme of multiple compaction of the soil from the impact of repeated loads of road rollers [4–6]. In this case, the time of impact of the road roller when compacting the soil in any area after n passes of road roller is determined by the formula:

$$t_n = nt_1 \tag{1}$$

where t_1, t_n - time of impact of the load on the ground base on the first and n- passages of the road roller; n - number of road roller passes.

The time of action of the load on the ground after the 1st pass of the road roller is determined by the speed of movement and the length of the contact:

$$t_1 = \frac{L}{V_{as}} \tag{2}$$

where L - length of contact of the roller with the ground in the direction of the road roller; V_{as} - average speed of the road roller. The description of the repeated compaction of bases with a significant thickness of the deformable layer is made by layer by layer summation similar to the repeated deformability of the soil under the wheels of a car [7, 8]. Taking into account the frequency of application of loads of the road roller on the soil base, the design scheme and the model of multiple compaction of the soil [5, 6, 9] the description of the possible compaction of the soil is made by layer-by-layer summation as follows:

$$\sum_{k=1}^{n} h_{rk} = \sum_{k=1}^{n} \frac{P_{gk} P_s aDarctg[(H_r - h_{rl}/aD)]}{E_k(P_s - P_{gk})} \tag{3}$$

where h_{rk} - absolute deformation of the soil after the k-th passage of the road roller; H_r - the thickness of the sealing layer; h_{rl} - linear deformation of the ground; a - attenuation coefficient of voltage in the ground; D - the diameter of the circle, equal to the footprint of the roller of road roller; P_s - bearing capacity of soil; P_{gk} - dynamic value of the subsequent specific load; E_k - soil deformation modulus corresponding to the k -th passage of the road roller. The obtained equations allow us to establish the value of compaction by the calculated stabilized value of soil deformation after the k- passage of the road roller.

3.2 Calculated Assessment of the Compaction Index of the Road Base

The results of calculations by (3) of compaction of sand, loam and clay on the number of passes of road rollers for 5 types of masses are shown in Figs. 1, 2 and 3.

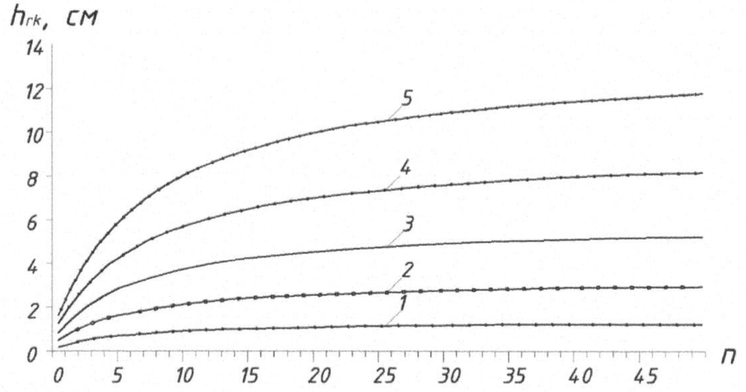

Fig. 1. Compaction dependence on the number of passes (for sand: 1 - P = 3t; 2 - P = 5t ; 3 - P = 7t; 4 - P = 10t; 5 - P = 14t).

In Fig. 1 the most stable seal base of sand is 12 cm from the maximum mass of the roll of road roller 14t, and when the mass of the roll of road roller 10t stable seal will be only 7 cm, which is heavier roll compacted sand to a depth twice as much. The average weight of the roll compacted the base by 5 cm and the lightweight of the rolls by 2 cm or less.

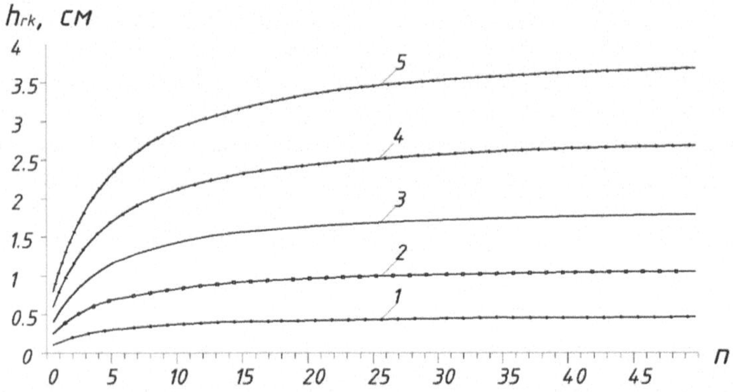

Fig. 2. Compaction dependence on the number of passes (for loam: 1 - P = 3t; 2 - P = 5t; 3 - P = 7t; 4 - P = 10 t; 5 - P = 14t).

According to Fig. 2, the largest stabilized compaction of the base consisting of loam is 3.5 cm from the maximum mass of the roll of the road roller 14t. With a 10t roll, the stabilized seal is 2.7 cm, which differs slightly from the previous one. The average weight of the roller compacted loam at 1.5 cm, and lightweight rolls compacted loam at 1 cm or less.

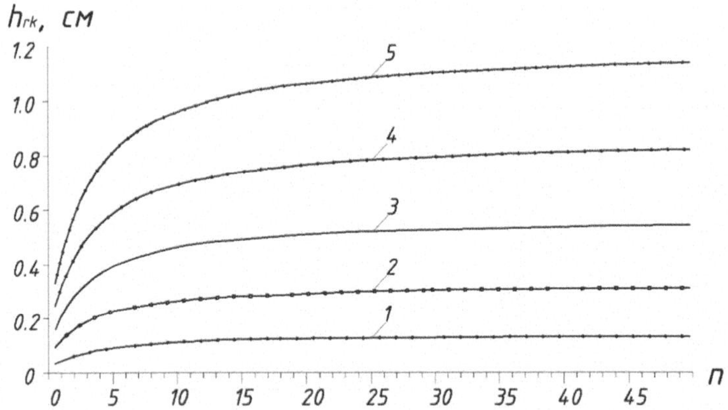

Fig. 3. Compaction dependence on the number of passes (for clay: 1 - P = 3t; 2 - P = 5t; 3 - P = 7t; 4 - P = 10t; 5 - P = 14t).

According to Fig. 3, the greatest compaction of the base consisting of clay is 1.2 cm from the maximum mass of the roll 14t. With the mass of the roll of the road roller 10t, the compaction is 0.7 cm, which differs slightly from the previous one. The average weight of the roll compacted the base by 0.5 cm and the lightweight of the rollers by 0.2 cm or less.

The analysis of graphic dependences shows that at increase in weight of a roller of a road roller increases, and the size of compaction of soil that as a result reduces quantity of passes of a road roller on one trace which is economically effective. For example, a road roller with a roll mass of 14t compacts the sand in 9 passes as much as a road roller with a roll mass of 10t compacts in 50 passes. Road rollers with masses of rollers 7t, 5t and 3t seal the sand to a stable state for 50 passes on one track, respectively, on the same sand road rollers with masses of rollers 10t, 7t and 5t seal the soil to a stable state for 5, 3 and 2 passes on one track.

3.3 Validation of the Theoretical Results

The results of calculations on the Eq. (3) in relation to the deformation of the soil from the impact of repeated loads of self-propelled wheeled vehicles, in order to assess their patency, are adequate to the experiments [8].

In this article, the scope of the Eq. (3) in relation to multiple compaction of the soil from the impact of roller loads is expanded in order to establish the maximum number of passes of the rink by the value of the stabilized soil sediment.

Therefore, any results of calculations by Eq. (3) to establish the value of stabilized soil sediment in assessing its multiple compaction from the impact of road roller loads will be adequate to the experimental results.

4 Conclusion

The review and analysis of compaction of road building materials from the impact of road roller loads are done in the article. There are justified the characteristic seal in the form of a unit load that represent the ratio of normal load to the contact area of drum road roller with the base and the indicator of the compactibility of the foundation in the form of a stabilized value of its precipitation from the effects of repeated specific loads of a road roller when it passes over the same track.

The necessity of assessing the compaction of the base from the impact of repeated loads of the road roller, taking into account the strength and deformation properties of the base, is substantiated.

The equations for estimating the compaction of the road base from repeated exposure loads of a self-propelled road roller are provided. The calculated assessment of the compaction index of the road base from repeated exposure loads of a self-propelled road roller is carried out. It is revealed that from repeated passes of road rollers soils are compacted and the intensity of compaction is significantly influenced by the contact time of the roller of the road roll with the soil.

The proposed approach allows to assess the multiple compaction of soil layers from repeated exposure to frequently repeated loads of a self-propelled road roller on the stabilized value of soil precipitation, taking into account its strength and deformation characteristics, while existing studies are based on changes in soil density.

The study shows that the results do not contradict the experimental data, confirm the known views of scientists and can be recommended for use in practice.

Further research will be carried out in terms of justification and selection of optimal indicators of compaction of road construction materials from repeated exposure loads of the self-propelled roller. Research will also be conducted to improve the design parameters of a self-propelled road roller, taking into account the compaction of defined road-building materials.

References

1. Cherkasov, I.: Mechanical properties of soils in road construction. p. 198. Transport, Moscow (1976)
2. Sidenko, V., Batrakov, O., Volkov, M.: Automobile road. Improvement of design and construction methods. p. 241. Budivelnik, Kiev (1973)
3. Vasilyev, A., Sidenko, V.: Road maintenance and traffic management. p. 300. Transport, Moscow (1990)
4. Tassybekov, Z., Zhumagalieva, G.: Model of soil base compaction. Int. Sci. J. Actual Probl. Mod. **3**(53), 63–66 (2010). Series Technical Sciences. Bolashak Publishing House, Karaganda
5. Bekenov, T., Tassybekov, Z., Nussupbek, Z.: To development of model of compaction of the soil basis of roads. Bulletin of the Kyrgyz State University of Construction, Transport and Architecture. Bishkek (2012)
6. Bekenov, T., Tassybekov, Z., Nussupbek, Z.: Substantiation of parameters and establishment of a model of multiple compaction between road rollers. Way Sci. **5**(39), 17–19 (2017). Volgograd
7. Ageykin, Y.: Passability of automobile. p. 231. Machine Building, Moscow (1981)
8. Bekenov, T.: Mathematical model of deformation of the medium under repeatedly applied quasi-static loads. No. 8, pp. 69–72. News of Universities Mechanical Engineering (1989)
9. Bekenov, T., Tassybekov, Z., Nussupbek, Z.: Justification of the road based seal model from the influence of road roller loads. Bull. L. N. Gumilyov Eurasian Natl. Univ. **125**, 44–50 (2018). Astana
10. Bekenov, T., Tassybekov, Z., Zhumagalieva, G.: Compaction of the soil base of roads taking into account the influence of the roll diameter of the road roller. In: Proceedings of International Scientific and Practical Conference Improvement of Operational Properties of Transport and Technological Machines and Complexes, pp. 28–33. Omsk (2012)

Railway-Ferry Crossings of Ukraine and Baltics as an Efficient Link of the Baltic-Black Sea Transport Corridor

Larysa Parkhomenko[1], Ruslan Viznyak[1], Dmytro Skurikhin[1(✉)], and Janis Eiduks[2]

[1] Ukrainian State University of Railway Transport, Kharkiv 61001, Ukraine
l.o.parkhomenko@ukr.net, viznyakruslan1977@gmail.com, skurikhin@i.ua

[2] Institute of Transport, Riga Technical University, Azenes Street 12-316, Riga 1048, Latvia
janis.eiduks_2@rtu.lv

Abstract. The historical circumstances of the emergence and development of rail-water communication in the European international directions were determined and clarified. The significance of functioning of existing and perspective railway-ferry crossings of Ukraine and the Baltic countries in the context of the general transport chain of container and contrailer transport of a unified logistics system is revealed. Effective long-term operation of such logistics projects as Viking and Zubr, which are used to transport containers between the ports of the Black and Baltic Seas combines developed portship infrastructure and railway-ferry crossings complexes in Odessa, Chernomorsk, Yuzhny, Riga, Ventspils, Liepaja, Klaipeda, Tallinn-Muuga, etc. The current "EU-Ukraine Association Agreement" in the transport sector confirms the significance of the planned joint intentions and measures, and causes intensive development of intermodality and interoperability of the Ukrainian-Baltic international transport systems.

Keywords: Railway-ferry crossings · Intermodal transport corridors · Stress-strain state

1 Introduction

Modern Ukraine and the Baltic States, as countries some of which have access to the Black and Azov Seas, and some – to the Baltic Sea, have all technical and economic capacities for further intensive development of combined multimodal transportation and the formation of new logistic links within the Transport Corridor Europe-Caucasus-Asia (TRACECA) of the Eastern Partnership Transport Panel. Due to their favourable geographical position for export and import, as well as transit commodity exchange – an unprecedented decades-long powerful concentration of port infrastructure and ferry complexes of the Black Sea and Baltic regions, which has been tending only to increase and expand in recent times - they have become the common transport gates for entry and exit, and the symbol of the combination of the North and South of Europe in the ancient historical ways. Thus, timely and effective measures to

E. Ginters et al. (Eds.): ICTE ToL 2019, LNITI, pp. 171–179, 2020.
https://doi.org/10.1007/978-3-030-39688-6_23

identify the most important components of modern transport policy and logistics, as well as their general integration into the European international transport system are required. Combined railway and ferry lines are such a technical achievement and symbiosis of the sectors of transport complexes of Ukraine and the Baltic states, namely Latvia, as a developed European country, in the strategic and free competitive struggle for successful overtake and operative use of the main transport routes. Efficiency of operation of ferry lines (so called *crossings*) as separate logistical chains of general transport routes thus determines reliability of interaction of the main modes of transport, namely rail and water, in the space of their joint development, technical improvement and continuous increase of volumes of international transportations through our countries.

2 Problem Definition

The existing ferries in northern Europe: Hanstholm – Kristiansand (Norway – Denmark), Trelleborg – Copenhagen (Sweden – Denmark) across the Skagerrak and Kattegat straits, Stockholm – Turku (Sweden – Finland) via the Gulf of Bothnia and analogues: Klaipeda – Sassnitz, Rügen (Mukran) (Lithuania – Germany) and the newly built by Russia Ust-Luga-Baltiysk (former Pillau, Kaliningrad region) crossing, besides Ventspils – Nynäshamn, Travemünde (Latvia – Sweden and Germany, respectively), Klaipeda – Karlsham (Lithuania – Sweden), Klaipeda – Kiel (Lithuania – Germany), Tallinn – Stockholm (Estonia – Sweden), Tallinn – Helsinki (Estonia – Finland), etc. are the result of centuries-old practice of cargo and passenger transportation. Located at the crossroads of trade routes, Ukraine and the Baltic States, as a north-eastern transit territory, have played an important role in forming large historic transport corridors with their part between the Baltic and the Black Sea. Therefore, the common rail and sea transportation as a whole and its ferry link is a natural and successfully implemented logistical solution with well-grounded technical, economic and geopolitical reasons.

It is known that Klaipeda in Lithuania from the common geographical and economic space of the Baltic Sea with its own developed railway and road ferry link is connected with almost all major ports of the Baltic Sea, namely: Baltiysk (ex-Pillau), Ust-Luga, St. Petersburg, Vyborg (Russia), Tallinn, Muuga (Estonia), Ventspils, Riga, Liepaja (ex-Libau) (Latvia), Gdansk (ex-Danzig) - Gdynia (Poland), Kiel, Lubeck, Rostock, Szczecin (Stettin), Sassnitz (Rügen) and others (Germany), Stockholm (Sweden), Copenhagen (Denmark), Helsinki, Turku (Finland), which undoubtedly brought independent Lithuania, as well as Latvia and Estonia, to the modern level of European countries of the transit zone, which adhere to the logistical principles of developing their own market and economic growth. These criteria are also characteristic for Latvia and Estonia with their priorities of transshipment to the Western European countries. Ukraine already has as a country seeking for the integration, but with orientation of the technical and economic specification of the new generation rolling stock for the European track gauge, a unified automatic coupler design, new body and undercarriage designs, etc., as well as other technical innovations, for example, those for ensuring reliable fastening and the stability of tank cars, gondola cars and container and contrailer flat cars as heavy dynamic systems to the ferry decks and their class with respect to the course of transportation [1, 2].

The main purpose of the article is a feasibility study for expanding the operation of railway ferry crossings as high-tech multimodal units in modern logistic combined transport systems of Ukraine and the Baltic Sea, as well as developing technical measures aimed at improving the efficiency of operation and interaction of railway and water facilities international transportation by means of the Baltic-Black Sea transport corridor. For example, these are the following solutions:

- Re-equipment and modernization of existing car designs with special devices for secure fastening on the decks of the rail ferries in both car-ship directions, gradual transfer of the post-Soviet 1520 mm track gauge from the territory of independent Ukraine to pan-European 1435 mm track gauge network, etc. for a successful inclusion into the transport network of the EU and Schengen countries [3, 4];
- Refinement of the technique and technology of railway ferry vehicles for the purpose of visa-free transportation of cars by sea by lighterage, which improves the dynamic performance and safety of the combined rail-sea transportation, significantly simplifies handling of cars by the existing technology when they are fastened with chain ties, lanyard stoppers and stop jacks. Utilization of the existing port facilities without major capital investment for constructing new specialized infrastructure (existing ports).

3 The New Method of the Cars Loading and Fastening

The Baltic states with their well-developed port infrastructure and the availability of recognized ferry services, such as Klaipeda, Ventspils, and Muuga, leave alone more intense development and use of the sea coast in completely new places, can significantly expand their functions by justifying the expansion of the organization of railway-ferry transportation, further survey, arrangement and construction, as well as successful integration into the common logistics system of operation with further integration into the Baltic-Black Sea corridor.

Figure 1 shows the most intense routes near the Black Sea, which can be used as a link in the Baltic-Black Sea international transport corridor.

The close connection of the ports and ferry complexes of the Baltic states with the European countries situated on the Baltic Sea coast, such as Germany, Denmark, Sweden, as well as the North Sea basin, including Norway, Finland, brings the prospect of increasing economic exchange of production technologies to a common denominator, saturating the common market of modern goods and transport services, developing substantiated in terms of logistics promotion of transit and export-import traffic towards Ukraine, as a country of the Eastern region with the developed chain of seaports and ferry crossings of the TRACECA corridor [5, 6], as well as in the reciprocal direction to the West and North through Lithuania, Latvia, Estonia.

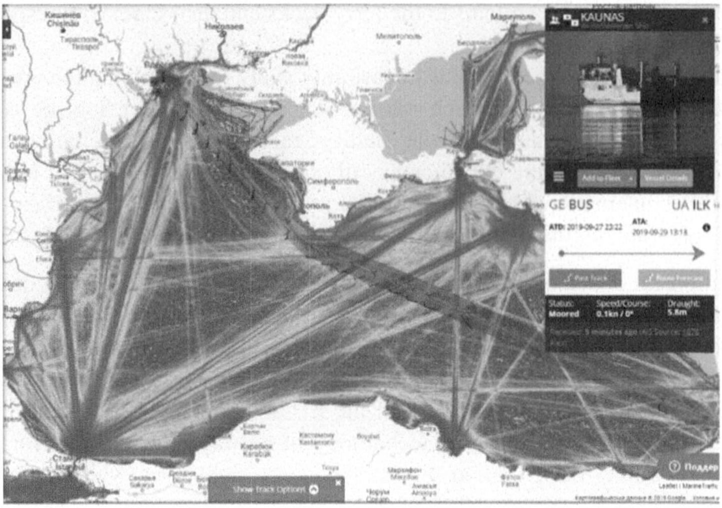

Fig. 1. Analysis of statistical data on the density and nature of ship traffic in the Black Sea with a route of KAUNAS (VILNIUS) ferry.

"UKRFERRY" shipping company transports from Ilyichivsk to Varna (Poti, Batumi, Georgia as well as Derince, Turkey) and has three ferry vessels. These are Ukrainian ferries "Heroes of Shipka" and "Heroes of Plevna" built in 1977, 1978/79 at the shipyard Uljanik, Yugoslavia, now Croatia. The Bulgarian Republic has vessel with virtually identical technical specification and appearance. These are "Heroes of Sevastopol" and "Heroes of Odessa". In addition, it owns the more modern ship Greifswald, built in 1988, and operates under the lease of the Georgian firm Instra and is under the classification supervision of the German Lloyd. These are vessels of classes RO-RO (roll on-roll off) and LO-LO (lift on-lift off), i.e. rolling car to the upper decks or lifting with the lifting mechanisms, respectively, can take 108 and 50 cars on board.

From the Baltic side, in Lithuania, the VILNIUS and KAUNAS SEAWAYS ferryboats built 1987/89, owned by DFDS (Norway), are operated, which are identical to GREIFSWALD. The corresponding ferry vessels are shown in Fig. 2.

The largest and most modern railway ferry on the Black Sea, capable of carrying passengers, as well as rolling stock, motor vehicles, containers and deck cargo, was built in 1988.

Ferryboat Greifswald has only two loading decks with five tracks. The upper deck is loaded via the main (lower) deck by the rolling ramp. In addition, there is a reserve space for loading road vehicles. The vessel has 2 cargo decks and can carry 50 universal cars with dangerous cargo in a closed cargo area (on the main deck) plus up to 50 to 60 heavy trucks TIR, as well as containers on the upper deck (half of the upper deck is covered with the topside, the other half is open). The upper deck is loaded from the main deck using the inner ramp. The vessel also has a stern ramp for motor vehicle loading.

Features of the dynamics of the car being transported by a railway ferryboat during heaving of the sea are the following. The dynamics of the sea waves have a significant effect on the magnitude of the forces acting on the car body being transported by ferry.

a) b)

c) d)

Fig. 2. Railway ferries with cars on board: (a) HEROES OF SHIPKI; (b) GREIFSWALD; (c) VILNIUS; (d) KAUNAS.

In assessing the navigability of a vessel, the formulas of the theory of progressive waves of relatively small amplitude are most commonly used to describe regular waves. The relation between the amplitude of a wave and its other kinematic parameters is established by the empirical formulas obtained by on-site observations [7].

Such waves are called trochoidal. The profile of a trochoidal is described by equations in parametric form:

$$\begin{aligned} x &= a + \mathrm{Re}^{kb} \sin(ka + \omega t) \\ z &= b - \mathrm{Re}^{kb} \cos(ka + \omega t) \end{aligned} \tag{1}$$

where a and b are coordinates of the center of orbit of the particle with coordinates x and z, but R is the radius of stationary circle. In addition, the following parameters were assumed, where λ is wave length, τ is wave period, ϑ is wave speed, ω is circular wave frequency, k is frequency of wave etc. These parameters are determined numerically and individually for the Black and Baltic Seas.

To compose the equations of motion of the car relative to the sea surface by a ferryboat, the D'Alembert-La-Grange principle of the second kind is used, and after certain transformations (2) is obtained.

To determine the force that will act on the car during sea rolling, the magnitude of acceleration would be calculated. The car body oscillates along the y and z axes in addition to other directions of combined movement. According to the laws of mechanics, acceleration of a material point, which is performing combined motion, is:

$$\begin{aligned} \frac{a_{11}\dot{q}_1^2}{2} + \frac{b_{11}\dot{q}_1^2}{2} + \frac{c_{11}q_1^2}{2} = F_i, \quad (a) \\ \frac{a_{22}\dot{q}_2^2}{2} + \frac{b_{22}\dot{q}_2^2}{2} + \frac{c_{22}q_2^2}{2} = F_j, \quad (b) \\ \frac{a_{33}\dot{q}_3^2}{2} + \frac{b_{33}\dot{q}_3^2}{2} + \frac{c_{33}q_3^2}{2} = F_k, \quad (c) \end{aligned} \tag{2}$$

where $q_i, \dot{q}_i, \ddot{q}_i$ are the generalized coordinates, velocities and accelerations, $a_{11}..a_{33}$- inertial coefficients, $b_{11}..b_{33}$- dissipative coefficients, $c_{11}..c_{33}$- stiffness coefficients, $F_{i..k}$- external forces. The geometric sum of the accelerations (relative and translational) can be specified as:

$$\ddot{q}(t) = (y + z)'' \tag{3}$$

Then, the equation of motion of the car body during rolling has the form:

$$a_{11}\ddot{q}_1 + b_{11}\dot{q}_1 + c_{11}q_1 = (y + z), \tag{4}$$

In real conditions free oscillations of the car will occur as a result of various deviations during fastening to the deck. Then the equations of motion that reflect the free-oscillation pattern then will have the following form:

$$\begin{aligned} a_{11}\ddot{q}_1 + b_{11}\dot{q}_1 - c_{11}q_1 = 0 \\ a_{22}\ddot{q}_2 + b_{22}\dot{q}_2 - c_{22}q_2 = 0 \\ a_{33}\ddot{q}_3 + b_{33}\dot{q}_3 - c_{11}q_3 = 0 \end{aligned} \tag{5}$$

The total acceleration of the car body can be determined:

$$a = \ddot{q}(t) + g\sin\theta \tag{6}$$

where θ - ferry roll angle.

Transportation of cars by sea using traditional fastening technology has an adverse effect on the preservation of their load-bearing structure, reliability and safety of the transportation process. During movement at high speeds this can lead to accidents. Therefore, to evaluate the strength of the car during its transportation by ferry and the real load in order to improve the interaction of railway rolling stock and sea ferry vessels, the analysis of the strength of the new generation of gondola car body was conducted (see Fig. 3). The analysis showed that the values of stress-strain state exceed the permissible values. In view of the identified shortcomings of the devices for interaction of cars with the decks of the railway ferry vessel, new car fastening schemes should be developed and implemented in order to automate the fastening process, ensure the reliability of operation and preservation of cars during heaving of the sea [8–10]. The results of stress-strain state of the modern gondola car body during rolling were obtained, taking into account the actual scheme of fastening to the deck.

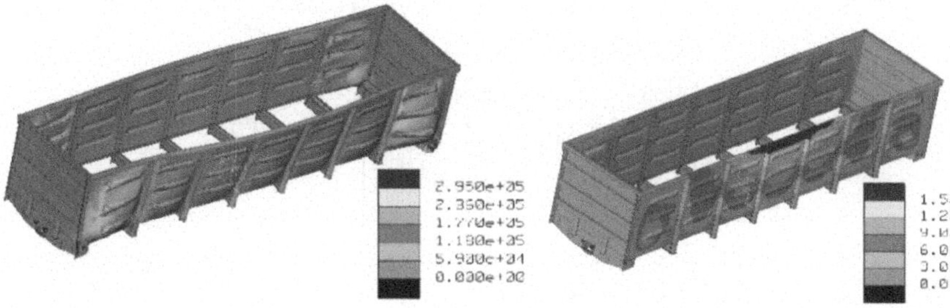

Fig. 3. The nature of the stress-strain state and movement in the elements of the car body during rolling.

An alternative option which will ensure the technical result of fastening of car bodies relative to the decks is the transportation of cars by sea without bogies, as shown in Figs. 4 and 5. Analysis of the statistics of ferry accidents showed that the main cause of loss of stability, and as a consequence, the wreck of ferryboats, is dis arrangement of cars relative to deck and damaged fastenings. Therefore, developing a new, improved system for the interaction of cars with the ferry decks, which would prevent displacement of its relative initial position during rolling is required [11–14].

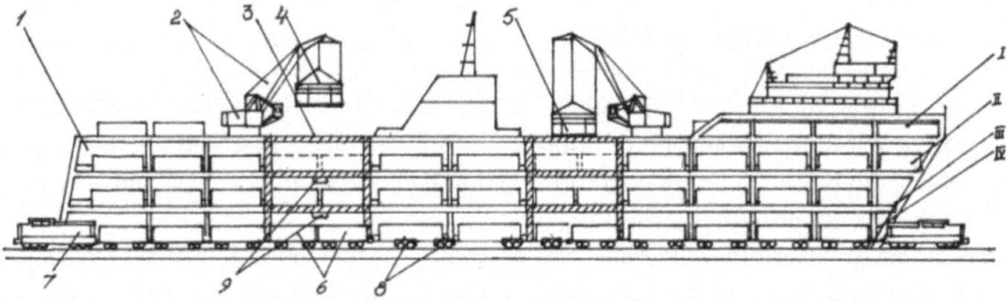

1- Ferry; 2 - Crane; 3 – Platform; 4 – Suspension system; 5 – Car body; 6 –Train car; 7 – Locomotive; 8 – Bogie; 9 – Horizontal movement device; 10 – Fastening devices; 11- Supporting devices; I-IV – Ferry decks

Fig. 4. Loading of the cars removed from the bogies to the ferry (side view).

The method can be used in rail-water communication with respect to international transport destinations not only at specially equipped ferry stations and complexes, but also at port berths, which greatly expands the operational capabilities of rail ferry services. In addition, the method can be successfully applied in the transportation of both wagons of various purpose and traction rolling stock or containers.

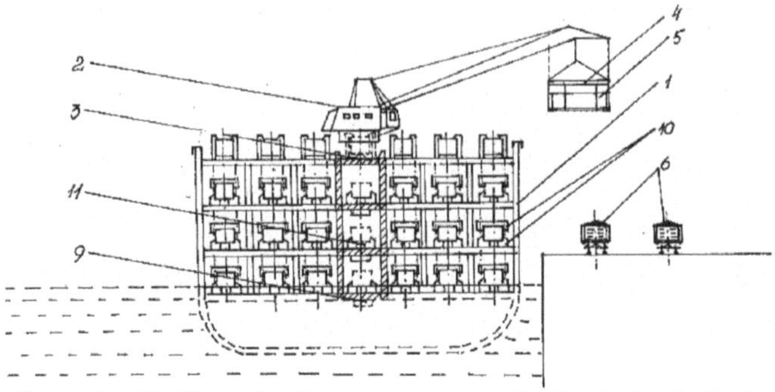

1- Ferry; 2 - Crane; 3 – Platform; 4 – Suspension system; 5 – Car body; 6 –Train car; 7 – Locomotive; 8 – Bogie; 9 – Horizontal movement device; 10 – Fastening devices; 11- Supporting devices; I-IV – Ferry decks

Fig. 5. Loading of the cars removed from the bogies to the ferry (front view).

The proposed method has been tested on a physical model that showed the feasibility of its application.

4 Conclusion

Ukraine and the Baltic States have great prospects for the development and efficient use of the railway and ferry lines of the unified logistic system of the Black Sea - Baltic International Transport Corridor, both from a geographic location and from a mutually developed transport infrastructure.

In order to ensure preservation of railway rolling stock, contrailers and containers during transportation in the international combined Ukraine-Baltic rail-sea routes, new alternative measures should be developed and implemented that will allow adjusting the vehicles to these operating conditions and reduce the number of damages occurring due to the fastening with standard devices.

In order to ensure the reliability and efficiency of transportation of loaded cars by sea by a railway ferry during sea rolling, a new sea ferry design for transportation of cars without bogies and a device for fastening cars to the ferry decks with fully automated fastening is proposed. The method helps to increase the payload of the vessel, increase cargo turnover and reliability of fixing of bodies of cargo units at increase of roll angles and ferry vessel differential in conditions of sea excitement.

The Railway Cars Department of the Ukrainian State University of Railway Transport has been conducting studies for a long time on the issues of car preservation during their operation in the international railway-sea routes, proposes new solutions in order to ensure the strength, reliability and preservation of passenger and freight cars during their transportation by sea which has sufficient business bases and can become a common creative discussion for further active and effective cooperation with our Baltic colleagues.

References

1. Fomin, O., Lovska, A., Masliyev, V., Tsymbaliuk, A., Burlutski, O.: Determining strength indicators for the bearing structure of a covered wagon's body made from round pipes when transported by a railroad ferry. Eastern-European J. Enterp. Technol. **1**(1–97), 33–40 (2019)
2. Fomin, O., Lovska, A., Kulbovskyi, I., Holub, H., Kozarchuk, I., Kharuta, V.: Determining the dynamic loading on a semi-wagon when fixing it with a viscous coupling to a ferry deck. Eastern-European J. Enterp. Technol. **2**(7–98), 6–12 (2019)
3. Vaiciunas, G.: Assessment of railway lines: an efficiency rating analysis for Baltic countries. Transp. Probl. **13**(1), 49–58 (2018)
4. Guler, N.: Black sea maritime transport corridors in relation with EU projects. WIT Trans. Built Environ. **79**, 293–300 (2005)
5. Davtyan, V.: Armenia's transport security within the framework of ITC TRACECA and "North-South". In: International Scientific Conference "Investment, Construction, Real Estate: New Technologies and Special-Purpose Development Priorities", Vol. 212, pp. 225–232 (2018)
6. Maciulis, A., Jakubauskas, G.: Challenges and opportunities for Lithuanian transport sector to become an integrated part of trans-Asian connections. In: Proceedings of the International Conference, pp. 95–98 (2007)
7. Golovko, V., Viznyak, R.: Railway-ferry crossings of Ukraine: historical past and modern ways of improving functioning efficiency. In: Collected scientific works of Ukrainian State Academy of Railway Transport 86, pp. 31–39 (2007)
8. Lovska, A., Fomin, O., Horban, A., Radkevych, V., Skok, P., Skliarenko, I.: Investigation of the dynamic loading of a body of passenger cars during transportation by rail ferry. EUREKA, Phys. Eng. **4**, 91–100 (2019)
9. Lovska, A.: Simulation of loads on the carrying structure of an articulated flat car in combined transportation. Int. J. Eng. Technol. (UAE) **7**(4), 140–146 (2018)
10. Lovskaya, A., Rybin, A.: The study of dynamic load on a wagon platform at a shunting collision. Eastern-European J. Enterp. Technol. **3**(7), 4–8 (2016)
11. Viznyak, R.: Increasing the Effectiveness of the Interaction between Rail and Maritime Transport in the Conditions of International Ferry Transportation. Grant Report No. 0108U088652 (2009)
12. Golovko, V., Viznyak, R., Golovko, T.: Ferryboat for the transport of railway rolling stock or containers. Patent No. 85438 Ukraine (2009)
13. Golovko, V., Viznyak, R., Golovko, T.: Method of loading and placing railway rolling stock units on a ferry vessel. Patent No. 85437 Ukraine (2009)
14. Golovko, V., Viznyak, R., Lovska, A.: Device for securing a railway car at the deck of a ferry vessel. Patent No. 91712 Ukraine (2010)

Perspective Directions for the Development of a New Generation Complex of Cargo Wagons

Ramis Zaripov[1], Nurbolat Sembaev[1], Ansagan Suleymenov[1], and Pavels Gavrilovs[2(✉)]

[1] Faculty of Metallurgy, Machine Building and Transport, Pavlodar State University, 140000 Pavlodar, Kazakhstan
ramis.zaripov@mail.ru, n.sembaev@mail.ru, ansar_muslim_91@mail.ru

[2] Institute of Railway Transport, Riga Technical University, Paula Valdena Street 1, Riga 1048, Latvia
pavels.gavrilovs@rtu.lv

Abstract. The article discusses the problems associated with the design of freight rail cars. The aim of the study is to determine methods for modernizing freight cars, as well as creating a whole complex of new generation cars. The results of the analysis of the state of the rolling stock of Kazakhstan, which can be used to judge the relevance of work, are reflected. It is determined that the basis for the development of rolling stock of a new generation should be the principle of a modular layout with a rational unification of units and systems that reduces the cost of these structures and their development. Some ways of developing the design of freight cars are considered in detail. The prospects for the use of composite material - fiberglass as the material of the body and elements of the car on the example of the boiler of the tank car are examined. The replacement of steel sheets of the car body of a freight car with sheets of aluminum alloys is considered. The parameters of articulated cars are considered using the example of a hopper car.

Keywords: New generation freight wagon · Composite materials · Railway transport

1 Introduction

Railway transport is an essential component of the industrial infrastructure of the Republic of Kazakhstan. The stability of the state economy, ensuring national security, improving the conditions and living standards of the population depend on its sustainable and effective work.

Currently, the work is underway to create innovative wagons with improved technical, economic, strength and dynamic performances. The main attention is paid to the development of wagons for the transportation of petroleum products, namely, an increase in their carrying capacity, a reduction in the metal consumption of the structure, a decrease of the center of gravity of the wagon to improve its dynamic

E. Ginters et al. (Eds.): ICTE ToL 2019, LNITI, pp. 180–187, 2020.
https://doi.org/10.1007/978-3-030-39688-6_24

characteristics, and more efficient use of net linear load which largely determines the efficiency of railway transport.

The aim of the study is to determine the methods of modernization of freight wagons, as well as the creation of a whole range of new generation wagons. The most significant contribution to the solution of the parameter optimization problems, the selection of a rational design scheme and the justification of the technical and economic indicators of freight wagons with the help of a specially developed method of selecting types and optimizing the main parameters, which allows determining the latter by the criterion of minimum reduced costs and being universal for all types of freight wagons were made by L. A. Shadur and V. V. Lukin.

In accordance with the developed method, the work was carried out to justify the feasibility of creating specialized freight wagon. A review of the work related to the selection of railway wagons parameters allowed to highlight the main trends in their development: increase in productivity (increase in loading capacity, full use of axial load), reduction of material consumption, reduction of operating costs, improvement of strength characteristics, indicators reliability and maintainability, ensuring the safety of goods and environmental safety.

2 The State of the Inventory Fleet of Wagons in Kazakhstan

In connection with the development of the mining industry, most of the freight traffic is made up of such cargoes as coal, ore, crushed stone, quartzite and oil products. In this regard, gondola wagons, tanks and special-purpose wagons occupy the vast majority of the fleet inventory of wagons (see Fig. 1) [1].

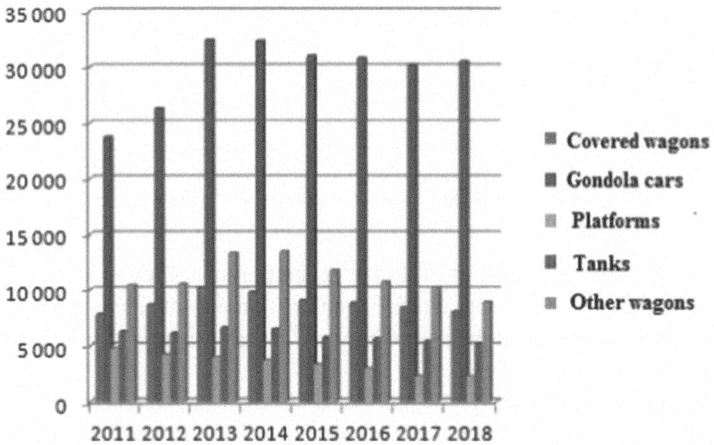

Fig. 1. Number of freight wagons in Kazakhstan.

Having made an analysis of the state of the inventory of wagons the following conclusions can be inferred:

- The average age of a freight wagon has decreased in recent years due to the active attraction of investments in the industry, but remains high and exceeds 75% of the designated service life;
- Gondola wagons have the highest wear, which is associated with a shorter service life compared to other types of rolling stock (22 years), the age of every third gondola wagon operated on the railway network exceeds the service life. Freight operators need high-capacity wagons with improved technical and economic parameters and a long service life.

To analyze possible modernization options, a freight wagon should be considered a system consisting of subsystems (see Fig. 2). Each subsystem is formed to obtain the technical and economic indicators specified for the wagon and in accordance with the restrictions imposed by the wagon as a whole, and from other subsystems.

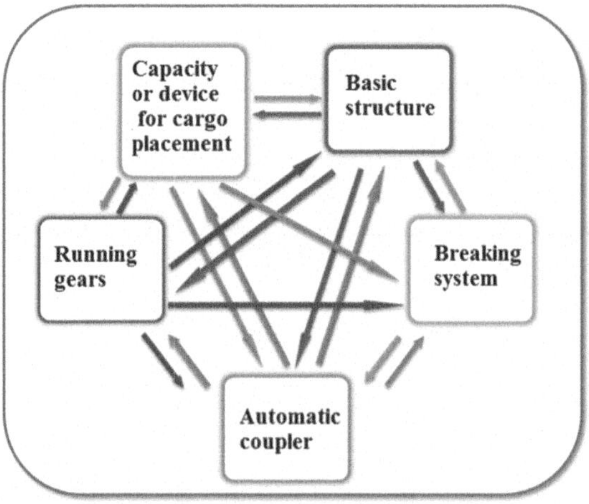

Fig. 2. Cargo wagon subsystems interrelated.

The basis for the development of a new generation of rolling stock should be the principle of a modular layout with a rational unification of units and systems, reducing the cost of these structures and their development. Freight wagons should be created on the base of unified basic projects.

The transition from the design of individual specialized wagons to the design of a family of wagons on the basis of the basic design will allow the creation of wagons with a given level of reliability and durability. With development of the basic modular components of freight wagon structures, unified procedures for their maintenance and repair should be provided. The efficiency of new generation wagons is achieved by:

- Compliance of designs with current regulatory documents of the customer and economic development forecasts during the assigned service life;

- Increase in load capacity (not less than 5%) compared with existing prototypes;
- Decrease in specific material consumption per unit of wagon capacity, body volume and floor area;
- The use of carts with improved systems for spring suspension reducing dynamic loads in the bearing nodes of the wagons and the elements of the track's upper structure;
- Increase the service life of freight wagons up to 32 years.

So, some of the possible options for upgrading the body structure of the new generation freight wagons are analysed in detail.

3 Wagon with a Body Made of Composite Material

Currently, the main materials used in wagon building are carbon and low alloy steels, and aluminum. The most common composite materials in railway engineering are coal, glass, organoplastics. Composites have several advantages and disadvantages over traditional car building materials. Comparative characteristics of the materials are presented in Table 1.

Table 1. Comparative characteristics of steel, aluminum and composite material.

Criterion	Steel	Aluminum	Composite material
Density, kg/m^3	7800	2640–2800	1800–1900
Elastic modulus, GPa	210	70–71	55
Tensile strength, MPa	240	50–440	1700
Thermal conductivity, W/Km	17,5–58	140–190	0,3–0,5
Conductivity, Ohm	Conductor	Conductor	$1 \cdot 10^{10}$
Corrosion resistance (external environment)	Low	High	Absolute
Resistance to aggressive environment	Low	Low	High
Hygroscopicity, %	No	No	0,5
Operating temperature, °C	from −60 to +300	from −70 to +210	from −60 to +600
Durability, years	About 40	About 50	More than 80

Thus, with respect to steel and aluminum, composites are lightweight, but at the same time they are not inferior to them in strength characteristics. In addition, composite materials have low thermal conductivity, biological and chemical resistance, durability, are dielectrics and can withstand temperatures up to 600 °C. However, the cost of composites is several times higher than the cost of steel and aluminum, which is explained by the complex technological process for the production of composites and the need to use expensive equipment and raw materials [2]. High prospects for the use of composite materials in the wagon building industry are due to:

- High resistance of composites to the chemical effects of transported goods and the environment;
- Compatibility with both chemical and food products;
- Ability to maintain mechanical properties throughout the entire service life when exposed to elevated and lowered temperatures;
- Lack of the need for expensive coatings;
- Reducing the frequency of repairs.

All these factors affect the reduction in the cost of the life cycle of an innovative product. It should be noted that the use of composite materials can reduce the tare weight of the wagon. The production of wagon parts from composites is less labor-intensive compared to the production of these parts from metals. As a rule, it takes more time to form a part from a composite material than to stamp a steel sheet, but metal processing includes a number of additional operations, such as welding and stripping, which are not typical for processing non-metallic materials [3].

The high cost of the composite material is compensated by the efficiency of its application, which, due to the strength characteristics, is higher than the efficiency of application in carriage building of metals. Despite the fact that the introduction of parts made of composite materials increases the initial cost of wagons, the properties acquired by the part ensure its proper operation throughout the entire life cycle of the wagon. Only a small proportion of Russian wagon-building plants carry out technical modernization of their products. The use of composites NPK Uralvagonzavod is considered in Table 2.

Table 2. Characteristics of innovative products of JSC "Uralvagonzavod" with components made of composite materials.

Product Name	Appearance	Purpose	Technical characteristics
Composite material hopper wagon with removable roof		Transportation of mineral fertilizers and other non-hazardous bulk cargo	Carrying capacity 71 t. Volume 86 m^3. Tare Weight 22.5 t
A hopper wagon with a body made of composite material		Transportation of mineral fertilizers and other non-hazardous bulk cargo	Service life - up to 100 years. Payload 74 t. Volume 125 m^3. Tare Weight 25.5 t
Tank container made of composite material		Transportation of dangerous goods (hydrochloric acid, sodium hydroxide, technical, etc.)	Service life - up to 100 years. Capacity 24 m^3. Tare Weight 4.5 t

The parts of the railway wagon, which can be made of composite material are following:

- Boiler tank wagon;
- Removable gondola roof;
- Adapters for the axle box of the trolley;
- Walls and roof of a covered wagon;
- Roof, manhole covers and walls of the hopper car;
- Elements of brake devices.

Consider the boilers of tank wagons. The use of fiberglass for the manufacture of tank boilers can significantly reduce operating costs due to the resistance of fiberglass to aggressive loads, lower thermal conductivity compared to steel and a sufficiently high mechanical strength with significantly lower dead weight. In turn, saving the weight of the boiler due to the use of fiberglass in its design can increase the payload capacity of each mobile unit by 3–5%.

According to the few researches available the first attempts to create fibreglass boilers for railway tanks abroad took place in the 60 s. So, in the USA in 1963, light tanks XP-1 were put into operation, with a capacity of 85 m^3, a tare weight of 23.6 tons, with fiberglass boilers based on epoxy resin [4]. It is mounted on standard trolleys and has a steel standard frame. Over six years of operation, no damage was found. Unfortunately, there are no similar tanks in operation in the CIS countries. For the production of such tanks it is necessary to choose the method of manufacturing fiberglass. A suitable method for manufacturing a fiberglass boiler is the winding method.

4 Aluminum Alloys in the Design of Wagons

The prospect of expanding the use of aluminum and aluminum alloys is determined by its large natural resources and a valuable range of properties.

The advantages of aluminum and its alloys include a large range of strength (σ_B = 100–800 MPa), low density (2.5–2.85 g/cm^3) and high specific strength, heat and electrical conductivity, reflectivity and corrosion resistance in various location.

Currently, the main technological process in the assembly of wagons is welding, more than 95% of steel structures. During installation, the application level of welding is about 60%.

Despite the fact that the mechanical connections of metalwork elements (rivets, bolts) have been known for a long time, in recent years, a type of mechanical connections, called "cold rivets", or pins with a crimping head (PwCH connections), has become widespread. Such a connection has a number of advantages compared to other types of mechanical and welded joints, since it allows to connect structural elements from dissimilar materials, for example, aluminum alloy - steel - composite material, maintains a predetermined tightness due to the lack of unscrewing the threaded nuts, does not lead to warpage and has a high tear, and shear resistance [5]. The use of PwCH connections with the correct choice of the type of pins and crimp heads, as well as the number of PwCH connections of individual elements of the body of a freight wagon made of aluminum alloy, allows for long-term operation of the body without damaging the base metal and joints.

The use of aluminum creates additional value for the carrier by reducing the weight of the wagon, its corrosion resistance, increased carrying capacity and increased service life. Alcoa has created aluminum semi-finished products that are successfully used for innovative solutions in the production of the 1520 Space freight train. Hopper wagons with aluminum structural parts were demonstrated by the Russian wagon building companies - NPK United Wagon Company and RM Rail at the 5th Anniversary International Railway exhibition of equipment and technologies EXPO 1520.

5 Articulated Freight Wagons

A promising direction for improving the efficiency of rail transportation is the development and commissioning of articulated six-axle freight wagons.

An objective assessment of articulated wagons will make it possible to find their place in the structure of the country's rolling stock and choose design solutions that meet the operating conditions on 1520 mm gauge railways. The development of scientifically based technical solutions for articulated wagons for 1520 mm gauge, which allows increasing their technical and economic efficiency and ensuring traffic safety, is a promising area for research. The articulated hopper wagon model 19-6978 (-01) has no analogues in the domestic market. In North America it is represented in Canada and unlike the Russian model it is not equipped with bunkers in the cantilever parts and has different axial load of the bogies (25 tf for the middle and 20 tf for the extreme), which reduces the linear load of the "Canadian".

The body volume increased to 160 m^3 allows you to efficiently transport the entire range of bulk cargo with full use of a carrying capacity of 113.5 tons. The linear load of the wagon is 7.68 t/m compared to 6.79 t/m for models on the market, which allows shippers to be provided and operators significant savings on the cost of transportation of goods, and also makes it possible to significantly increase the throughput of railways [6]. For further modernization, it will be effective to replace steel cladding with aluminium or fibreglass. Such an implementation will reduce the tare factor and the load on the path.

6 Conclusion

Based on the results of the study, the following conclusions can be drawn:

- The development of a complex of new-generation freight wagons with improved technical and economic indicators is a promising direction for the development of transport and the country's economy;
- Currently, wagon-building plants produce wagons with increased axial load, which is achieved through the use of innovative bogies;
- The possibility of increasing axial load contributes to studies to increase the carrying capacity of freight wagons, which is currently necessary. The use of light

materials in the construction of wagon bodies allows to reduce its own weight and thereby increase the carrying capacity.

Research conducted in the work can be used in the design of new types of cars and the modernization of existing ones. In the future, the authors plan to develop the design of a new generation of freight cars with increased carrying capacity and the use of new materials. These data will increase the efficiency of freight transportation by wagons.

References

1. Committee on Statistics of the Republic of Kazakhstan: Information portal, http://www.stat.gov.kz. Accessed 21 Oct 2019
2. Karakayev, A.K., Zaripov, R.: Composite materials in freight wagon building science and technology of Kazakhstan, vol. 1, pp. 4–9 (2016)
3. Gavrilov, P., Zaripov, R.: Innovative technologies in the production and repair of railway wagons. Monograph, pp. 116–122. Pavlodar (2018)
4. Zaripov, R., Gavrilov, P.: Research opportunities to improve technical and economic performance pf freight car through the introduction of lightweight materials in their construction. J. Procedia Eng. **187**, 22–29 (2017)
5. Shurtakov, A.: Optimizatsia sostava i mekhanicheskih svoystv svarnih i krepezhnih soedineniy aluminievih splavov dlya sozdaniya kuzovov zheleznodorozhnih vagonov novogo pokoleniya. Dis. on the competition uch. step. cand. tech. sciences (in Russian), pp. 151–157. Moscow (2017)
6. Boronenko, Yu.P.: Innovatsionniy gruzovoy podvizhnoy sostav zheleznih dorog i ego visokotechnologichnoe proizvodstvo (in Russian). Science and transport, No. 3, pp. 18–21 (2012)

Information Support of the Task of Rolling Stock Health Management

Sergii Panchenko[1], Sergii Zmii[1(✉)], Ivan Siroklyn[1], Vladislav Sahaidachnyi[1], and Ilja Korago[2]

[1] ICST Department, Ukrainian State University of Railway Transport, Kharkiv 61050, Ukraine
panchenko074@ukr.net, zmii.serhii@kart.edu.ua, seroklin.iv@gmail.com, vlad.sagaidachniy@gmail.com
[2] Institute of Transport, Riga Technical University, Azenes Street 12-316, Riga 1048, Latvia
ilja.korado@ldz.lv

Abstract. The existing automation systems in railway transport provide a sufficient amount of redundant information. The comprehensive analysis of that allows controlling the wide range of operational characteristics for the tasks of optimizing the assets and their resources management. The goal of the research is to develop a method for the automatic identification of railway cars for the check-out system of operation life expenditure by means of the data integration of the existing commercial and diagnostic systems. The paper proposes a method for determining the parameters of railway cars in the process of movement based upon calculating the statistical characteristics of the acceleration and kinematic relations between the state coordinates. The probabilistic decision-making method and set theory are used for reconcile the data obtained to determine the railway cars with the data of commercial information systems.

Keywords: Railway transport · Automatic identification · Probabilistic decision-making

1 Introduction

Existing automation systems in rail transport provide a sufficient amount of redundant information. A comprehensive analysis of this information allows you to control a wide range of operational characteristics, including the tasks of optimizing asset and resource management.

Currently, all technological operations that are carried out by rolling stocks in railway transport are subject to registration processes in commercial and diagnostic systems. However, due to the lack of technical means that allow automatic registration, control, processing and transmission of information, most of the registration data of trains is entered manually by the operator. Because of this, all information about the rolling stock has low reliability and is not delivered on time.

One of the solutions to this problem is the introduction of an automatic train identification system. This solution allows to:

E. Ginters et al. (Eds.): ICTE ToL 2019, LNITI, pp. 188–195, 2020.
https://doi.org/10.1007/978-3-030-39688-6_25

- Increase the reliability and efficiency of control systems for the consumption of technical resources;
- Increase the safety of the car;
- Optimize the use of the rolling stock.

The aim of the research is to develop a method for automatic identification of trains for prognostic and health management systems.

2 Review of the Existing Approaches

One of the tasks of prognostic and health management systems is the identification of individual rolling stocks on the track or their identification in the train. Nowadays, many different methods have been developed to solve such a problem. The most common method in the world is radio frequency identification [1]. The main advantages of this system is a non-contact identification method, ease of implementation in the process control system. The disadvantages of this system include the need to install passive sensors on each mobile unit, as well as the need to install readers in the identification zone. Another way is video analysis of types of rolling stocks [2]. With this method of identifying rolling stocks, it occurs by recognizing the wagon number. The number is read using a digital video camera. In addition to the need for installing cameras, this method has problems recognizing the number of the car. The problems are associated with frequent fuzzy image fixation, especially at sufficiently high speeds, as well as contamination of the inscriptions on the cars. These shortcomings lead to the impossibility of accurate identification of wagons. The next method of identifying a rolling stock number is real-time train tracking using satellite navigation systems [3]. The features of this method include monitoring the movement of rolling stocks using a satellite navigation system, as well as the need for using GSM-communication to exchange information. Another identification method is to determine the type of rolling stock based on its static characteristics [4]. To determine the static characteristics of rolling stocks, a control section is utilized using point track sensors or optical cable [5] as primary sources of information.

The paper offers an improvement of the existing method for obtaining the static characteristics of rolling stocks for their subsequent identification.

3 The Method of the Rolling Stock Types Identification in a Train

In general, the procedure for identifying a rolling stock type in a train on the basis of the static characteristics of the rolling stock consists of the following steps:

- Measurement of static parameters of rolling stock;
- Generation of a train model;
- Identification of the train based on the model generated using the train database [4].

The process of measuring the motion parameters is accompanied by errors related to the vibration of the rail, the wear of the flange of the wheel, the discretion of information processing and more. For further processing of measurement results, it is advisable to use an algorithm for estimating primary information.

The potentialities of the algorithm for estimating the state of an object are determined by the degree of adequacy of the model of its motion, which is chosen for the construction of the algorithm and real conditions. With this in mind, it is proposed to use the optimal linear algorithm (Kalman algorithm), which is based on a stochastic Markov model of rolling stock motion in the state space [6, 7]. This model is based on a statistical description of the processes of acceleration or deceleration of the moving unit and the kinematics equations that bind the coordinate (range, distance to a certain point), its first and second derivatives (velocity, acceleration). The optimal linear algorithm is determined by the recurrence relations [6, 7]:

$$\hat{z}_e(t_n) = \Phi(\Delta t) \cdot \hat{z}(t_{n-1}) \tag{1}$$

where $\hat{z}_e$ is mathematical expectation of the state assessment until the next measurements; $\hat{z}$ is assessment of the condition with the next measurements; $\Phi(\Delta t)$ is transition state matrix:

$$\hat{z}(t_n) = \hat{z}_e(t_n) + K(t_n) \cdot [y(t_n) - C \cdot \hat{z}_e(t_n)] \tag{2}$$

where K is matrix of optimal coefficients; y is vector-column of observation results; C is matrix of observations. Calculation of matrix K is done taking into account the measurement accuracy and the intensity of the rolling stock acceleration change at recurrent ratios [7]. The covariance matrix of a priori errors:

$$M(t_n) = \Phi(\Delta t) \cdot P(t_{n-1}) \cdot \Phi(\Delta t)^T + Q(\Delta t) \cdot \sigma_w^2 \cdot Q(\Delta t)^T \tag{3}$$

where σ_w – white sequence with known covariance structure, this is the input white noise contribution to the state vector; Q is excitation matrix. The covariance matrix of a posteriori errors:

$$P(t_n) = M(t_n) - \Delta M(t_n) \tag{4}$$

A matrix that characterizes the reduction of a priori errors is determined as:

$$\Delta M(t_n) = M(t_n) \cdot C^T(t_n) \cdot \frac{M(t_n) \cdot C(t_n)}{C(t_n) \cdot M(t_n) \cdot C^T(t_n) + \sigma_\eta^2} \tag{5}$$

where σ_η – is rms error of measurements.

However, the matrix of optimal coefficients is following:

$$K(t_n) = \frac{M(t_n) \cdot C^T(t_n)}{C(t_n) \cdot M(t_n) \cdot C^T(t_n) + \sigma_\eta^2} \tag{6}$$

Thus, the formation of algorithm evaluation results consists of two components - measurement results and extrapolation of motion parameters based on the already available data. The relationship between the significance of the extrapolated data and the newly obtained data depends on the matrix of relative coefficients. Such a matrix over time accumulates information and increases the priority of extrapolated data. This allows to reduce the measurement noise received from the primary sources.

To identify the rolling stock type based on the static data estimation, Bayesian decision theory is used [8]. This method of analysis is only used to compare distances for each railway wagon separately. This is because the distance between the wheel pairs of adjacent wagons can vary widely. This situation leads to the deterioration of further comparisons and, as a consequence, the erroneous identification of trains. Based on the analysis of the static characteristics of rolling stock used on the railway, it can be concluded that most wagons have similar static characteristics. Thus, when identifying a single wag-on, static data identifies the set to which the wagon belongs. Identification of a particular type of wagon is made only after identification of the entire train based on the comparison of data from commercial and diagnostic systems.

4 Calculation of the Probability of a Train Identification

The identification of the type of rolling stock of a train is performed by a consistent comparison of the static results of the measurement of the rolling stock parameters with the static characteristics of the passport data of the rolling stocks. However, during the formation of an ordered array of static characteristics of rolling stocks from the commercial and diagnostic systems, it may happen that two types of rolling stocks will have the same arrays, because of the coincidence of the passport data of the axle distances of each car. This makes impossible to identify.

In order to be able to use this method, it is necessary that there would be no rolling stocks with identical static characteristics in the original base of the mobile units after identification. Let us introduce the concept of the uniqueness coefficient of static characteristics of a train ζ_i. It is necessary to analyze the ability to identify mobile units based on data from commercial and diagnostic systems.

The coefficient of uniqueness of the static characteristics of the rolling stock will be understood to be a numerical value, which indicates how much the static characteristics of the mobile unit are comparable to other static characteristics in the database of commercial and diagnostic systems on a specific section of the railway track:

$$\xi_i = \frac{\sum_{k=1}^{N} P(O_{i/k})}{N} \tag{7}$$

where $P(O_{i/k})$ is the probability of the coincidence of the type of the i-th rolling stock with the type of the k-th rolling stock; N – is number of rolling stocks in the database of commercial and diagnostic systems on a specific section of the railway track.

The value $P(O_{i/k})$ can be found by the following formula:

$$P(O_{i/k}) = \prod_{j=1}^{n} P[a_i(j) \rightarrow a_k(j)] \tag{8}$$

where $P[a_i(j) \rightarrow a_k(j)]$ is the probability of static coincidence of the set of the j-th car for the i-th and k-th rolling stock; n is number of cars of rolling stocks compared.

Thus, the higher the coefficient of uniqueness of the static characteristics of the rolling stock with the same number of mobile units, the more likely it is to detect identification errors. It should be borne in mind that the method of sequential comparison of the measured static characteristics of each wagon with the passport static characteristics of the corresponding wagons obtained from the commercial and diagnostic systems is used to identify the rolling stocks. Then, the probability of identifying the k-th rolling stock is:

$$P(D_k) = \frac{P(D_{k/k}) \cdot \xi_k}{\sum_{i=1}^{N} P(D_{i/k}) \cdot \xi_i} \tag{9}$$

where $P(D_{i/k})$ is the probability of choosing the i-th train in case of choosing the k-th train; I is number of trains with a similar number of wagons in the database.

The probability of choosing the i-th train in case of choosing the k-th train

$$P(D_{i/k}) = \frac{\sum_{j=1}^{n} \beta_j \cdot p\left[v_{i/k}(j)\right]}{n} \tag{10}$$

where β are coefficients which indicate the priority of identifying a single car for the whole train; n is number of train rolling stocks; $p[\upsilon_{i/k}(j)]$ – is probability of belonging of static characteristics of the j-th car of the k-th train to the set to which the j-th car of the i-th rolling stock belongs.

5 The Results of the Experiment

A field experiment was conducted to test the possibility of using an optimal linear algorithm in systems for determining the static characteristics of rolling stocks. The experiment was conducted on the site where floor sensors were installed to collect primary information (see Fig. 1).

These sensors are connected to an existing system (Hot Axle Box Detection). One of the tasks of the system is the determination of the types of moving units by static characteristics. In the course of the experiment, a controller is connected to the primary sensors of the existing socket state control system. The data obtained were processed in MatLab using an optimal linear estimation algorithm.

Fig. 1. Connecting equipment for the experiment.

At the same time, this data was collected and processed by the system based on existing algorithms (see Fig. 2) [4].

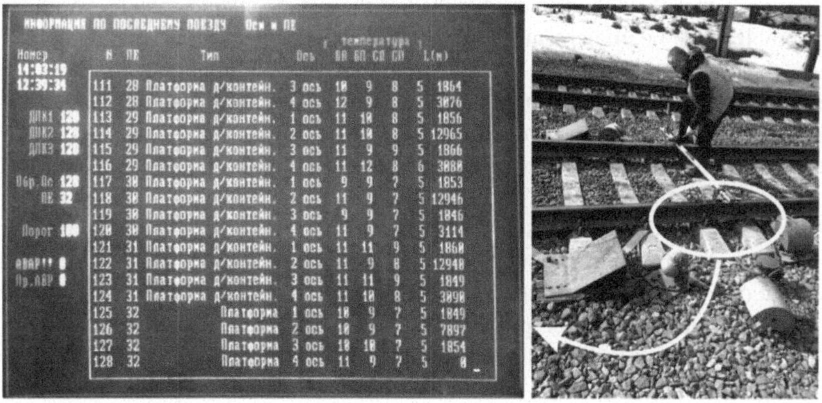

Fig. 2. The results of the algorithm of the tested system.

The results of the algorithm of the tested system were also compared with the data of commercial systems of control movement of rolling stocks and goods. For a visual representation of the results of the existing algorithm and the optimal linear algorithm, the results of the standard deviation from the passport data of the results of the algorithms are presented (see Fig. 3).

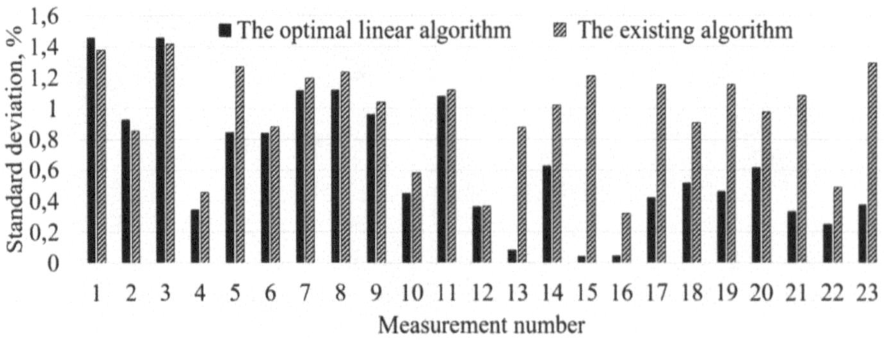

Fig. 3. Comparison of the results of using algorithms.

The value of the estimation error obtained by use of an optimal linear algorithm is almost 4 times smaller than calculated for previous algorithms used.

6 Conclusion

In existing commercial and diagnostic systems, a large amount of information is collected. This information is entered manually by the operator, which leads to errors. At the same time, part of this information can be obtained on the basis of processing existing information from floor systems.

The types of rolling stocks are identified of their static characteristics. The solution of the problem can be implemented using the Bayesian theory of decision making. This theory gives a minimum error of determination, according to the criterion of the minimum standard deviation. In turn, the static characteristics can be obtained using track sensors. It is possible using the optimal linear Kalman algorithm. As it can be seen from the results of the experiment, the proposed method allows to increase the accuracy of the estimation data by 19% compared to existing approaches.

However, this solution has several limitations, namely: lower accuracy of estimation at the initial moment of time (low accuracy of estimation at the stage of data accumulation); modeling the state equation; setting the correct noise profiles (σ_w, σ_η); capturing the true sensor noise profile.

Thus, the use of an optimal linear algorithm for estimating static parameters allows one to increase their accuracy. The method can be useful to improve the quality of determining the types of moving units in Hot Axle Box Detection systems.

The approach to identifying moving units is not only an alternative to traditional identification systems, but also allows to find and timely eliminate errors in manual input of information in commercial systems of the movement of railway cars and goods control.

In the future, it is proposed to improve the identification method by integrating the locomotive tracking subsystem. That in turn will allow to track operations leading to a change in the types of rolling stock in the train in automatic mode.

References

1. Cheng, G., Wenjun, F.: A new automatic train identification system based on RFID. In: Web Information Systems Engineering, WISE 2010, pp. 463–473 (2010)
2. Panchenko, S., Siroklin, I., Lapko, A., Kameniev, A., Zmii, S.: Improvement of the accuracy of determining movement parameters of cuts on classification humps by methods of video analysis. East.-Eur. J. Enterp. Technol. **4**(3), 25–30 (2016)
3. Ye, J., Shi, L., He, H.: Design of wireless intelligent train identification system based on GPS/GPRS. In: International Conference on Information Engineering and Computer Science, ICECS, Wuhan, vol. 1, pp. 1–4 (2009)
4. Zhukovitsky, I., Egorov, O.: The procedure for the identification of trains using information from ASK VP UZ E. Railway Transp. Inf. Control Syst. **6**(13), 61–66 (2015)
5. Iele, A., Laudati, A., Mazzino, N., Cutolo, A., Cusano, A.: A novel fiber optic sensing system for weighing in motion and wheel flat detection. In: Wheel Detection Forum, Viena, p. 1 (2017)
6. Rail, R., Sengupta, M.: Alternative and innovative approach for train tracking technology. In: Wheel Detection Forum, Viena, pp. 1–21 (2017)
7. Hismatulin, V., Volichenko, I.: The optimal linear algorithm for estimating the coordinates of the state of a moving unit. Collect. Sci. Works DonIZT **37**, 10–14 (2014)
8. Lavine, M.: Frequentist, Bayes, or other? Am. Stat. **73**, 312–318 (2019). Statistical Inference in the 21st Century: A World Beyond "p < 0,05"

Analysis of Highly Defective Rails with Horizontal Splitting in the Head

Pavels Gavrilovs[1], Viktors Ivanovs[2(✉)], Pavels Stankevics[2], and Fjodors Mihailovs[2]

[1] Institute of Railway Transport, Riga Technical University, Paula Valdena Street 1, Riga 1048, Latvia
pavels.gavrilovs@rtu.lv

[2] Institute of Railway Transport, Riga Technical University, Azenes Street 12, Riga 1004, Latvia
viktors.ivanovs_2@edu.rtu.lv, {pavels.stankevics, fjodors.mihailovs}@rtu.lv

Abstract. Safe movement of the rolling stock is the major responsibility of every country with an operating railway sector. This type of major defect under code XX 30.1, once detected, poses a threat to further operation of the rolling stock. Horizontal splitting in the head reaching deeper than 8 mm from the tread surface can result in chipping of the rail head, which in its turn can cause the rolling stock to run off the track. This article discusses statistics of highly defective rails under code 30.1 in Latvian railway between 2011 and 2018. A piece of chip from the rail head of type R-65 was examined. This fragment of rail was examined for metal hardness, chemical composition and metal microstructures. Full evaluation and main conclusions were made about the quality of rail steel, and also versions of occurrence and development of this type of major defect were discussed.

Keywords: Railway transport · Highly defective rails · Steel composition

1 Introduction

The purpose of this study is to analyze highly defective rails under code XX30.1 and also to determine the numerical and mass fractions of hardness, chemical composition and structure of metal of rail steel 76 HAF using advanced digital methods and tools. The research process consisted of four stages: Stage 1: analysis of horizontal splitting in the head, under code XX 30.1 on the Latvian railway was carried during years of 2011, 2012, 2013, 2014, 2015, 2016, 2017, 2018. Stage 2: determination of metal hardness according to the Brinell scale (Brinell Hardness Number - HB) with a modern device "Krautkamer MIC10". The obtained results should be compared with the data of the manufacturer's factory – Tahil. Stage 3: determination of the chemical composition of the rail steel of R-65, *76HAF* (in the rail head) using the "PMI – Master PRO". The obtained data should be compared with the manufacturer's data. Stage 4: determination of the rail steel structure. Drawing of main conclusions about the quality of the rail steel of *76HAF*. Stage 5: drawing of main conclusions and summary of the cause and

E. Ginters et al. (Eds.): ICTE ToL 2019, LNITI, pp. 196–202, 2020.
https://doi.org/10.1007/978-3-030-39688-6_26

development of the fracture from the horizontal splitting in the head, under code XX 30.1 (see Fig. 1).

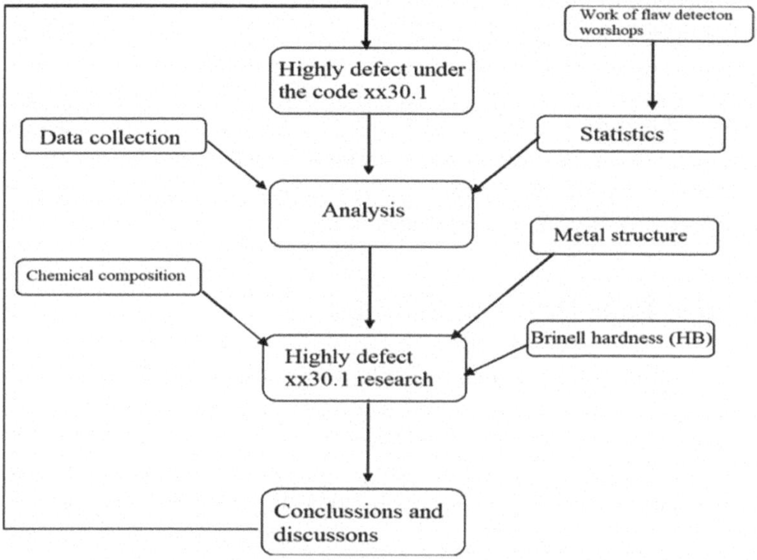

Fig. 1. Summarized algorithms of stages on rails analysis.

During the last eight years defect detection equipments detected larger number of highly defective rails under code XX 30.1 in the railway tracks of Latvia. This type of major defect is classified as horizontal splitting of the rail head, leading to a progressing crack originating inside of the head. Digit 1 shows that this type of major defect is located in end-to-end joint area. In these areas cracks usually form not further than 6.35 mm from the tread surface and spread horizontally in all directions [1]. Such cracks almost always cause uneven tread surface of the head. When the defect reaches lateral edge of the head, it looks like a longitudinal split (see Fig. 2).

Fig. 2. Chip from the head of rail of type R-65.

Such cracks are caused by rolled blisters inside the rail head, segregation and foreign inclusions. This type of defect was detected with an ultrasound probe PC at ray penetration angle 0^0. Length of horizontal split of the head is 450 mm. Depth of bedding of this type of major defect varies between 25–27 mm [3]. This study is aimed at an highly defect metallographic study of *76HAF* rail steel, which will help defectoscopy workshops in detecting defects under the code XX30.1.

Defectoscopes of type RDM - 22, 23, 24, manufactured in Moldova, are widely used to detect this type of defect in the railway tracks of Latvia. These defectoscopes detect horizontal extentional cracks going deeper than 8.0 mm and, regardless of their length and area of location in depths of the rail, classify them as highly defective that must be replaced within 5 h of their detection [2].

2 Defective Rails Detection Statistics

At the Stage 1 of the study on the Latvian Railway, statistics of highly defective rails under code XX30.1 were collected and analysed. Between 2011–2018, 619 highly defective rails under code XX 30.1 were detected in Latvian railway. In 2011, 88 cases were registered. The largest number of highly defective rails, 98 cases, were detected in 2012. Meanwhile 84 cases were detected in 2013. In 2014, 90 highly defective rails were detected, making an increase by 6 cases in comparison to 2013. Between 2015 and 2016, number of highly defective rails fell from 57 to 46. In 2017, number of highly defective rails steeply increased to 87 cases and in 2018 decreased down to 69 cases (see Fig. 3) [4].

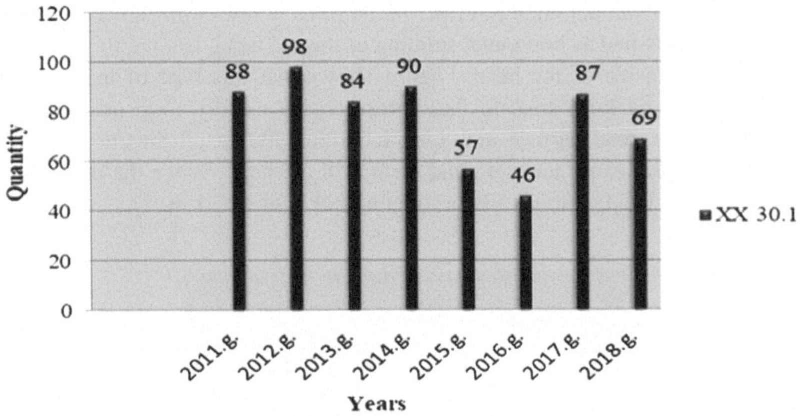

Fig. 3. Quantitative diagram of highly defective rails under code 30.1 in the railway of Latvia between 2011 and 2018.

3 Hardness of Highly Defective Rail of Type R-65

At the Stage 2 of the research, highly defective rail from Tahil factory was tested in the laboratory of Riga Technical University for its Brinell hardness (HB), and the obtained results were compared with the Russian standard GOST R 51685-2013. Measurements were taken with measuring device "Krautkamer MIC10". Hardness of the rail steel was tested according to standard ISO/17025 – Certificate of Calibration. Research results are shown in Fig. 4.

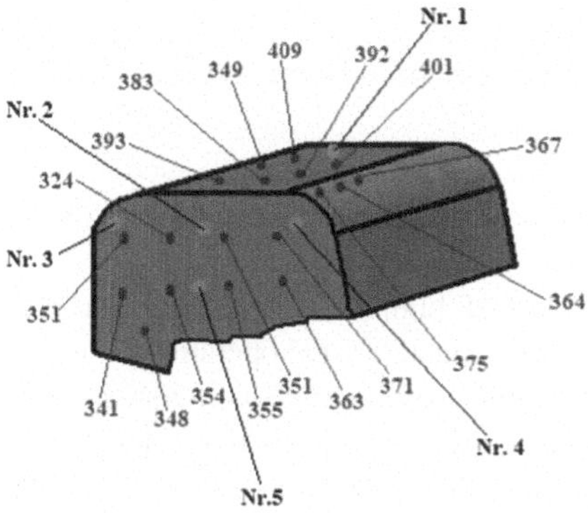

Fig. 4. Determination of hardness of a chip from the head of rail type R-65.

On the tread surface of the head (Point Nr. 1) (HB): 393, 383, 375, 349, 392, 364, 409, 401, 367. In 10 mm depth from the tread surface of the head along the vertical axis of the rail (Point Nr. 2), not less than (HB): 324, 351. In 10 mm depth from the surface of fluting of rail (Points Nr. 3 and Nr. 4) (HB): 351, 371. In 22 mm depth from the tread surface of the head along the vertical axis of the rail (Point Nr. 5) (HB): 341, 354, 355, 363, 348 (Table 1).

Table 1. The point value of the hardness of the rail type R-65.

Hardness determination spot	Rail make hardness 76HAF	
	Hardness according to manufacturer (HB)	Average determined hardness (HB)
On the tread surface of the head (Point Nr. 1):	350–390	381
In 10 mm depth from the tread surface of the head along the vertical axis of the rail (Point Nr. 2) no less than:	341	349
In 10 mm depth from the surface of fluting of rail (Points Nr. 3 and Nr. 4) no less than:		
In 22 mm depth from the tread surface of the head along the vertical axis of the rail (Point Nr. 5) no less than:	321	352

Average hardness on tread surface of the head of rail (Point Nr. 1) was 381 (HB), corresponding to the hardness data provided by the manufacturer. Hardness at 10 mm depth from the tread surface of the head along the vertical axis (Point Nr. 2) and surface of rail fluting (Points Nr. 3 and Nr. 4) was 349 (HB), which is by 8 HB more than the stated parameters [5]. In 22 mm depth from the tread surface of the head along the vertical axis of the rail (Point Nr. 5), HB was 352. Hardness testing of highly defective rail under code XX 30.1 evidences compliance of the hardness parameters specified by the manufacturer (GOST R 51685-2013) with mean values of data established in metallographic laboratory of Riga Technical University.

4 Chemical Composition of Major Defect Under Code XX 30.1

At Stage 3 phase of the research the chemical composition of the chip from a highly defective rail is examined.

The chip from the head of highly defective rail was examined with an optical emission spectrometer "PMI – Master PRO" operating on the principle of local burning of sample's surface to determine the chemical composition and afterwards sending the obtained data to the printer [6]. The obtained results of chemical composition are shown in Table 2. While the chemical composition of rail steels produced in Japan, France, Austria and Poland differs from the chemical composition of domestic steels: the silicon content in the steels produced in France, Japan and Poland is 1.5–2 times higher than in domestic ones, all the steels are imported manufactures are alloyed with chrome, rail steel made in France is microalloyed with titanium. Russian-made steels are microalloyed with vanadium, the aluminum content in them is higher than in imported ones. Steel of Polish manufacture was deoxidized with aluminum, the aluminum content in it was 0.020% of the mass. The total content of sulfur and phosphorus in steels of domestic and foreign production does not differ.

Afterwards the obtained chemical composition data were compared with the factory standard GOST R 51685-2013. Research conducted regarding the chemical composition of the chip from a highly defective rail allowed providing full estimation of mass fractions of chemical elements [7]. Main mass fractions in the chemical composition consisted from: (C) carbon with mass friction exceeding the specified ratio by 0.34%, (Mn) manganese within the permissible ratio and equalling 0.798%, (Si) silicon content by 0.003% less than the minimum permissible values specified in standard GOST R 51685-2013. Equally important element in the chemical composition is (V) vanadium, which increases hardness and durability and its mass fraction was by 0.0924% less. (Cr) chrome content was by 0.1036% less, and chrome is commonly known to increase hardness and durability without compromising plasticity of steel. (N) nitrogen content in the examined sample of rail steel was not detected. Content of another important element, (P) phosphorus, was by 0.0121% less. Mass fraction of (S) sulphur was in the stated range and equalled 0.0217%. (Al) aluminium contained in the rail steel is a good oxidising agent, neutralising harmful admixtures of phosphorous, and it also increases impact resistance of rail steel, and content of said element was increased by 0.0072%, being a good indicator in the chemical composition of rail steel *76HAF* [8].

Table 2. Average permissible values of chemical composition in % ratio in the examined highly defective rail (head) under code XX 30.1 and of type R-65.

Value	Fe	C	Si	Mn	P	S	Cr	Mo	Ni
Min	–	0,950	0,150	0,400	0,0000	0,0000	0,0000	–	0,0000
Max	–	1,05	0,350	0,600	0,0350	0,0350	0,200	–	0,200
Head	97,40	1,15	0,223	0,793	0,0091	0,0231	0,0727	0,0030	0,0897
	97,40	1,12	0,228	0,788	0,0052	0,0224	0,0806	0,0030	0,0941
	97,30	1,19	0,231	0,812	0,0094	0,0196	0,0761	0,0030	0,0950
Average	97,36	1,16	0,227	0,798	0,0079	0,0217	0,0764	0,0030	0,0930
Value	Al	Co	Cu	Nb	Ti	V	W	Pb	Zr
Min	–	0,0000	0,0000	0,0000	0,0000	0,0000	0,0000	0,0000	–
Max	–	0,100	0,400	0,0700	0,0700	0,100	0,100	0,0700	–
Head	0,0120	0,0257	0,0158	0,0030	0,0020	0,0071	0,0250	0,0100	0,0030
	0,0119	0,0268	0,0161	0,0030	0,0020	0,0069	0,0250	0,0100	0,0030
	0,0127	0,0270	0,0163	0,0030	0,0020	0,0088	0,0250	0,0100	0,0030
Average	0,0122	0,0265	0,0161	0,0030	0,0020	0,0076	0,0250	0,0100	0,0030

On the basis of chemical composition analysis one can arrive at the following conclusion: mass fractions of chemical elements correspond to the standard GOST R 51685-2013 and they cannot be the underlying cause of non-compliance of the established mass fractions of chemical elements.

5 Metal Structure of the Rail Chip

At Stage 4 of the research structure of metal was analysed with electronic microscope Carl Zeiss Axiovert 40 MAT, and data were transferred to the computer. The sample under examination was corroded with 5% solution of the nitric acid $HN0_3$ [9]. Metal structure was explored under (x200) magnification. The rail steel has pearlite structure and ferrite admixtures. Area of pearlite plate did not exceed 100 nm, and ratio of thickness of the iron carbide and thickness of the ferrite in the pearlite was no less than 0.15%. Summing up the ferrite admixtures in metal structure, it was proved that rails of type R-65 [10] with steel grade *76HAF* have the highest level of cyclic crack resistance K_{fC} (56–59 MPa), characterized by a smaller grain size obtained as a result of recrystallization during repeated furnace heating under quenching.

6 Conclusion

The horizontal crack 25–27 mm below the tread surface of the rail head was identified by defectoscope RDM-22. Obtained statistical data from the Latvian railway confirm topicality of the problem of highly defective rails under code XX 30.1. Research conducted to determine hardness compliance with the manufacturer's standard GOST R 51685-2013.

Chemical composition of the highly defective rail steel also confirmed compliance with the manufacturer's standard. Build-ups of ferrite and non-metallic inclusions as black spots were detected in the metal structure.

Researches performed by the metallographic laboratory of Riga Technical University allows concluding that horizontal cracking and chipping from the rail head was caused by con-compliance of manufacturing of rail steel that resulted in hairlines. Hairlines are caused when rolled linear non-metal impurities emerge on the surface of the bars during hot and cold rolling, and also when metal deforms during the manufacturing process. Formation of hairlines was enhanced by presence of non-uniform areas on the surface of slabs, rolled stock that facilitates oxidization of steel along grain edges during heating of metal while rolling followed by hot rolling.

In further studies of this type of defect, it is planned to study samples of rail steel from the initial stage of development of the defect. Analysis of highly defective rails reveals that full estimation of series of factors that directly influence quality of rail steel *76HAF* must be examined. It creates the need for additionally control of rail steel manufacturing.

References

1. Siamak, H., Alfredo, N., David, M.J.: Semi-supervised rail defect detection from imbalanced image data. IFAC-Papers OnLine **49**(3), 78–83 (2016)
2. Federalnoje Agenstvo po Tehniceskomu Regulirovaniju i Metrologii. Nacionalnij Standart Rossijskoj Federacii GOST R 51685-2013. Relsi zeleznodoroznije Obwije tehniceskije uslovija. (EH 13674-1:2011, NEQ) (in Russian), pp. 1–30. Standartinform, Moskava (2014)
3. Jagodkins, M.: Valsts Akciju Sabiedriba "Latvijas Dzelzcels" Instrukcija C-010 Sliezu defekti, to pazimes, klasifikacija, katalogs, markesana, un vilcienu kustibas nosacijumi pa defektivam un loti defektivam sliedem (in Latvian), p. 28. Riga (2017)
4. Glotovs, V., Makedons, V.: Work analysis of defectoscopy in 2011–2018, Jelgava, pp. 1–14 (2019)
5. Baiko, S.: Latvian railway SJSC Service of railway equipment. Norms and specifications for defective rails NTD-C-97, Riga, pp. 51–65 (1997)
6. Dijs, S., Tipainis, A., Gavrilovs, P.: Restoration of railway turnout elements with manual metal arc welding and flux-cored arc. In: Proceedings of the 9th International Scientific Conference TRANSBALTICA, Vilnius, vol. 134, pp. 353–358 (2015)
7. Dijs, S., Tipainis, A.: Assessment of railway turnout element restoration using MMA and FCAF welding. In: 15th International Scientific Conference "Engineering for Rural Development", vol. 15, pp. 606–611. Latvia University of Agriculture, Jelgava (2016)
8. Korostelev, P.P.: Reagents for technical analysis, Moscow, p. 104 (1988)
9. Matvecov, V., Miroshnikov, N., Kratenok, V.: Damages of railway rails in different operation conditions. Bull. Ternopil Nat. Techn. Univ. **3**(71), 182–189 (2013)
10. Mezitis, M., Vasiljevs, A.: Feasibility of IIoT application in railway signaling and interlocking systems. In: Proceedings of 23rd International Scientific Conference "Transport Means", Palanga, pp. 169–175 (2019)

Using the New Car Braking Systems in Marshalling Yards

Oleksandr Ohar[1], Igor Berestov[1], Maxim Kutsenko[1], Julia Smachilo[1](✉), and Ilja Kogaro[2]

[1] Ukrainian State University of Railway Transport, Feyerbakh Square 7, Kharkiv 61001, Ukraine
smachilo.julia@gmail.com
[2] Riga Technical University, Azenes Street 12, Riga 1048, Latvia

Abstract. The purpose of the work is substantiating conditions for use of the target gravity braking technology for railway cars. To implement this technology, a marshalling system with a specific plan and profile design is proposed. Unlike conventional marshalling humps, the proposed system does not have any braking positions on the descending slope. The speed of cars rolling down from the hump is reduced with the help of the brake position in the classification bowl and elevations before it. Sufficient intervals between the groups of cars on switches are ensured only with the longitudinal profile. The main advantages of the marshalling system with target gravity braking technology include the significantly simplified rolling speed regulation without need for automated braking on the descending slope, and reduced effect of human factor and random parameters. Conditions of use of the target gravity braking technology are justified on the basis on the analysis of the calculation data of the rolling parameters of estimated cars with different values of uncoupling speed, wind speed and direction, permissible trailing-point movement speed and exceeding the permissible speed of entering the parking retarder position.

Keywords: Marshalling system · Target gravity car braking · Longitudinal profile · Plan of the neck · Uncoupling of cars

1 Introduction

One of the important problems in railway transport is increasing the efficiency of train uncoupling on marshalling humps. Today, the known ways of solving this problem include the optimization of the plan and profile of the marshalling system, choosing rational types of car retarders and braking modes, using modern automation devices, expanding tasks solved by complex automated train uncoupling systems, improvement of rolling speed regulation for cars, increasing the degree of consideration of random parameters in the design and technological calculations, as well as designing and developing decision support systems for operators of the hump.

Since the parameters of the plan of the yard neck are the input data for the calculation of the height and longitudinal profile of the hump, the choice of car retarder types and influence the choice of rolling speed regulation modes for cars, optimization of these parameters will result in better energy and resource saving of marshalling systems.

E. Ginters et al. (Eds.): ICTE ToL 2019, LNITI, pp. 203–210, 2020.
https://doi.org/10.1007/978-3-030-39688-6_27

By optimizing the longitudinal profile, the nominal uncoupling speed can be increased, which will result in reduced idle time which trains spend waiting for uncoupling in the receiving yard. In addition, increased uncoupling speed will also contribute to reducing diesel fuel consumption for train humping and uncoupling.

The choice of rational car retarder types, characterized by a wide range of technical and operational indicators, the main of which include braking power, permissible entry speed, energy consumption for one activation, duration of braking and duration of braking release, will have positive effect on the quality of car braking control on both mechanized and automated marshalling humps, and will reduce the cost of electricity for braking.

Optimization of braking modes has also the energy saving effect. In addition, braking modes are the main tool in maintaining the set up uncoupling speed, the permissible speed of collision of cars and the minimum average "gap" on the tracks of the marshalling yard. Thus, the application of rational modes will have an additional effect of cutting costs for compensation of damage to cars and cargo and the extra costs of idle time when trains wait in the yard to be uncoupled. It should also be noted that the problem of choosing a rational braking mode is one of the most complicated optimization problems because of the necessity to find a solution for conflicting problems of interval and target regulation of rolling speed of cars.

One of the most powerful measures to increase the efficiency of train uncoupling is the introduction of complex automation systems for hump technological processes. These systems provide high accuracy of calculation and realization of control parameters for cars rolling down the hump in the conditions of uncertainty and presence of multiple random parameters. Different scientific approaches are used in comprehensive automation systems to solve complex technological problems. As a rule, they have different techniques of formation and implementation of controlling actions. In most cases, these techniques are more important in solving the problems of increasing the efficiency of the train uncoupling process than the technical implementation of management actions.

Improvement of the plan and profile design of the marshalling hump, the means of regulation of the rolling speed of cars, the automation systems for the hump technological processes and car braking modes with the aim of increasing the efficiency of the process of train uncoupling have been considered in details in [1–15]. For example, in [1, 2] this goal is achieved by optimizing the shunting neck design, in [3–6] – by determining the optimal parameters of the longitudinal profile of marshalling humps, in [7, 8] – by developing new and upgrading the design of the existing roll-off speed controllers, in [9–11] – by forming approaches to automated brake control and in [12–15] – by optimizing the car braking modes.

Thus, many researchers have theoretically proved that the efficiency of train uncoupling can be increased, provided that the proposed measures are used on the marshalling hump. Some scientists have proved this possibility not only in theory, but also in practice – through the actual implementation of the results of their studies in the industrial facilities.

It should also be noted that the effectiveness of the train uncoupling is significantly influenced by a number of factors that are quite difficult to take into account, predict or formalize. These factors include the technical condition of the roll-off speed controllers and the automation devices, occasional braking performance of retarders, unpredictable behavior of the air masses, condition of the wheel pairs of the cars, the response time, and the level of professional training of operators.

Thus, the marshalling hump is a very complicated system consisting of multiple interrelated and interacting elements that is exposed to influence of numerous external and internal random factors. In view of this, the problem of formation of mathematical models of hump processes, which take into account all features of functioning of the hump and provide high accuracy of design and technological calculations is rather difficult. It should be noted that this problem has not been finally solved so far.

The purpose of this work is determining conditions of use of the target gravity braking technology on marshalling systems. The studies performed may be of interest for institutions involved in railway station design and to researchers engaged in the calculation and optimization of design and technological parameters of marshalling humps.

2 Design of the Marshalling System

Introduction of the target gravity braking technology is one of the ways to improve the efficiency of train uncoupling. This technology can be implemented by using a marshalling system with a specific plan and profile design (Fig. 1).

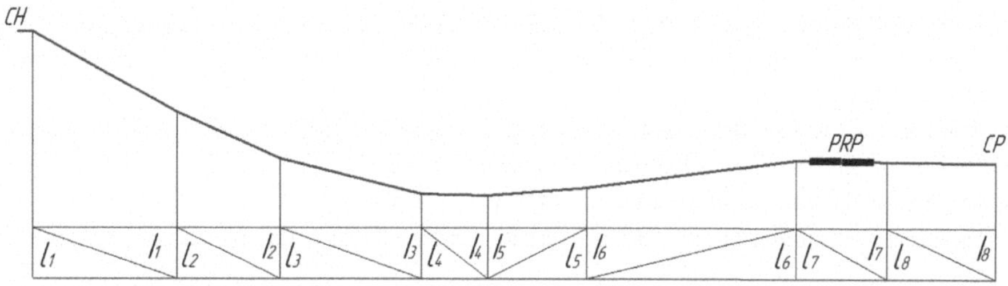

Fig. 1. Longitudinal profile of the marshalling system with target gravity car braking technology.

The specific feature of this design is location of a part or the whole switching area together with the beginning of the marshalling tracks (to the parking retarder position (PRP)) on the ascent. Other elements in the section from the crest of the hump (CH) to the calculation point (CP) are located on the descent (CP is located at a distance of 50 m from the end of PRP).

The design parameters of the marshalling system should ensure fulfilling the following conditions:

- Rolling of the slow light car (SL) in unfavorable winter conditions from CH to CP of the track with the highest resistance;
- Sufficient intervals at the switches in combination SL − FH (δ_{SL-FH}), where FH is a fast heavy car. An interval is the time difference between the moment when FH enters the switch and the moment when SL leaves this switch. The interval is considered sufficient provided the condition $\delta_{SL-FH} \geq 1$ is fulfilled, where 1 is the time required to switch the point.

The use of the proposed system is limited if the speed of cars passing the switches located on the descending part and/or the permissible speed at which cars enter the PRP retarders are exceeded. Symmetric switches 1/6 railway frogs are laid on the descending part of the system. The permissible speed of trailing-point movement for these switches is not more than 40 km/h or 11.1 m/s. The permissible entry speed for modern retarders is 8 m/s and 8.5 m/s. For further studies, the speed of 8.5 m/s is taken as the upper speed limit of cars entering the PRP retarder.

3 Analytical Calculation Data

To justify the use of the marshalling system with the target gravity braking technology, analytical calculation of rolling down of estimated cars (SL, FH) in the combination SL − FH in favorable and unfavorable weather conditions with the initial uncoupling speed (V_o) 1.4 m/s, 1.6 m/s and 1.8 m/s and wind speed (V_w) 4 m/s, 6 m/s, and 8 m/s has been performed.

The above process was calculated using the following input data:

- Number of tracks in the marshalling yard – 32;
- Weight of FH – 85 t;
- Main specific resistance of FH – 0.5 N/kN;
- Weight of SL – 22 t;
- Main specific resistance of SL – 4.5 N/kN;
- Estimated ambient temperature in unfavorable conditions of rolling down – −15 °C, in favorable conditions – +30 °C;
- Estimated wind speed in favorable conditions of rolling down – 8 m/s;
- Angle between the axis of the track on CH and the longitudinal axis of the marshalling yard – 0.41°;
- Number of the longitudinal profile elements from CH to CP – 8;
- Incline of the first (speed) profile element (I_1) – 50°/oo;
- Incline of the profile element with PRP (I_7) – 1.5°/oo;
- Incline of the marshalling tracks from the end of PRP to CP (I_8) – 0.6°/oo.

Height (H) and the longitudinal profile of the marshalling system with the target gravity braking technology for the specified operating conditions and assuming that the difference of inclines of the second (I_2) and third (I_3), third and fourth (I_4) profile elements is maximum (25°/oo), is calculated using the following optimization model:

$$H = (I_2 \cdot l_2 + (I_2 - 25) \cdot l_3 + (I_2 - 50) \cdot l_4 + I_5 \cdot l_5 + I_6 \cdot l_6) \cdot 10^{-3} + \\ + C \rightarrow H_{min}, \tag{1}$$

$$\begin{cases} 25 \leq I_2 \leq 105; \\ I_4 - I_5 \leq 25,\ I_5 - I_6 \leq 25; \\ h_{\omega}^{SL}(I_2, I_5, I_6) = H + h_0^{SL}; \\ \delta_{SL-FH}^{SW}(I_2, I_5, I_6) = 1, \end{cases} \tag{2}$$

where I_5, I_6 are inclines of the fifth and sixth longitudinal profile elements, °/oo;

l_1, …, l_8 are the length of the first, …, eighth longitudinal profile elements, m;

$$C = (50 \cdot l_1 + 1.5 \cdot l_7 + 0.6 \cdot l_8) \cdot 10^{-3} \tag{3}$$

H_{min} is minimum possible height of the marshalling system, m;

$h_{\omega}^{SL}(I_2, I_5, I_6)$ is energetic height of SL expended against all types of resistance during rolling down in unfavorable weather conditions from CH to CP of the track with heavy resistance, m. en. h.;

h_0^{SL} is the initial energetic height of SL on CH, m. en. h.;

$\delta_{SL-FH}^{SW}(I_2, I_5, I_6)$ is the interval on the last switch when SL rolls down to the track with high resistance and FH rolls down to the adjacent track, s.

To reduce the calculations, the angle between the wind direction and the longitudinal axis of the marshalling yard (*B*) was determined for the maximum and minimum resistance to movement of cars from the environment and wind when they roll down from CH to CP. According to the conducted study, this angle is 36.5° for unfavorable weather conditions (Fig. 2), and 200° for favorable ones (Fig. 3).

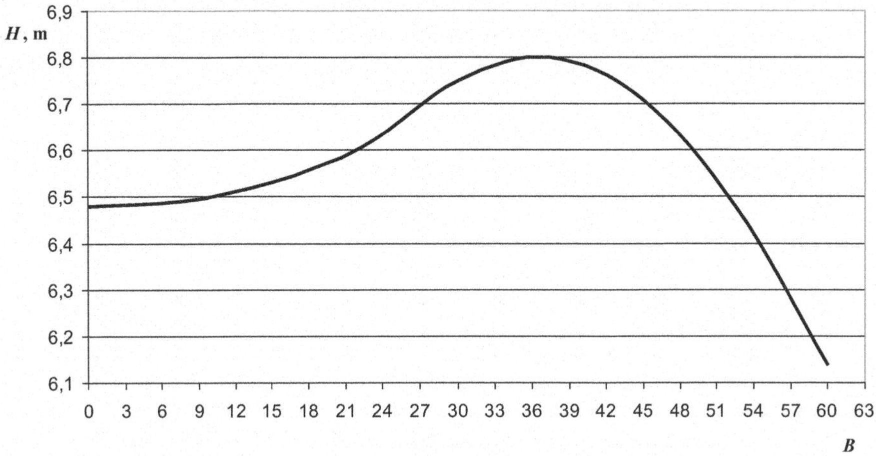

Fig. 2. Dependence of the height of the marshalling system on the angle between the wind direction and the longitudinal axis of the marshalling yard.

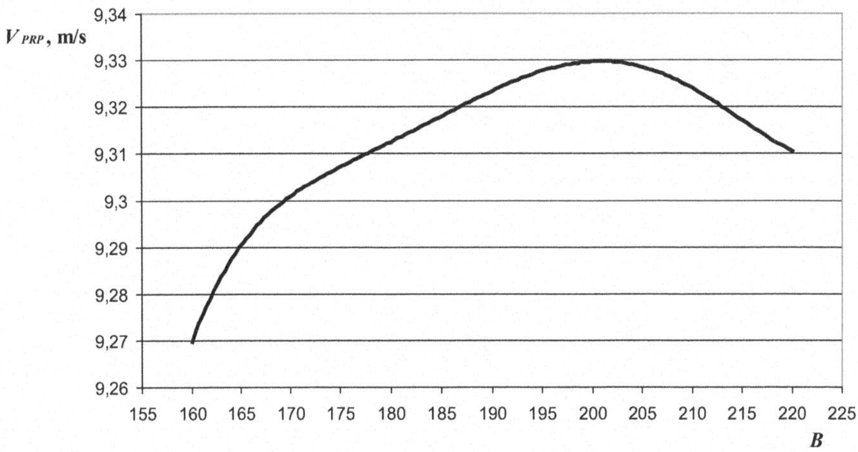

Fig. 3. Dependence of the entry speed of FH to PRP (V_{PRP}) on the angle between the wind direction and the longitudinal axis of the marshalling yard.

The analytical calculation data of rolling of the estimated cars (Figs. 4, 5 and 6) prove that the scope of application of the marshalling system with the target gravity braking technology depends significantly on the estimated weather conditions. E.g., rolling speed of FH down the descending part in favorable weather conditions does not exceed the permissible speed of trailing-point movement on symmetric switches at initial uncoupling speed up to 1.53 m/s and the height of the system calculated supposing SL rolls down from CH to CP at the ambient temperature −15 °C, $B = 36.5^{\circ}$ and $V_w = 8$ m/s (Fig. 4). With the heights of the marshalling system calculated at the headwind 6 m/s and 4 m/s, permissible speed of movement of FH on the switches is ensured if initial uncoupling speeds are up to 1.79 m/s and 2 m/s, respectively.

Similar is the situation with the speed of entry of cars to PRP (Fig. 5). With the heights of the marshalling system calculated at the headwind speed 4 m/s, 6 m/s and

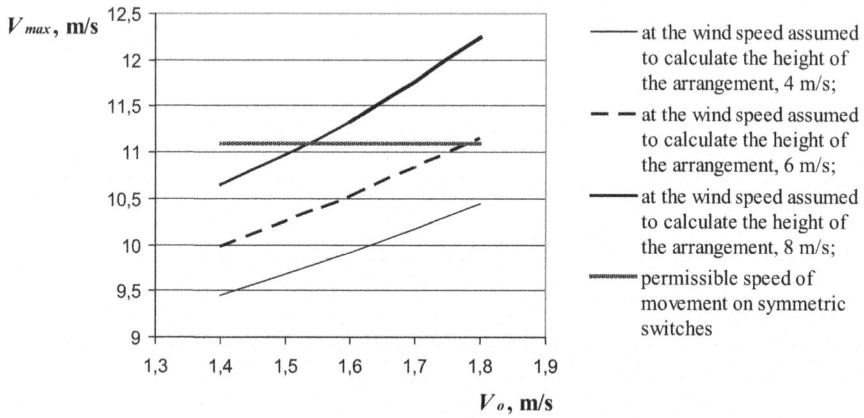

Fig. 4. Dependence of the maximum rolling speed of FH on the descending part in favorable weather conditions (V_{max}) on the initial uncoupling speed.

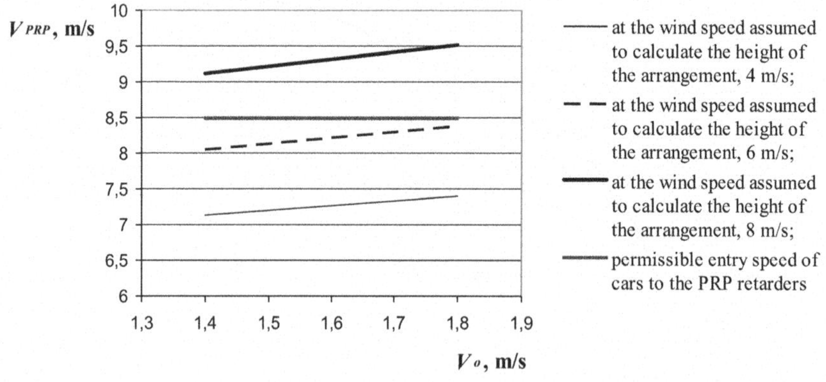

Fig. 5. Dependence of the entry speed of FH to PRP in favorable weather conditions on the initial uncoupling speed.

8 m/s, permissible entry speed of FH in favorable weather conditions is ensured at uncoupling speed up to 2.8 m/s, 1.9 m/s and 0.75 m/s, respectively.

The dependences shown in Fig. 6 indicate that with the height of the marshalling system calculated at wind speed up to 6.2 m/s, the scope of its application is limited by the permissible speed of trailing-point movement of cars on symmetric switches, and with the height of the marshalling system calculated at wind speed more than 6.2 m/s – by the permissible speed of entry of cars on the PRP retarder.

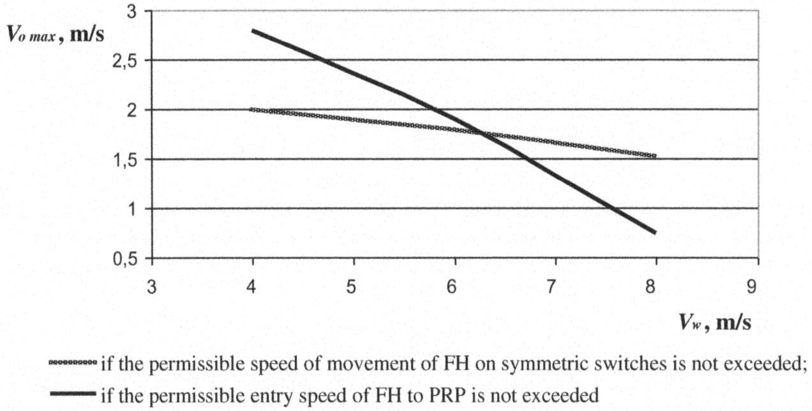

Fig. 6. Dependence of the maximum uncoupling speed of cars (*Vo max*) on the wind speed taken for calculation of the height of the system.

In view of the obtained results, it can be concluded that if the marshalling yard has more tracks, the proposed system can only be used in areas with slow winds during the seasons that are unfavorable to rolling.

4 Conclusion

The conducted studies have shown that the marshalling system with target gravity braking technology can definitely be used if the sorting yard has up to 32 tracks and the estimated wind speed is up to 6 m/s, while providing the uncoupling speed of cars up to 1.8 m/s and, respectively, high handling capacity of the system. If the marshalling yard has more tracks and/or in case of stronger winds, the marshalling system can be used provided that the uncoupling speed is reduced. In this case, the feasibility of using the system should be verified by feasibility studies. This is due to the fact that when the uncoupling speed decreases, the diesel fuel consumption for train humping and uncoupling and additional costs caused by the idle time of trains waiting for uncoupling in the yard increases.

The main benefit of the use of the marshalling system with target gravity braking technology is the significantly simplified rolling speed regulation without need for automated braking only on YBP, reduced impact of multiple external and internal random factors, as well as human factor on the efficiency of train uncoupling.

In view of the advantages of the new car braking technology, the application of the results of the study in designing by the relevant institutions and educational establishments will allow to increase their competitiveness in the design market and to gain additional profit from the development of new and redesign of the existing marshalling systems.

Further studies will focus on improving the hump neck design. Such improvement will be aimed at ensuring the minimum distance that the car runs from CH to CP. This will contribute to reducing the height of the marshalling system with the target gravity braking technology, the capacity of PRP and expand the scope of application of the new braking technology.

References

1. Bobrovsky, V., Kolesnik, A., Dorosh, A.: Improving the design of the road development plan of the yard neck. Transportation systems and transportation technologies. Bulletin of Dnipropetrovsk National University of Railway Transport Named After Academician V. Lazaryan **1**, 27–33 (2011)
2. Rozsokha, O.: The efficiency of gravity humps in the application of new yard neck. East.-Eur. J. Enterp. Technol. **5/3**(35), 51–59 (2008)
3. Wei, Q., Yin, G.-D., Lai, L.: Research on the length optimization of hump accelerate slope based on humping safety of vehicle. J. Railw. Eng. Soc. **7**(33), 78–82 (2016)
4. Zhang, H., Yang, J., Yang, T.: Multiobjective optimization model for profile design of hump distributing zone. Math. Probl. Eng. **2017**, 1–11 (2017)
5. Bobrovsky, V., Kolesnik, A.: Determination of the rational design of the longitudinal profile of the gravity hump. Transportation systems and transportation technologies. Bulletin of Dnipropetrovsk national university of railway transport named after academician V. Lazaryan **4**, 19–24 (2012)
6. Bessonenko, S.: Optimization of the main parameters of the gravity humps. Improving rail operations: a collection of scientific papers, Novosibirsk (2008)
7. Kobzev, V.: New wagon retarders. Autom. Commun. Inform. **10**, 10–12 (2009)
8. Kobzev, V.: The development of technical means of mechanization of gravity humps. Autom. Commun. Inform. **2**, 13–14 (2011)
9. Shelukhin, V., Malyshev, I.: Universal module for controlling brake positions. Autom. Commun. Inform. **5**, 12–14 (2000)
10. Shabelnikov, A., Olgeizer, I., Rogov, S.: Management of brake means of gravity humps: improving quality and efficiency. Bull. Rostov State Transp. Univ. **2**, 74–79 (2015)
11. Shabelnikov, A., Lyabakh, N.: Intellectualization of sorting processes control on the basis of instrumental determination of analogies. In: 2nd International Conference on Intelligent Information Technologies for Industry, pp. 138–145. IITI, Varna (2017)
12. Bessonenko, S.: Modes of braking of cuts at hump retarder. Improving the operational work of railways: a collection of scientific papers, Novosibirsk (2007)
13. Bobrovsky, V., Kozachenko, D., Bozhko N.: Optimization of braking modes of cuts on gravity humps. Makovetsky, Dnepropetrovsk (2010)
14. Shi, W., Li, X., Hu, M., Gao, L.: Research on rail car rolling speed regulating mode of hump yard under heavy load conditions. J. China Railw. Soc. **6**(40), 1–10 (2018)
15. Kozachenko, D., Bobrovskyi, V., Demchenko, Y.: A method for optimization of time intervals between rolling cuts on gravity humps. J. Mod. Transp. **3**(26), 189–199 (2018)

Fastening Device of the Railway Rolling Stock

Anastasia Pinchuk[1(✉)], Pavel Dadoenkov[1], Andrejs Baranovskis[1], Aleksandrs Nikolajevs[1], and Assylzhan Mukasheva[2]

[1] Institute of Transport, Riga Technical University, Kalku Street 1, 1658 Riga, Latvia
midzun@gmail.com
[2] L.N. Gumilyov Eurasian National University, Munaitpasov Street 5, Nur-Sultan, Kazakhstan

Abstract. In the paper the fastening device of the railway rolling stock is described. The main idea of the construction was its simplicity and safety. The device consists of brake pads, which are mounted on the running rails with typical rail linings. Brake pads are put and removed from the heads of the running rails, using a track switch. Additional elements, namely laser and Parktronic, are built into the brake pads. Elements send the signal through the radio transmitter into the driver's cabin, about the distance left to the complete rolling of the wheel pair on the brake shoe. Due to this device, the accuracy of the rolling stock fastening is increased. When the train departs, the driver receives a signal through the radio transmitter about whether the shoes remained under the wheel pairs.

Keywords: Railway transport · Fastening device · Efficiency assessment

1 Introduction

In railway transport a constant problem is the safety of train traffic and the reduction of manual labor, namely, related to the fixing and controlling the speed of cars on the sorting yard tracks when they freely roll down the slope, where brake devices are used in the form of shoes that are manually placed under the wheel pair of a wagon. For this purpose, calculations were made in MatLab software to determine problem of fixing of the brake shoe under the pair of wheels. It was also compared to currently used method of fixing of the brake shoe and improvement of the brake shoe. The aim of this work is to develop approach for partial rolling assessment of a pair of wheels on a brake shoe, which results from a lot of external factors.

2 Rolling onto Brake Shoe

Safety of train traffic and reduction of manual labor partially related to fixing and controlling the speed of cars on rail yards to prevent them from rolling down the slope by using brake devices such as shoes, manually placed under a wheel pair of a car, is a constant problem in railway transport [1].

E. Ginters et al. (Eds.): ICTE ToL 2019, LNITI, pp. 211–217, 2020.
https://doi.org/10.1007/978-3-030-39688-6_28

Calculations allowed to build a model of rolling onto the brake shoe in existing operational conditions of the rolling stock. In order to model the rolling onto the brake shoe, software tool Fuzzy Logic Toolbox with extension package MatLab that contains tools for projecting fuzzy logic systems was used [9, 10].

In order to depict selected fuzzy subsets, the following factors affecting the rolling onto the brake sue were used as linguistic variables:

- Human factor (alert locomotive driver; moderately tired locomotive driver, tired locomotive driver);
- Weather (not slippery, slippery);
- Track profile (downward slope, normal, upward slope);
- Train weight (empty, loaded);
- Pressing force (pressing on automatic coupling, moderate pressing on automatic coupling, no pressure on automatic coupling).

Mamdani algorithm was used to arrive at fuzzy logic inference. Processing of fuzzy inference rules in expert system consists of:

- Calculation of degrees of truth of left-hand side of rules (between "if" and "then")—determination of degree of membership of input data to fuzzy subsets, shown in the left-hand side of rules of inference;
- Modification of fuzzy subsets, shown in the right-hand side of rules of inference (after "then"), according to truth values obtained in the first stage;
- Grouping (superposition) of modified subsets;
- Scaling of superposition result—shift from fuzzy subsets to scalar values (see Fig. 1).

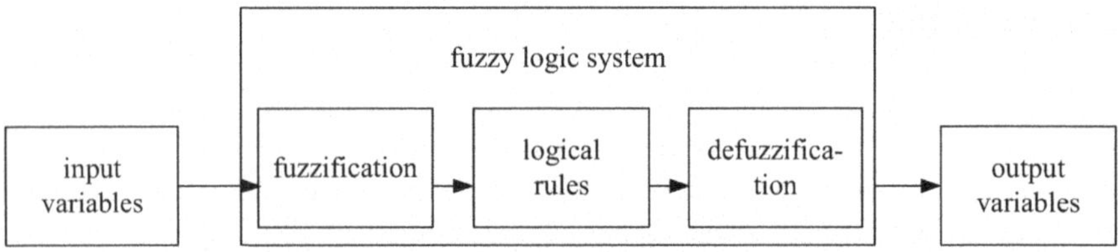

Fig. 1. Block diagram of the output signal of the fuzzy controller.

This system allows evaluating the rolling onto the brake shoe on the basis of entered linguistic variable output.

Figure 2 shows Gaussian membership function of distribution. It is membership function for the variable "rolling onto brake shoe", where condition values for rolling onto a brake shoe are put at axis X, and membership function values for "rolling onto a brake shoe" are put at axis Y. As the values of 0.7 and higher are presented as high probability of rolling onto a shoe, then membership function takes the value of 1 that corresponds to membership truth of complete rolling onto a brake shoe. When the values are from 0 to 0.3 (low probability of rolling onto brake shoe), the value of membership function is 0. The point of crossing of these two membership functions is the midvalue of 0.5. As the research cannot be based on two values only (there is rolling/there is no rolling), a middle value (not complete rolling) is needed. So there was introduced the middle value that proportionally divides membership functions. In the range of 0.3–0.7 the rolling is considered medium (the wheel pair does not completely rest on the head of the brake shoe). But in the range of 0.7–1 membership function increases and therefore the credibility of statement raises when the value approaches the value of complete rolling onto brake shoe.

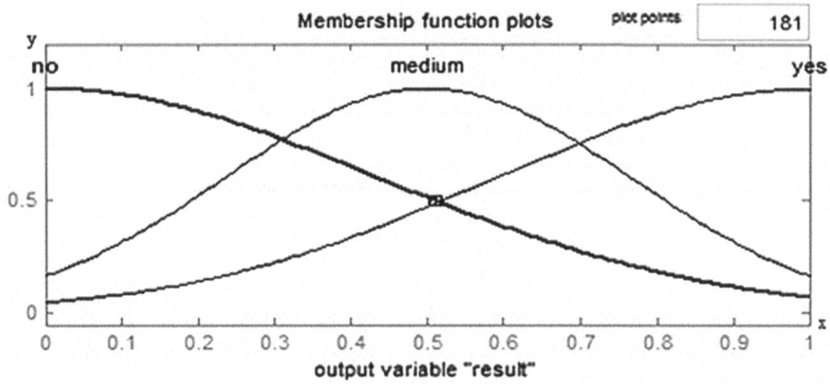

Fig. 2. Parameters of linguistic output variable in MatLab (rolling onto the brake shoe).

A feature of the fuzzy expert system is the simultaneous operation of all given rules, with varying degrees of their influence on the output value. The risk level is assessed in the process of fuzzy output, using a lot of fuzzy rules, which together constitute the knowledge base of this subject area, which are presented as the following:

$$if\ (x_1 = A)\ and\ (x_2 = B)\ then\ (y = F), \tag{1}$$

where $A, B, \ldots, F$ - these are linguistic values identified in a fuzzy way through the corresponding membership functions for the variables x_1, x_2 and y, and depend on human factor, weather, track profile, train weight and pressing force on automatic coupling. Variable y is Boolean (rolling onto brake shoe is present or absent). In the article 39 rules were introduced for the purpose of calculation. One of the rules as an example is specified bellow:

$$\begin{gathered} if\ (human\_factor\ is\ peppy)\ and\ (weather\ is\ not\_glides)\ and\ (track\_profile\ is\ slope)\ and \\ (train\_weight\ is\ empty)\ and\ (pressing\_force\ is\ no)\ then\ (result\ is\ yes) \end{gathered} \tag{2}$$

where x_1 - *human_factor*, x_2 - *weather*, x_3 - *track_profile*, x_4 - *train_weight*, x_5 - *pressing_force*. *A* - *peppy*, *B* - *not_glides*, *C* - *slope*, *D* - *empty*, *E* - *no*, and the result *F* - *yes*. Each parameter, that is a part of fuzzy logic, has its own change form and parameters. For example, (see Fig. 3) the characteristics of parameter change is shown:

when the driver is peppy (value 1), weather is not glides (0.5), track profile is slope (value −0.5), train is empty (value 29.5) and there is no pressing on the automatic coupling (value 0.2). As a result the rolling onto brake shoe will be medium, because its value is 0.58 that makes more than a half. However as it is obvious from the Fig. 2, the value of 0.7−1 is considered complete rolling and up to value 0.7 rolling onto brake shoe is medium (wheel pair does not fully rest on the head of brake shoe).

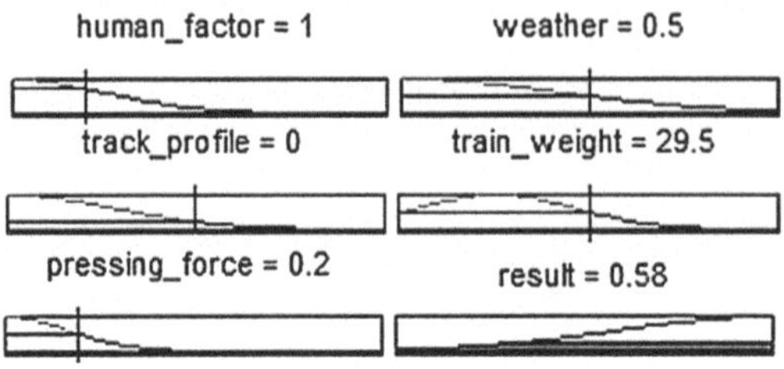

Fig. 3. A part of results of calculation of fuzzy inference for 39 given rules.

According to the calculation, rolling of the rolling stock onto the brake shoe, is not efficient when determining linguistic variable inputs. See calculation results below, which were obtained with the MatLab applying the above 39 rules. Figure 4 represents two input variables, namely human factor and track profile.

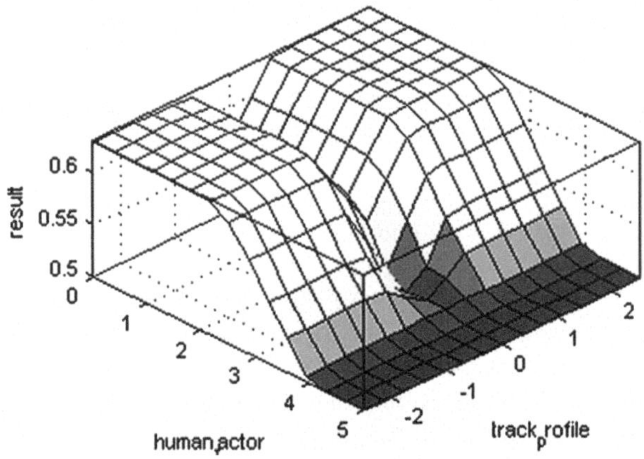

Fig. 4. Results of calculations human factor – track profile (rolling onto the brake shoe).

As it is clear from the outcome of the research, the value of human factor from 0 to 3 and the value of track profile from −2.5 to −1 gives complete rolling of wheel pair onto brake shoe, the same when the value of track profile is from 1 to 2.5. Complete rolling is guaranteed by driver's vivacity. However as the condition of driver worsens, his vivacity decreases, gradually the driver is getting tired which is obvious from human factor values (3−5), consequently the value of rolling onto brake shoe drops. It is clearly seen when analyzing the value of track profile from −2.5 to 2.5. It does not matter if there is slope or descent, rolling onto brake shoe starts decreasing as it is obvious from the figure after calculations. Consequently only when the driver is peppy he is capable of complete rolling of wheel pair onto brake shoe. However if the value of track profile is −0.75–0.75, rolling onto brake shoe differs from 1 to 0.75, that means complete rolling onto brake shoe to above medium which is actually rolling onto brake shoe as it was described above (from 0.7 to 1 rolling is complete) (see Fig. 1). According to the calculation, the level of rolling of the wheel pair on the brake shoe, with the measured parameters, is low, which corresponds to the initial input data. The fuzzy model for evaluating the wheel set rolling onto the brake shoe is adequate to the current process on the railway and allows to assess the risks of cars rolling off the brake shoe, which results from a lot of external factors. For this reason, an improved brake shoe has been developed, which allows the driver to initially see the wheel pair rolling on the brake shoe. This helps to reduce the risks described above in the current state of railway operation. The drawing below illustrates the operating principle of improved brake shoe (see Fig. 5).

The magnet, on which the shoe with a stop is placed, is mounted along the rail grid on the tracks, where the cars fixing is provided in the technical distribution act of the station. Additional elements are laser and parktronics. These elements are mounted on a portable device that is attached to a regular shoe. Elements allow for the attendant in the park, as well as the driver to safely fix the rolling stock. The driver, backing the train with the cars on the brake shoe, receives a signal via the repeater about the distance from the wheel pair to the brake shoe [4, 6–8]. In the driver's cab all data are shown on the display. Parktronic and laser rangefinder measure the distance from the brake shoe to the wheelset.

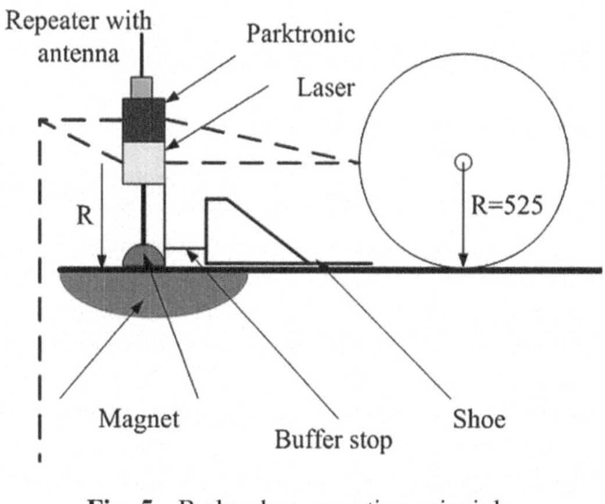

Fig. 5. Brake shoe operating principle.

Also, if it is necessary to remove the brake shoe from under the wheel pair, the driver will receive a signal via the repeater about its presence or absence under the wheel pair. Parktronic and laser rangefinder complement each other, and also this equipment makes the work of the yard operator on-duty (yardmaster) much safer. An improved brake shoe makes it impossible for the train to set off with a brake shoe under the wheel set, which excludes the damage of: wheelset, brake shoe, departure of the train with a brake shoe, arrow cut, rolling stock and railway infrastructure [1].

3 Conclusion

The presented improved model of the brake shoe allows rolling of the rolling stock onto the brake shoe at all times, since the engine driver always sees the distance from the wheel pair to the brake shoe. Unlike the existing fixation, it ensures precise rolling onto the brake shoe. Fixation of the rolling stock with the improved brake shoe is not affected by human factor and pressing force on automatic coupling. Also train weight do not affect rolling of the rolling stock, because the engine driver always sees the remaining distance until the brake shoe. Weather conditions and track profile do not have any impact either, because the shoe is equipped with a magnet.

Improvement of the brake shoe makes it possible to back the train more safely and to roll it on the brake shoe. It increases the work safety of the personnel involved in fixing the stock and reduces the time of train fixing. For this reason, the trains processing time in the technological processes of stations decreases, as well as the time at shunting; appears the possibility to reduce the staffs who are engaged in trains fixing. This reduces operating costs, cars downtime, delays of train and shunting locomotives before uncoupling and after attaching them to the train, as well as the additional time to perform these operations. But most importantly, it increases the level of safety on the railway. Further hardware and software development of an improved brake shoe is planned with the aim of assessing the roll of the wheel pair on the brake shoe in real time.

References

1. Pinchuk, A., Dadoenkov, P., Halutin, D., Korago, I.: Improved rolling system of railway stock on brake shoe. Procedia Comput. Sci. **149**, 258–263 (2019)
2. Karevs, V., et al.: Digital signal processing in real time in multi signal monitoring systems with high risk of application. In: Proceedings of 57th International Scientific Conference on Power and Electrical Engineering of Riga Technical University, Riga (2016)
3. Mezitis, M., et al.: Object controller of electric drive of point machine with expanded set of diagnostic tools. Procedia Comput. Sci. **104**, 393–399 (2016)
4. Sładkowski, A.: Rail Transport—Systems Approach. Part of the Studies in Systems, Decision and Control, vol. 87. Silesian University of Technology Katowice, Poland (2017)
5. Lu, S., Liu, Z., Nan, G., Zhang, G.: Automated visual inspection of brake shoe wear. In: Proceedings of SPIE, Beijing, China, p. 9675 (2015)
6. Yin, J., Tang, T., Yang, L., Xun, J., Huang, Y., Gao, Z.: Research and development of automatic train operation for railway transportation systems: a survey. Transp. Res. Part C: Emerg. Technol. **85**, 548–572 (2015)
7. Kurek, A., Modelski, J.: Application of the repeaters for indoor localization. In: Proceedings of the 9th European Conference on Wireless Technology (ECWT 2006), Poland (2007)

8. Solonnikov, Y., Ivanov, V.: Application of algorithms of the fuzzy logic on the basis of Siemens S7-300 for the electric drive control system. Int. Res. J. **8**(3), 140–150 (2017). Pacific National University, Khabarovsk
9. Aldjendi, I., et al.: Using fuzzy logic for the optimal placement of the compensating devices. Int. Res. J. **9**(40), 6–9 (2015). Belgorod State Technological University

On the Issue of Motion of Compensating Masses of an Automatic Balancing Device in Rotary Machines with a Horizontal Axis of Rotation

Guntis Strautmanis[1(✉)], Alexander Gorbenko[2], Mareks Mezitis[1], Valentina Strautmane[3], and Pavels Stankevics[1]

[1] Riga Technical University, 1, Kalku Street, Riga 1658, Latvia
{guntis.strautmanis,mareks.mezitis, pavels.stankevics}@rtu.lv

[2] Kerch State Maritime Technological University, 82, Ordzhonikidze Street, Kerch 298309, Russia
gan0941@yandex.ru

[3] Riga Technical University, Daugavpils Branch, 90, Smilshu Street, Daugavpils 5410, Latvia
valentina.strautmane@rtu.lv

Abstract. In mechanical engineering, including transport, rotary machines with a horizontal, vertical or inclined axis of rotation are widely used. They have an imbalance that causes vibration during their operation. The use of automatic balancing devices (ABD) can reduce the imbalance of unbalanced rotor systems operating in the super resonance velocity range. However, when accelerating the compensating masses in the rotor-ABD system, a side mode, different from the auto-balancing mode, may occur. In this paper, based on the mathematical model of a horizontal rotor with an ABD, the possible modes of motion of compensating masses are considered and the attraction regions of the main auto-balancing mode are constructed depending on the value of the viscous friction coefficient in the ABD. It is shown that at the decrease of the coefficient of viscous friction in the ABD, the attraction region of the auto-balancing mode decreases, while the rotor system switches to another stable mode.

Keywords: Railway transport · Rotor automatic balancing devices · Mathematical modeling

1 Introduction

The operation of rotor machines with a horizontal or vertical axis of rotation, having an imbalance, is accompanied by the excitation of increased vibrations in the super resonance zone of rotor velocity ranges. Installation of an automatic balancing device (ABD) makes it possible to significantly reduce the level of vibrations, which increases the reliability of rotary machines. In this paper, ball-type ABDs are considered, in which metal balls are used as compensating masses (CM) [1–3].

E. Ginters et al. (Eds.): ICTE ToL 2019, LNITI, pp. 218–226, 2020.
https://doi.org/10.1007/978-3-030-39688-6_29

In the auto-balancing mode, which is the main mode for the rotor system, the CMs are located on the opposite side of the imbalance and thereby reduce it in the automatic mode. At the same time, in a number of works [2, 4, 5] it was shown that additional motion modes that can occur during acceleration of a rotor with an ABD can be implemented in an ABD. Possible modes of motion can be periodic stable modes, or almost periodic ones [6]. Also in [6], a horizontal rotor with an ABD with one CM was considered, and the region of attraction for the auto-balancing mode was constructed for it. It is of interest to construct an auto-balancing region of attraction with a large number of CMs to determine their influence on this region.

The aim of this work is to construct and analyze the region of attraction of the auto-balancing mode for a horizontal rotor with an ABD containing two CMs.

2 Mathematical Model of a Rotor with an ABD

The work considers a symmetrical horizontal rigid rotor on an elastic dissipative suspension with an ABD.

The design model of the rotor system is shown in Fig. 1. The motion of the rotor axis is considered in a fixed coordinate system 0xyz, the motion of the CM is considered in a movable coordinate system $0_1x_1y_1z_1$, rotating with a constant angular velocity of the rotor (ω = const). Parameters of the rotor system with an ABD:

- ω – angular velocity of the rotor rotation, rad/s;
- m_r –the mass of the rotor, kg; m – the mass of one CM, kg;
- e = AC – rotor eccentricity, m;
- c – the stiffness of the elastic suspension of the rotor, N/m;
- b – the coefficient of external viscous damping of the rotor, kg/s;
- b_1 – coefficient of viscous friction in ABD, kg/s;
- p – the critical rotor velocity, rad/s;
- x_A, y_A – current coordinates of the geometric center of the disk and the axis of the rotor in a fixed coordinate system, m;
- R – the circle radius along which the centers of the CM, m.

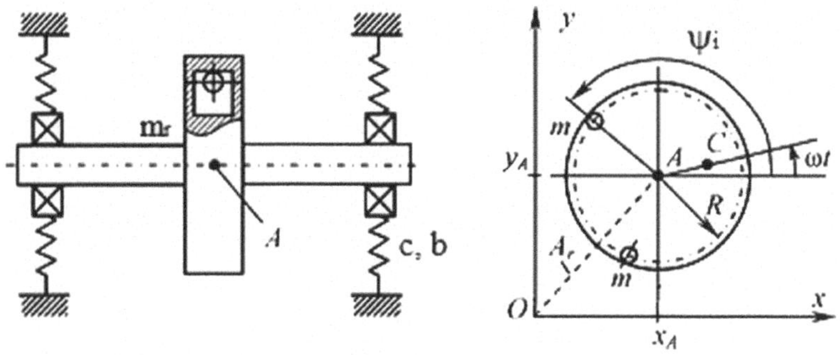

Fig. 1. The design model of the rotor system with an ABD.

The rotor system has four degrees of freedom, and its motion is described by four independent coordinates:

- x_A, y_A – current coordinates of the geometric axis of the rotor in a fixed coordinate system, m;
- Ψ_1, Ψ_2 – current angular coordinates of the CM in the movable coordinate system, rad.

The equations of motion of the presented rotor system are given [6]:

$$\begin{aligned} &M\ddot{x} - mR\ddot{\psi}_1 \sin(\varphi+\psi_1) - mR\ddot{\psi}_2 \sin(\varphi+\psi_2) \\ &\quad = -b\dot{x} - cx + mR\left(\dot{\varphi}+\dot{\psi}_1\right)^2 \cos(\varphi+\psi_1) + mR\left(\dot{\varphi}+\dot{\psi}_2\right)^2 \cos(\varphi+\psi_2) \\ &\quad + m_r e\dot{\varphi}^2 \cos\varphi - m_r g; \end{aligned} \tag{1}$$

$$\begin{aligned} &M\ddot{y} + mR\ddot{\psi}_1 \cos(\varphi+\psi_1) + mR\ddot{\psi}_2 \cos(\varphi+\psi_2) \\ &\quad = -b\dot{y} - cy + mR\left(\dot{\varphi}+\dot{\psi}_1\right)^2 \sin(\varphi+\psi_1) + mR\left(\dot{\varphi}+\dot{\psi}_2\right)^2 \sin(\varphi+\psi_2) \\ &\quad + m_r e\dot{\varphi}^2 \sin\varphi; \end{aligned} \tag{2}$$

$$\begin{aligned} &\frac{7}{5} mR^2\ddot{\psi}_1 - mR[\ddot{x}\, sin(\varphi+\psi_1) - \ddot{y}\, cos(\omega t+\psi_1)] \\ &\qquad = -b_0R^2\dot{\psi}_1 + mgR\, sin(\varphi+\psi_1); \end{aligned} \tag{3}$$

$$\begin{aligned} &\frac{7}{5} mR^2\ddot{\psi}_2 - mR[\ddot{x}\, sin(\varphi+\psi_2) - \ddot{y}\, cos(\omega t+\psi_2)] \\ &\qquad = -b_0R^2\dot{\psi}_2 + mgR\, sin(\varphi+\psi_2); \end{aligned} \tag{4}$$

and during the study were presented in a dimensionless form. For this, Eqs. (1) and (2) were divided into Mp^2e, and Eqs. (3) and (4) were divided into mp^2R^2. After transformations, the system of Eqs. (1) … (4) in dimensionless form, written in matrix form, has the following form:

$$\begin{pmatrix} 1 & 0 & -b1\cdot sin(\varphi+\psi_1) & -b1\cdot sin(\varphi+\psi_2) \\ 0 & 1 & b1\cdot cos(\varphi+\psi_1) & b1\cdot cos(\varphi+\psi_2) \\ 0 & 0 & 1 & -b3 \\ 0 & 0 & 0 & 1 \end{pmatrix} = \begin{pmatrix} d1 \\ d2 \\ d3 \\ \frac{d4+d3\cdot b3}{c1} \end{pmatrix}, \tag{5}$$

where

$$b1 = \frac{m}{M}\frac{R}{e} = m_0 e_2; \quad b3 = \frac{m}{M}\frac{cos(\psi_2-\psi_1)}{\left(\frac{7}{5}-\frac{m}{M}\right)} = m_0\frac{cos(\psi_2-\psi_1)}{\left(\frac{7}{5}-m_0\right)};$$

$$c1 = (1 - b3 * b3).$$

$$d1 = -n\dot{\bar{x}} - \bar{x} + b1\left(\omega_0 + \dot{\bar{\psi}}_1\right)^2 \cos(\varphi + \psi_1) + b1\left(\omega_0 + \dot{\bar{\psi}}_2\right)^2 \cos(\varphi + \psi_2) + (1 - 2m_0)\omega_0^2 \cos\varphi - (1 - 2m_0)g_0; \tag{6}$$

$$d2 = -n\dot{\bar{y}} - \bar{y} + b1\left(\omega_0 + \dot{\bar{\psi}}_1\right)^2 \sin(\varphi + \psi_1) + b1\left(\omega_0 + \dot{\bar{\psi}}_2\right)^2 \sin(\varphi + \psi_2) + (1 - 2m_0)\omega_0^2 \sin\varphi; \tag{7}$$

$$d3 = \frac{1}{7/5 - m_0}[1/e_2(n\dot{\bar{y}} + \bar{y})\cos(\varphi + \psi_1) - 1/e_2(n\dot{\bar{x}} + \bar{x})\sin(\varphi + \psi_1) - m_0\left(\omega_0 + \dot{\bar{\psi}}_2\right)^2 \sin(\psi_2 - \psi_1) + (1 - 2m_0)1/e_2\,\omega_0^2 \sin\psi_1 - n_0\dot{\bar{\psi}}_1 + g_0/e_2\, 2m_0\, sin(\varphi + \psi_1)]; \tag{8}$$

$$d4 = \frac{1}{7/5 - m_0}[1/e_2(n\dot{\bar{y}} + \bar{y})\cos(\varphi + \psi_2) - 1/e_2(n\dot{\bar{x}} + \bar{x})\sin(\varphi + \psi_2) + m_0\left(\omega_0 + \dot{\bar{\psi}}_1\right)^2 \sin(\psi_2 - \psi_1) + (1 - 2m_0)1/e_2\omega_0^2 \sin\psi_2 - n_0\dot{\bar{\psi}}_2 + g_0/e_2 2m_0\, sin(\varphi + \psi_2)]. \tag{9}$$

Dimensionless parameters of the rotor system:

$$n = \frac{b}{Mp};\; n_0 = \frac{b_0}{mp};\; e_2 = \frac{R}{e};\; \dot{\bar{\varphi}} = \bar{\omega} = \frac{\omega}{p};\; m_0 = \frac{m}{M};\; g_0 = \frac{9{,}81}{ep^2};\; p^2 = \frac{c}{M}.$$

The differential equations of motion of the rotor system with an ABD were studied using the SPRING software [7]. The results of the calculation of the system of differential equations were obtained in the form of dimensionless velocities and coordinates:

$$\ddot{\bar{x}} = \frac{\ddot{x}}{ep^2};\; \ddot{\bar{y}} = \frac{\ddot{y}}{ep^2};\; \ddot{\bar{\psi}}_i = \frac{\ddot{\psi}_i}{p^2}; \qquad \dot{\bar{x}} = \frac{\dot{x}}{ep};\; \dot{\bar{y}} = \frac{\dot{y}}{ep};\; \dot{\bar{\psi}}_i = \frac{\dot{\psi}_i}{p};$$
$$\bar{x} = \frac{x}{e};\; \bar{y} = \frac{y}{e};\; \bar{\psi}_i = \psi_i.$$

The calculation of the motion of the rotor system was carried out at p = 1 s^{-1}. To obtain the actual values of the motion parameters, they should be recalculated taking into account the specific values of the eccentricity of the rotor "e" and the critical rotational velocity of the rotor "p".

3 Results of the Calculations of the Mathematical Model

The paper considers the influence of the coefficient of viscous friction in the automatic balancing device on the region of attraction of the auto-balancing mode in the rotor system with a horizontal axis of rotation. In this case, the initial velocity of the CM in the circumferential direction of the balancing device $\dot{\bar{\psi}}_1$ and the original coordinate $\bar{\psi}_1$ were changed. The initial conditions of the rotor system in this case were as follows: $\tau = 0; \bar{x} = -96{,}661129; \dot{\bar{x}} = -0{,}008786; \bar{y} = 0{,}002342;\ \dot{\bar{y}} = 0{,}290157; \bar{\psi}_2 = 3{,}546597;$ $\dot{\bar{\psi}}_2 = -0{,}273826$, under which the condition of automatic balancing of the rotor is ensured. In this case, the first CM should also take the necessary position $\bar{\psi}_1 = 2{,}734986; \dot{\bar{\psi}}_1 = -0{,}274189$. When constructing the region of attraction of the auto-balancing mode, the dimensionless angular velocity of the rotor was assumed to be constant and equal to $\bar{\omega} = 3$. The studies were carried out at various values of the relative coefficient of viscous friction in the ABD: $n_0 = (0{,}05;\ 0{,}0075;\ 0{,}005)$. The attraction region of the auto-balancing mode was constructed at the following parameters of the rotor system: $n = 0{,}1;\ e_0 = 2{,}1;\ e_2 = 76{,}230003; g_0 = 98{,}1;\ m_0 = 0{,}007$.

The calculation results are shown in Figs. 2 and 3.

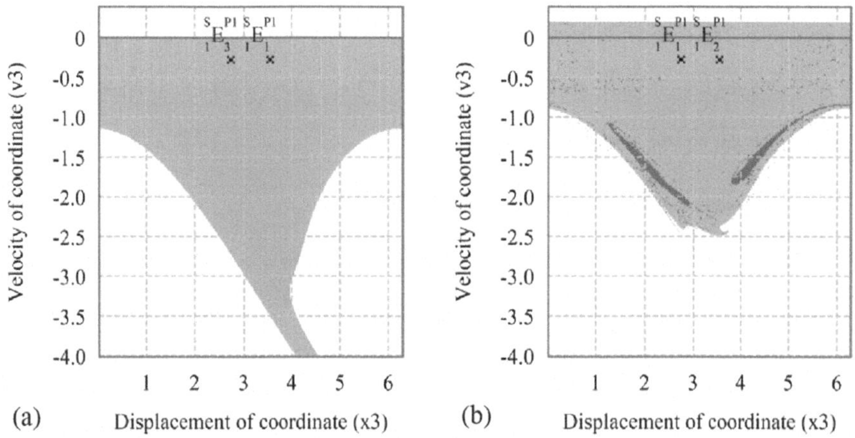

Fig. 2. Region of attraction of the auto-balancing mode (light gray color) at the value of the relative coefficient of viscous friction in the ABD (a) $n_0 = 0{,}05$; (b) $n_0 = 0{,}0075$; ${}^{S}_{1}E^{P1}_{n}$ – coordinate of the fixed point of the auto-balancing mode.

The regions of attraction are constructed under an unstressed condition for the motion of compensating masses. The masses seem to slip through each other, or move along parallels, closely placed paths, without hitting each other.

The fixed point of the auto-balancing mode (stable operation mode of the rotor system) is indicated by symbol ${}^{S}_{1}E^{P1}_{n}$ (where *S* is stable mode; *P1* is main mode; n is point number). From a comparison of the regions of attraction (see Figs. 2 and 3), it is obvious that with a decrease in the relative viscous friction coefficient in the ABD n_0, the attraction region of the auto-balancing mode decreases.

At the same time, it can be seen from Fig. 3 that there is an area around the attraction region (let us call this area a "cloud"), in which the auto-balancing mode is not secured. In this space (cloud), a slight change in the initial conditions of the CM motion leads to the loss of the auto-balancing mode.

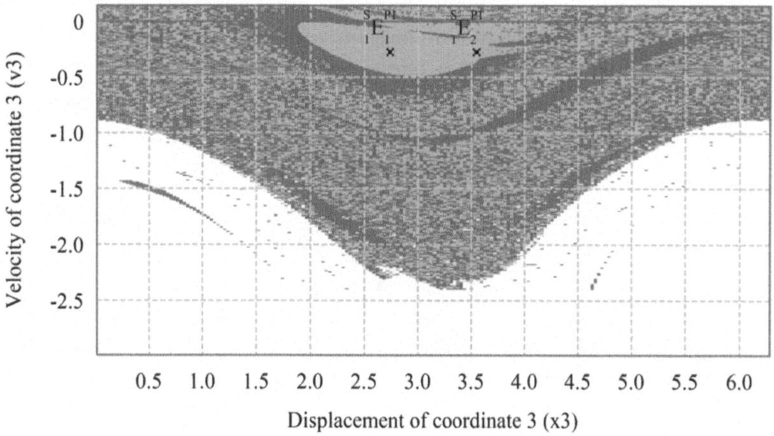

Fig. 3. Region of attraction of the auto-balancing mode (light gray color) at the value of the relative coefficient of viscous friction in the ABD $n_0 = 0{,}005$.

In the auto-balancing mode (in Fig. 3 point ${}^S_1E^{P1}_1$) the compensating masses, as follows from Fig. 4a and b, are placed opposite the imbalance and oscillate with respect to the equilibrium position. The higher the rotor velocity (the graphs are constructed at dimensionless angular velocities of the rotor $\bar{\omega} = 3$ and $\bar{\omega} = 10$) is, the closer the compensating masses are located to the optimal position, and their oscillations decrease. In this case (see Fig. 4c), the rotor is not completely balanced and it also makes small oscillations, which tend to decrease when increasing the rotor velocity.

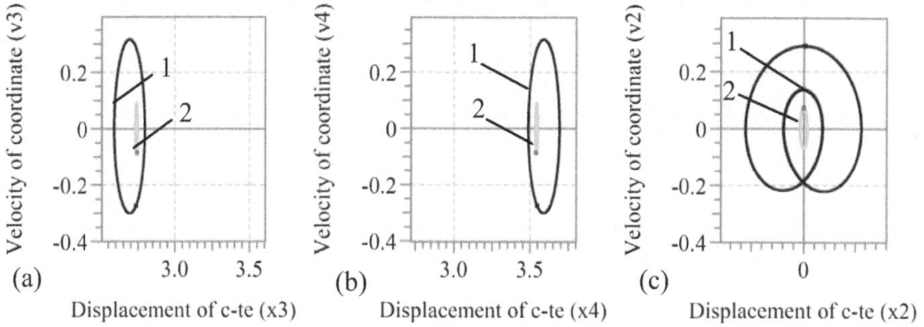

Fig. 4. Phase trajectories of the auto-balancing mode at the value of the relative coefficient of viscous friction $n_0 = 0{,}005$ (1 – at $\bar{\omega} = 3$, 2 – at $\bar{\omega} = 10$); (a) designations along the coordinate axes $v3 = \dot{\bar{\psi}}_1$; $x3 = \bar{\psi}_1 = \psi_1$; (b) designations along the coordinate axes $v4 = \dot{\bar{\psi}}_2$; $x4 = \bar{\psi}_2 = \psi_2$; (c) designations along the coordinate axes $v2 = \dot{\bar{y}}$; $x2 = \bar{y}$.

If for the first CM the initial velocity and the initial coordinate are chosen outside the region of attraction, for example, $\dot{\bar{\psi}}_1 = -0{,}6; \bar{\psi}_1 = \psi_1 = 3{,}1415$, the rotor system will not return to the main auto-balancing mode, but will continue to move with a non-periodic or almost periodic mode. Figure 5a shows the oscillations of the rotor axis and Fig. 5b shows a graph of the CM motion near the equilibrium position, which acquire a complex character. Figure 5c shows a phase portrait of the motion of the first compensating mass. The second compensating mass has a similar phase portrait of the motion.

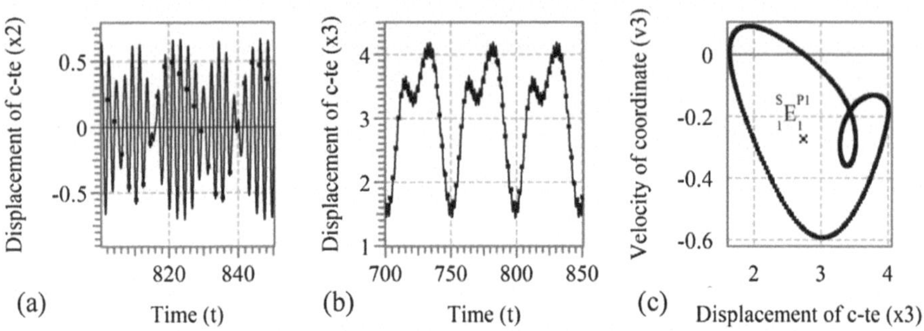

Fig. 5. Graphs of the motion of the rotor system at the value of the relative coefficient of viscous friction $n_0 = 0{,}005$ at the initial conditions for the first CM $\dot{\bar{\psi}}_1 = -0{,}6; \bar{\psi}_1 = \psi_1 = 3{,}1415$; (a) graph of the motion of the rotor axis along the coordinate $2 = \bar{y}$; (b) graph of the motion of the first CM $x3 = \bar{\psi}_1 = \psi_1$; (c) phase portrait for the first CM $v3 = \dot{\bar{\psi}}_1; x3 = \bar{\psi}_1 = \psi_1$.

If for the first CM the initial velocity and the initial coordinate are chosen with even a large deviation from the attraction region, for example, $\bar{\psi}_1 = -1{,}1; \bar{\psi}_1 = \psi_1 = 3{,}1415$, then in this case a mode will be implemented in which one CM will rotate relative to the axis of rotation of the rotor, and the second CM will make oscillatory movements without rotation relative to the horizontal axis. The indicated mode of motion of the rotor system is shown in Fig. 6. The rotor axis in this case will oscillate with increased amplitudes.

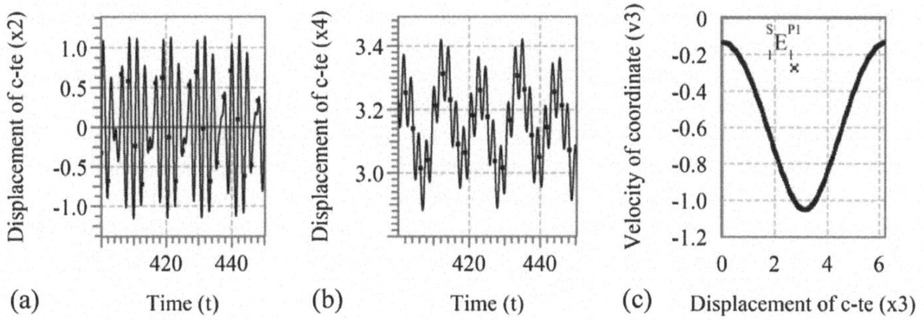

Fig. 6. Graphs of the motion of the rotor system at the value of the relative coefficient of viscous friction $n_0 = 0{,}005$ at the initial conditions for the first CM $\dot{\bar{\psi}}_1 = -1{,}1; \bar{\psi}_1 = \psi_1 = 3{,}1415$; (a) graph of the motion of the rotor axis along the coordinate $x2 = \bar{y}$; (b) graph of the motion of the first CM $x4 = \bar{\psi}_2 = \psi_2$; (c) phase portrait for the first CM $v3 = \dot{\bar{\psi}}_1; x3 = \bar{\psi}_1 = \psi_1$.

4 Conclusion

The calculation of the mathematical model of a horizontal rotor with an automatic balancing device with two compensating masses made it possible to state that the rotor system has several stable modes of motion.

The analysis of the constructed regions of attraction of the auto-balancing mode showed that the size of the region depends on the value of the coefficient of viscous friction in the automatic balancing device. The lower the coefficient of viscous friction is, the smaller the attraction region of the auto-balancing mode is.

In case the initial conditions for compensating masses go beyond the area of attraction of the auto-balancing mode, almost periodic or chaotic modes of motion are realized in the rotor system. In this case, the compensating masses make continuous motion relative to the body of the automatic balancing device and the oscillation amplitudes increase in the rotor.

The results obtained will be useful for the practical design and operation of rotary machines with an automatic balancing device. In addition, the work serves the basis for the development and further study of the automatic balancing device, taking into account the collisions of the compensating masses.

References

1. Gorbenko, A.N.: On the stability of self-balancing of a rotor with the help of balls. Strength Mater. **35**(3), 305–312 (2003)
2. Filimonikhin, G.B.: Balancing and Vibration Protection of Rotors by Autobalancers with Solid Corrective Weights. Kirovograd, KNTU, vol. 352 (2004)
3. Strautmanis, G., Grinevich, I., Strautmane, V.: The influence of automatic equalizer and rotor parameters on the ball's motion mode. In: Mechatronic Systems and Materials 2014, Selected Papers, pp. 135–141 (2015)

4. Artjunin, A.I.: The study of the movement of the rotor with auto-balancer. In: Proceedings of Universities Engineering, vol. 1, pp. 15–19 (1993)
5. Gorbenko, A., Strautmanis, G., Filimonikhin, G., Mezitis, M.: Motion modes of the nonlinear mechanical system of the rotor autobalancer. Vibroeng. Procedia **25**, 1–6 (2019)
6. Strautmanis, G., Mezitis, M., Strautmane, V., Gorbenko, A., Shchukins, I., Gromova, M.: Study of the area of attraction of the auto-balancing mode in a ball-type automatic balancing device with a horizontal axis of rotation. Vibroeng. Procedia **24**, 63–67 (2019)
7. Scukins, I., Zakrzevskis, M., Ivanov, Y., et al.: Application of software SPRING and method of complete bifurcation groups for the bifurcation analysis of nonlinear dynamical system. J. Vibroeng. **10**(4), 510–518 (2008)

Efficiency of Cellular Networks of Mobile Communication in Transport

Valentin Popov(✉)

Riga Technical University, Azenes Str. 12a, 1048 Riga, Latvia
popovs@latnet.lv

Abstract. The paper provides a theoretical analysis of the effectiveness of cellular mobile networks, which are used by operators in both public and transport networks of the GSM-R standard, several options of approaches to determining efficiencies of cellular networks in mobile communications (CNMC) in transport, including of GSM-Railway. The formulas and numerical estimates of the above CNMC efficiencies are provided. The article will contribute to the improvement of mobile network in transport and help in the development of new efficient network applications in the transport sector.

Keywords: Cellular networks of mobile communications (CNMC) · Energetically efficiency · Spectral efficiency

1 Introduction

The growth in the number of cellular networks of mobile communication (CNMC) subscribers, including Global System for Mobile Communications – Railway (GSM-R) (see Fig. 1), and the need of introducing new technologies related to the increase in transmission rates of digital signals and bandwidth of radio frequencies are the most important factors that will influence the development of CNMC in the coming years. As is known [1–8], the effectiveness of CNMC is understood as a function that depends on the main technical and economic parameters of CNMC, which determines the optimal energy, spectral and user resources of CNMC. Increasing the efficiency of the CNMC allows [4, 5]:

- Reduce costs for the use of the allocated frequency range;
- Reduce the cost of acquisition, installation, power supply and maintenance of base stations;
- Increase the capacity of the network by means of cell expansion, which in turn increases the capacity of the entire system as a whole and directly affects the quality of service to end users, especially in urban areas.

Thus, the aim of the work is to implement a theoretical analysis of the effectiveness of cellular mobile networks, which are used by operators in both public and transport networks of the GSM-R standard.

Further the energy and spectral resources of the CNMC and related 4 approaches to evaluating its effectiveness are discussed:

E. Ginters et al. (Eds.): ICTE ToL 2019, LNITI, pp. 227–235, 2020.
https://doi.org/10.1007/978-3-030-39688-6_30

- Analysis of energy and spectral efficiency of CNMC;
- Assessment of efficiency of CNMC in terms of subscribers, accounting for the selected radio frequency band;
- Calculation of spectral efficiency of CNMC as a function of the cluster traffic per one MHz;
- Determination of spectral efficiency of CNMC as a [(bps)/Hz] and {[(bps)/Hz]/cell}.

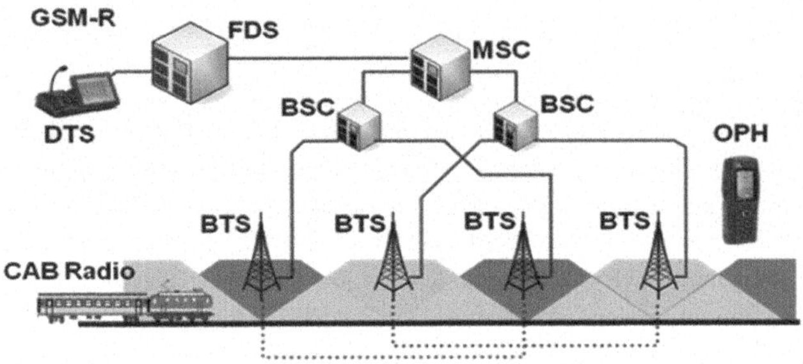

Fig. 1. Public CNMC and GSM-R network (FDS - function switch of dispatching subsystem, DTS - dispatcher terminal station, CAB Radio - train radio station, OPH - specialized wearable terminal, BTS - base transceiver station, BSC - base station controller, MSC - mobile switching center).

2 Energy and Spectral (Band) Efficiency of CNMC

As is known [3], the energy resource is understood as the power of radio signals propagating from BTS to MS and back in the conditions of interference and noise. In this case, the energy efficiency of CNMC is defined as the ratio of the average square value of the power of the useful radio signal P_S to the average square value of the power of noise P_N at the receiver input, which is necessary to ensure a given quality of radio communication, i.e.:

$$E_{eff} = \frac{P_s}{P_N} \tag{1}$$

At present only digital radio communication is used in the CNMC, therefore, to determine the value of E_{eff}, by the following estimates:

- When transmitting 1 bit of the message, the normalized signal-to-noise ratio is [3]:

$$h_{b^2} = \frac{E_b}{N_0} \tag{2}$$

where E_b is the energy of radio signal spent for transmitting 1 bit of the message, N_O - spectral density of all noises at radio circuit recounted at the receiver input. It is necessary to ensure the required error probability value h_b^2 (Eq. (2)), which is a universal indicator of the energy efficiency for a certain method of transmission of digital information [3]. Regardless of the type of signals used, the types of modulation and demodulation, the types of encoding and decoding, the best option for using the energy potential is the one for which the required value of h_b^2 is minimized, since this means that at equal radio channel noises (by spectral density) the radio line uses less energy of the transmitter to transmit each bit of the message.

- Taking into account that at an average speed of digital information transmission υ_b, the value of the transmission time per one bit is:

$$T_b = \frac{1}{\upsilon_b} \tag{3}$$

the signal energy E_b can be defined as:

$$E_b = P_s \cdot T_b = \frac{P_s}{\upsilon_b} \tag{4}$$

- Assuming that the information bandwidth F_{inf} is numerically equal to the digital information transmission rate $F_{inf} = \upsilon_b$, the value of h_b^2 can be considered as the ratio of the radio signal power to the noise power in the information band ($P_N = N_0 \times F_{inf}$):

$$h_b^2 = \frac{P_s}{P_N} = \frac{P_s}{N_0 \cdot F_{inf}} \text{ or } E_{eff} = h_{b^2} = \frac{P_s}{P_N} = \frac{P_s}{N_0 \cdot F_{ch}} \tag{5}$$

- The efficiency of using the radio frequency working band by the digital information transmission system is characterized by the ratio of the digital signal transmission rate implemented in this system to the bandwidth of radio frequencies occupied by the radio channel F_{ch}:

$$\gamma = \frac{\upsilon_b}{F_{ch}} \tag{6}$$

this value is called the specific bit rate of digital information and is measured in $\frac{bps}{Hz}$.

- The Shannon theorem can be used to obtain a relation between the quantities h_b^2 and γ:

$$\mathrm{v_b = F_{ch} \cdot \log_2(1 + \frac{P_s}{P_N}) = F_{ch} \cdot \log_2\left(1 + \frac{P_s}{N_0 \cdot F_{ch}}\right) = F_{ch} \cdot \log_2\left(1 + \frac{P_s \cdot v_b}{N_0 \cdot v_b \cdot F_{ch}}\right)} \tag{7}$$

From which follows:

$$\frac{v_b}{F_{ch}} = log_2\left(1 + h_b^2 \cdot \frac{v_b}{F_{ch}}\right) \text{ as a result, } \gamma = \frac{v_b}{F_{ch}} = log_2\left(1 + h_b^2 \cdot \gamma\right) \tag{8}$$

the final form of the functional dependence $h_b^2(\gamma)$ is written:

$$E_{eff}(\gamma) = h_b^2(\gamma) = \frac{2^\gamma - 1}{\gamma} \tag{9}$$

Thus, formula (9) determines the relationship between the energy and spectral efficiency of the CNMC for the case of the best use of the digital radio link. The numerical estimates of the expression (9) are presented in the Table 1.

Table 1. Numerical estimates of the expression $h_b^2(\gamma)$.

$\gamma = v\_b/F\_ch,\ bps/Hz$	$h_b^2(\gamma) = 10 \cdot \lg h_b^2 = 10 \cdot \lg\left(\frac{2^\gamma - 1}{\gamma}\right), dB$
−2	−4,25
−1	−3
0	−1,59
1	0
2	1,8
5	7,8
10	29
$\gamma \gg 1$	$h_b^2(\gamma) = 3.01 \cdot \gamma - 10 \cdot \lg \gamma, dB$

As follows from Table 1:

- At a specific transmission rate tending to zero, the value of $h_b \to -1{,}59$ dB, arises with orthogonal coding (i.e., this condition is the limiting one established by the information theory [3]), which corresponds to the extremely low spectral (band) efficiency;
- At $\gamma = v\_b/F\_ch = 1$, the value of $h_b^2(1) \to 0$ dB, i.e. with the same radio channel noise in the radio link, the minimum energy of the transmitter is used to transmit each bit of the message;
- In the area where the spectral efficiency lies within the limits of $1 < \gamma < 2$, the value $0 \le h_b^2 \le 1{,}8$ dB [3];
- With increasing specific transfer rate from 2 to 4 (i.e., spectral efficiency), the value of $h_b^2\ (\gamma)$ increases, which means that the energy efficiency falls, since for the transmission of digital signals with increasing radio signal power at a fixed noise power the transmitter output power increase is required.
- For $\gamma \gg 1$ Eq. (9) takes the form:

$$h_b^2(\gamma) = 3.01 \cdot \gamma - 10 \cdot lg\,\gamma, dB \tag{10}$$

The Numerical Cases Related. The numerical examples are provided as following:

- For digital voice transmission systems of the PDH standard, the AMI transmission rate for one channel is 64 kps, with γ = 64000/3100 = 20.6 (bps)/Hz (with the standard voice channel band 3.1 kHz), the energy efficiency in this case will be equal to: $h_b^2(\gamma) = 3.01 \cdot \gamma - 10 \cdot \lg \gamma, dB$ = $h_b^2(\gamma) = 3.01 \cdot 20.6 - 10 \cdot \lg 20.6 =$ 50.66 dB, which conforms with norms in respect of signal-to-noise ratio for voice channels - about 50 dB [3];
- For the GSM standard, the transmission rate for one radio channel is 270.833 kbps with a band of 200 kHz, with γ = 270.833/200 = 1.354 (bps)/Hz, the energy efficiency will be: $h_b^2(\gamma) = 10 \cdot \lg h_b^2 = 10 \cdot \lg\left(\frac{2^\gamma - 1}{\gamma}\right) = 10 \cdot \lg\left(\frac{2^{1.354} - 1}{1.354}\right) \approx 0.6\ dB$, which indicates a high degree of energy efficiency.

3 The Efficiency of CNMC in Terms of the Number of Subscribers per Dedicated Band of Radio Frequencies

The following assumptions are made [8]:

- Let the mobile stations be evenly distributed in the service area of the operator, which has approximately the form of a circle with radius R_O, while the coverage area will be equal to:

$$S_0 = \pi \cdot R_0^2 \tag{11}$$

- Let the entire service area be divided into cells, forming a flat regular hexagonal lattice, each cell being a hexagon with the radius of the circumscribed cell R_C, with the cell area equal to:

$$S_c = \frac{3\sqrt{3}}{2} \cdot R_c^2 \approx 2,6 \cdot R_c^2 \tag{12}$$

- However, if the cells are represented as circles with radius R_C, then:

$$S_c = \pi \cdot R_c^2 \tag{13}$$

- Then the number BTS in the service area will be (see Fig. 1):

$$N_{BTS} = \frac{S_0}{S_c} \approx \frac{\pi \cdot R_c^2}{2.6 \cdot R_c^2} = 1,21 \cdot \left(\frac{R_0}{R_c}\right)^2 \quad (14)$$

Taking into account the cluster structure of the CNMC [6, 8] with the cluster order $N_{cl} = (1/3) \cdot q^2$, the radio frequency band for the CNMC, and the set of frequencies for each BTS consists of n_{BTS} channels having a bandwidth of each - F_{ch}:

$$F_S = F_{ch} \cdot n_{BTS} \cdot N_{cl} \quad (15)$$

The number of active subscribers in the service area will be:

$$N_i = N_{BTS} \cdot n_{BTS} \quad (16)$$

Then the efficiency of using the allocated radio frequency band is defined as:

$$E_f(F_{ch}, N_{cl}, (R_0/R_c)) = \frac{N_i}{F_S} = \frac{N_{BTS} \cdot n_{BTS}}{F_{ch}.N_{cl} \cdot n_{BTS}} = \frac{1,21}{F_{ch} \cdot N_{cl}} \cdot \left(\frac{R_0}{R_c}\right)^2, \frac{subscribes}{MHz} \quad (17)$$

It follows from (17) that:

- The value of E_f does not depend on the number of channels n_{BTS};
- The value of E_f decreases with the growth of the order of the N_{cl} cluster and the associated cell radius R_c, which reduces co-channel interferences, but also reduces the number of subscribers per 1 MHz;
- With regard to the sufficient value of the efficiency at using the allocated radio frequency band for the CNMC, it is necessary to reduce the order of the N_{cl} cluster when meeting the requirements for the level of co-channel interference.

4 Spectral Efficiency of the CNMC as a Function of Cluster Traffic

In [7] the following expression of the effectiveness of CNMC was proposed:

$$E_f(A_{cl}, S_{cl}, F_{ch}, N_{cl}, n_{BTS}) = \frac{A_{cl}}{S_{cl}} \cdot \frac{1}{2F_{ch} \cdot N_{cl} \cdot n_{BTS}} \frac{Erl}{km^2 \cdot MHz} \quad (18)$$

where A_{cl} is the traffic per cluster, S_{cl} is the cluster area, $2F_{ch}$ is the double value of the radio channel band that takes into account the frequency bands allocated in both directions of the radio link, and the calculation algorithm E_f can be expressed in the following procedures. Given the area of radio coverage of the operator and the electrodynamics calculation of the maximum radius of individual cells [8], the area of one cell (according to formulas (12) or (13)) is sought. In this case, the number of cells in the service area is:

$$N_{cel} = \frac{S_0}{S_{cel}} \tag{19}$$

The coefficient of co-channel recurrence q is determined from the expression [6, 8]:

$$\left(\frac{D - R_{max}{}^{n}}{R_{max}}\right) = \left(\frac{\mathrm{D}}{\mathrm{R}_{\max}} - 1\right)^{\mathrm{n}} = (\mathrm{q}-1)^{\mathrm{n}} \leq \left(\frac{\mathrm{S}}{\mathrm{I}}\right)_{\mathrm{per}} \tag{20}$$

where $\left(\frac{S}{I}\right)_{per}$ is given value of the permissible signal level to the interference level, but n is the decay exponent (determining the influence of the environment on the propagation conditions of radio waves), determined from Table 2:

Table 2. Values of attenuation exponent n.

Propagation media	Attenuation exponent n
Free space (Line of Sight - LOS)	2
Suburban (LOS/NonLOS)	2.7–3.5
City (urban) (NonLOS)	3–5

The cluster order is defined as a rounded integer:

$$N_{cl} = \frac{1}{3}(q)^2 \tag{21}$$

The area of the cluster is determined:

$$S_{cl} = N_{cl} \cdot S_{cel} \tag{22}$$

Given the standards for the F_{ch} band and the number of n_{BTS} channels for one cell, we determine the total occupied band of radio frequencies in one direction of the CNMC (16):

$$F_S = F_{ch} \cdot n_{BTS} \cdot N_{cl} \tag{23}$$

The effectiveness of CNMC (18) can be found in the form:

$$E_f = \frac{A_{cl}}{S_{cl}} \cdot \frac{1}{2F_{ch} \cdot N_{cl} \cdot n_{BTS}} \frac{Erl}{km^2 \cdot MHz} \tag{24}$$

Thus, the spectral efficiency of CNMC will depend not only on the cluster's traffic and area, but also on the doubled band of the radio channel, the number of clusters and the number of base stations.

The Numerical Case Related. A numerical example is provided bellow of calculating the effectiveness of CNMC by formula (18) [7]. For the GSM-900 standard, with the following basic data: $\frac{A_{cl}}{S_{cl}} = 0,5\frac{erl}{km^2}$, F_{ch} = 0.2 MHz, N_{cl} = 3.8 - TDM slots, p_B = 2% - failure probability and number of channels n_{BTS} = 20, $E_f = \frac{A_{cl}}{S_{cl}} \cdot \frac{1}{2F_{ch} \cdot N_{cl} \cdot n_{BTS}} = 0,5 \cdot \frac{1}{2 \cdot 0,2 \cdot 3 \cdot 20} = 0,02\frac{erl}{km^2 \cdot MHz}$. In [7], numerical estimates of the dependence of the CNMC efficiency on the cell radius (assuming that all cells in the service area of the operator are the same in size and base stations have the same directional pattern), is shown that with increasing radius of the cell efficiency decreases (for example, for the GSM standard at R_c= 0.5 km - E_f= 0.416 [erl/(km^2 ·MHz)], and for R_c = 8 km E_f = 0.0347 [erl/(km^2 ·MHz)].

5 Spectral Efficiency of the CNMC

The spectral efficiency of the CNMC, as shown above, is an indicator calculated as the ratio of the rate (in bps) of the transmitted digital data per 1 Hz of the radio frequency band used: {(bps/Hz}- formula - (6). This value characterizes the speed of information transfer in a given band of radio frequencies. Spectral efficiency has a huge impact on the efficiency of using frequency resources, dedicated network and quality of services (QoS). The spectral efficiency can also be expressed as:

$$\gamma = \frac{v_b}{F_{ch}} \cdot \frac{1}{N_{cel}} \{[(bps)/Hz]/cell\} \tag{25}$$

Considering the need to provide higher bandwidth and the possible increase in the available radio spectrum for 3G networks, the development of 4G LTE was designed to use a bandwidth wider than 5 MHz. At the same time, to provide greater flexibility in the development of the standard LTE (Release 8 E-UTRAN), both the use of a bandwidth of 5 MHz and less than 5 MHz was envisaged. The main task facing the LTE developers was the task of significantly improving the spectral efficiency (bringing it up to 5 bps per Hertz), throughput on the direct channel (downlink) to 100 Mbps with a bandwidth of 20 MHz and 50 Mbps on the reverse (the line "up"), i.e. 2.5 bps per 1 Hz, and also in reducing the transmission delay of data packets to 10 ms compared to 80 ms in HSDPA (Release 5) and simplifying the network architecture. As follows from the CNMC comparative analysis of spectral efficiency, depending on the bandwidth of the channel, the largest value $\gamma = \frac{v_b}{F_{ch}}$ is in the LTE standard at F_{ch} = 20 MHz γ> 4.5 (bps)/Hz.

6 Conclusion

The article considers various methods of determining the efficiencies of CNMC, which allows to evaluate the perspectives of the corresponding generation of CNMC and use these estimates to optimize the CNMC. It should be noted that (other factors (range, bandwidth, cell spacing, etc.) being equal) the spectral efficiency of a particular technology is the parameter on which the provision of the greatest number of users to CNMC services depends. The analysis of the efficiency and selection of the CNMC generations allows the designers of mobile systems on the railway transport: to reduce the costs for the use of the allocated frequency range, lower the costs of acquisition, installation, power supply and maintenance of base stations, to increase the capacity of the network through the expansion of the cell, which in turn increases the bandwidth capacity of the network as a whole and has a direct impact on the quality of service to end users in the transport sector, especially in urban areas. This problem will be quite sharp with upcoming development and implementation of new mobile communication standards, such as 5G, LTE-R, etc. to which usability in railway communication will be aimed further research activities.

References

1. Zyuko, A., Falco, A., Panfilov, I.: Noise immunity and efficiency of information transmission systems. Radio Communications. Moscow (1985)
2. Hurley, S.: Planning effective cellular radio networks. IEEE Trans. Veh. Technol. **51**(2), 243–253 (2002)
3. Volkov, L., et al.: Digital radio communication systems. Eko-Trends, Moscow (2005)
4. Tikhvinsky, V., et al.: Spectral efficiency of mobile communication systems. Eko-Trends, Moscow (2007)
5. Tikhvinsky, V., Terentyev, S., et al.: Spectral efficiency of UMTS and LTE mobile communication systems. Russian Acad. Nat. Sci. **1**, 21–26 (2008)
6. Popov, V.: Foundations of cellular communication GSM standard. Eko-Trends (2005)
7. Sukachev, E.: Mobile networks of radio communication with mobile objects. ONAC name of A.S. Popov, Odessa (2013)
8. Popov, V., Skudnov, V.: Foundations of cellular networks of mobile communication designing. Gorjatchaja Linija-TELECOM, Moscow (2017)

Bio-Inspired Optimization of Rateless Codes for Reliable Intelligent Transportation Systems

Sergii Prykhodko[1], Mykola Shtompel[1(✉)], Oleksandr Sievierinov[2], Viacheslav Tretiak[3], Andrii Vlasov[3], and Vitalii Martovytskyi[2]

[1] Ukrainian State University of Railway Transport, 61001 Kharkiv, Ukraine
{prihodko, shtompel.mykola}@kart.edu.ua

[2] Kharkov National University of Radio Electronics, 61166 Kharkiv, Ukraine
oleksandr.sievierinov@nure.ua, martovytskyi@gmail.com

[3] Ivan Kozhedub Kharkov National University of Air Force, 61023 Kharkiv, Ukraine
slava_tr@ukr.net, vav_and@i.ua

Abstract. The quality of service the users of intelligent transportation systems highly depends on the methods of information processing in telecommunication networks. In particular, forward error correction codes are an important element of many telecommunication technologies. In last decade a lot of studies have been carried out and attempted to improve the performance of different codes. The study indicates that rateless codes can be used to increase the reliability of data transmission in telecommunication networks of intelligent transportation systems. In the article bio-inspired optimization method of degree distributions of rateless codes is discussed. The main stages and features of this method are presented. The simulation results approve that proposed approach significantly improves error performance of rateless codes. Due to this, the effectiveness and robustness of network-based intelligent transportation systems are increased.

Keywords: Intelligent transportation systems · Telecommunication · Bio-inspired optimization

1 Introduction

The intelligent transportation systems play major role in development of modern cities around the world. These systems consist of advanced traffic management and information systems, advanced vehicle control systems, advanced public transportation systems, etc. The various telecommunication networks are a cyber-physical basis for organizing information exchange in intelligent transport systems. An important aspect of designing these systems is ensuring the specified quality of service the users. The different information processing methods are applied in telecommunication technologies to fulfill this requirement [1].

One of the most important index of quality of service is the reliability of information transmission. The channel coding is a typical approach to improve this parameter. Therefore different forward error correction codes are a part of most telecommunication technologies. It is advisable to use packet-switched networks for

E. Ginters et al. (Eds.): ICTE ToL 2019, LNITI, pp. 236–242, 2020.
https://doi.org/10.1007/978-3-030-39688-6_31

transmitting multimedia information and connecting heterogeneous equipment in modern intelligent transportation systems. Rateless codes have shown excellent performance for information transmission in these networks [2].

Luby transform (LT) codes were the first practical realization of rateless codes but they have relatively large the encoding and decoding complexity [3]. Raptor codes are another example of this class of codes. They are characterized improved performance due to serially concatenate structure with some fixed-rate code as an outer code and LT code as an inner code [4]. Therefore Raptor codes are the preferred scheme for increasing the reliability of information transfer in modern network-based intelligent transportation systems.

2 Problem Definition

The performance of Raptor codes with fixed outer code is largely determined by the characteristics of LT codes. In the encoding process of LT codes, the original message is divided into k source symbols, the code symbols are then formed by modulo addition several random information symbols. The encoding and decoding processes of these codes can be represented using Tanner graphs, which consist of source and code vertices. The main feature of LT codes is to use of a special degree distribution $p(d)$ of code vertices. In fact, $p(d)$ is the probability distribution of a random variable d, which determines the number of source symbols that are used to form some code symbol. A typical approach to decoding LT codes is the belief propagation algorithm which has low computational complexity. However, this algorithm is sensitive to the used degree distribution and requires n code symbols for successful decoding.

With this in mind in [3], ideal and robust soliton degree distributions were proposed for the classical LT codes but they are not suitable for Raptor codes. These codes requires modified LT codes with specific degree distributions with some very different properties. Based on theoretical analysis, several degree distributions for long LT codes with certain particular sizes of source symbols were obtained [4]. By long LT codes are understood as codes for which the number of source symbols $k > 10000$.

On the other hand, the implementation of modern telecommunication technologies requires the design of LT codes for fewer source symbols. For decision this problem, the several optimization approaches based on evolutionary computation to the degree distribution search were proposed.

In [5] the approach with using covariance matrix adaptation evolution strategy to directly optimize the degree distributions of LT codes was presented. The feasibility of customizing degree distributions for different purposes was demonstrated. The simulation showed LT codes with the optimized degree distributions have good characteristics for a wide range of small source symbol sizes, but the results are not available for LT codes with number of source symbols $k > 1000$.

In [6] the features to employ different types of evolution strategies, namely, covariance matrix adaptation evolution strategy, natural evolution strategy, differential evolution, in optimization the degree distributions of small-length LT codes was considered. Experiment results confirmed that covariance matrix adaptation evolution strategy and natural evolution strategy produce near-optimal degree distributions of rateless codes which suitable for different applications.

From a practical point of view, short LT codes, for which number of source symbols $1000 \leq k \leq 10000$, play an important role in network-based intelligent transportation systems. Thus, the actual problem is a generalization of these optimization approaches for search degree distributions for these LT codes.

The main task of the LT code as a part of Raptor code is to provide the minimum required number of unrecovered source symbols for a given number of received code symbols, i.e. for a given overhead $R = n/k$. This indicator is called the failure rate F.

Formally, this problem can be represented as the following optimization problem

$$f(p^*(d)) = \min_{p(d) \in p'(d)} f(p(d)), \tag{1}$$

$$f(p(d)) = F, \tag{2}$$

$$R = const,\ p'(d) = \left\{ p(d) \middle| \begin{array}{l} \sum_{d=1}^{k} p(d) = 1 \\ 0 \leq p(d) < 1 \end{array} \right\}, \tag{3}$$

where $p(d)$ is the trial degree distribution of given LT code; $p'(d)$ is the set of admissible degree distributions of given LT code; $p^*(d)$ is the best degree distribution of given LT code.

From the analysis of objective function (2) and constraints (3) it follows that the formulated minimization problem (1) is a nonlinear programming problem, for the solution of which it is proposed to use a bio-inspired approach.

3 Proposed Optimization Method of LT Codes

In the general case, the degree distribution of LT code consists of k elements, which leads to a large dimension of the optimization problem (1) for codes with number of source symbols $1000 \leq k \leq 10000$. By analogy with [5], to reduce the dimension of the optimization problem, it is proposed to use a truncated degree distribution

$$w(d) = \begin{cases} p(d) \text{ if } d \in D \\ 0 \text{ if } d \notin D \end{cases}, \tag{4}$$

where D is the selected subset of degrees d.

In the paper, it is proposed to use the cuckoo search procedure as a bio-inspiration search mechanism for optimization degree distribution (4). This procedure is characterized by high efficiency and has a small number of tuning parameters [7]. During optimization, the degree distribution of given LT code is converted to a real-number decision vector. The erasure channel model is used for submission to the information transmission over the packet-switched network.

The proposed optimization method of given LT code based on cuckoo search procedure consists of the following stages.

Stage 1. Selecting of parameters of LT code and characteristics of erasure channel model.

At this stage, the number of source symbols k, the formation of a subset D with non-zero probability values in the degree distribution, the erasure probability in the channel q are selected.

Stage 2. Setting the failure rate F for given LT code as an optimization criterion.

Stage 3. Optimizing the degree distribution of given LT code based on the cuckoo search procedure (while current number of iteration less maximum number of iterations, i.e. $t < T$).

Step 3.1. Forming the set of trial decision vectors.

At this step, initial population of N host nests $w_i^{(t)}$, $i \in [1, N]$ is randomly generated and evenly distributed in the search area.

Step 3.2. Converting trial decision vectors to appropriate degree distributions of LT code.

Step 3.3. Modeling the process of information transmission through an erasure channel using LT code with formed degree distributions.

At this step, the encoding of source symbols and decoding of received code symbols based on belief propagation algorithm are performed.

Step 3.4. Calculating the failure rate of LT code given by each of the degree distributions.

Step 3.5. Modification of trial decision vectors by the use of specific operators of the cuckoo search procedure.

Step 3.5.1. Finding a new cuckoo position based on Levy flights, which are realized by the formula

$$w_i^{(t+1)} = w_i^{(t)} + a \oplus L(x), \tag{5}$$

where $w_i^{(t)}$ is the vector of the cuckoo's starting position; a is the vector of step size for the corresponding components of the vector $w_i^{(t)}$; $L(x)$ is the random vector consisting of independent real numbers distributed by Levi law; x is the random step length in Levi distribution.

Step 3.5.2. Selecting the nest $w_j^{(t)}$ and replacing the egg in the nest with the cuckoo's egg if $f(w_i^{(t+1)}) < f(w_j^{(t)})$.

Step 3.5.3. Removing some of the worst randomly selected nests from the population with the probability b, and forming the same number of new nests according to step 3.1.

Step 3.5.4. Ranking the current decision vectors $w_i^{(t+1)}$, finding the current best decision vector $w_i^{*(t+1)}$ and switching to the Step 3.2.

Stage 4. Determining the best decision vector w^* and forming of appropriate degree distribution $w^*(d)$ of given LT code (when the maximum number of iterations is reached, i.e. $t = T$).

4 Experiments and Results

To evaluate the performance of the proposed optimization method of degree distributions of LT codes, a series of experiments were carried out.

The simulation was conducted for LT codes with several length, among which k is set as 1000, 5000 and 10000, and for channel with erasure probability $q = 0.2$. The parameter D was set as 8, 9 and 10, respectively. Each subset D contains $d = 1$, $d = 2$, $d = 3$, which play a key role in any degree distribution, and empirically selected additional degrees. In addition, experiments were carried out for fixed overhead values, namely, $R = 1.05$, $R = 1.1$, $R = 1.15$, $R = 1.2$, $R = 1.25$, for each value k.

For cuckoo search procedure, such parameter values were used: the population size N is 100; the maximum number of iterations T is 100; the probability of detecting a cuckoo egg by the host b is 0.3; the vector of step size a is 1.

The simulation was executed 50 times independently for each experiment to achieve statistical results.

The convergence process of existing evolutionary [5] and proposed bio-inspired optimization method for LT code with $k = 1000$ and $R = 1.1$ are shown in Fig. 1.

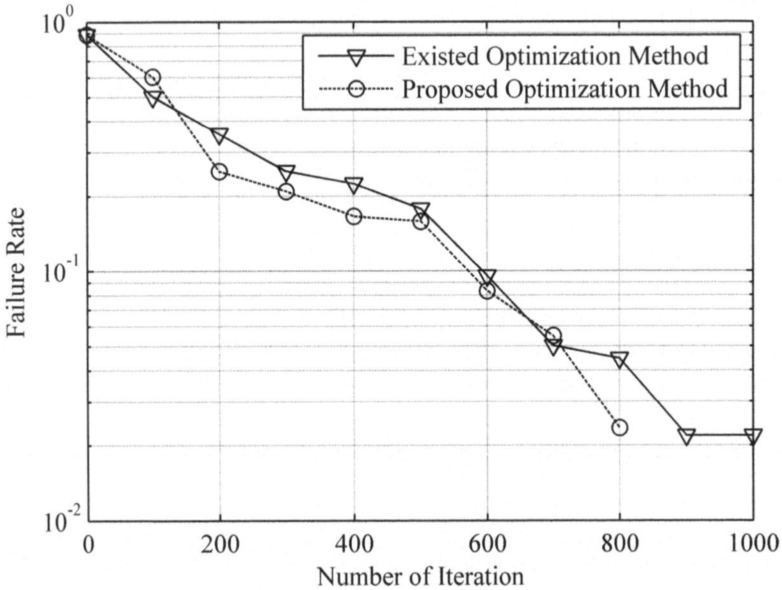

Fig. 1. The convergence process of optimization methods of LT codes.

It follows from Fig. 1 that the proposed optimization method has better convergence and near-optimal value of failure rate was determined after 800 iterations. As a result, the computational complexity of finding optimized distributions for this case decreases by 20%.

The best optimized degree distributions of several short LT codes with $R = 1.1$ are presented in Table 1.

Table 1. The best optimized degree distributions of short LT codes with $R = 1.1$.

$k = 1000$		$k = 5000$		$k = 10000$	
Degree	Probability	Degree	Probability	Degree	Probability
1	0.127	1	0.134	1	0.145
2	0.364	2	0.382	2	0.371
3	0.271	3	0.217	3	0.183
4	0.166	7	0.121	5	0.134
12	0.035	21	0.112	14	0.121
37	0.028	62	0.013	41	0.018
111	0.007	185	0.007	123	0.009
333	0.002	556	0.011	370	0.012
–	–	1667	0.003	1111	0.006
–	–	–	–	3333	0.001

From the analysis of Table 1 it follows that in the obtained degree distributions the greatest probabilities have degrees $d = 1$, $d = 2$, $d = 3$. This is because successful belief propagation decoding of LT codes with low overhead is mainly determined by the presence of small degrees.

The failure rate obtained for optimized degree distributions for LT codes with a fixed length, among which k is set as 1000, 5000 and 10000, and for various overhead is shown in Fig. 2.

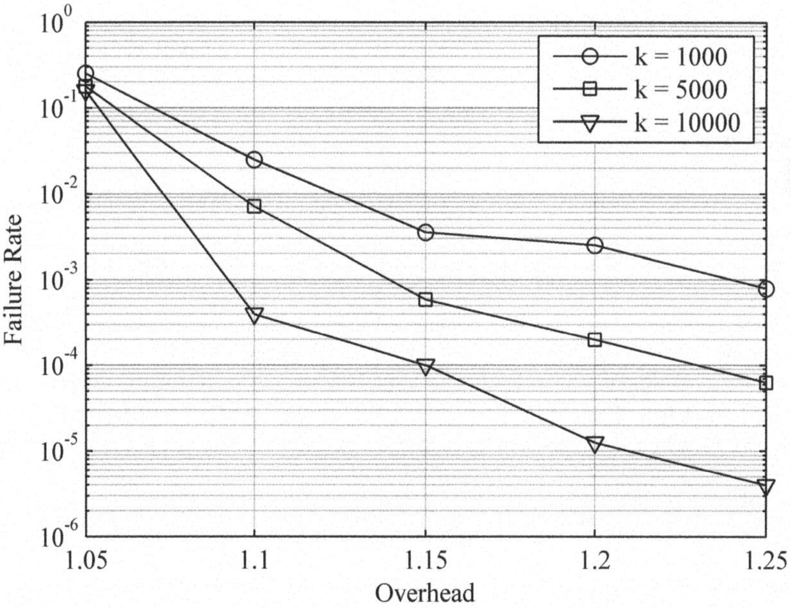

Fig. 2. The failure rate for optimized degree distributions of LT codes.

It follows from Fig. 2 that by increasing the number of information symbols of LT code for fixed overhead the failure rate is significantly reduced. For example, for $R = 1.1$, optimization of the degree distribution of LT code with $k = 10000$ ensures a decrease in the failure rate by almost 2 orders of magnitude compared to LT code with $k = 1000$. It should be noted that the obtained failure rate values for small-length LT code practically coincide with the results presented in [5], but the proposed approach has less computational complexity.

5 Conclusion

The use of Raptor codes allow to increase the quality of service the user of network-based intelligent transportation systems. The key element of Raptor codes are modified LT codes with specific degree distributions to ensure low failure rate with fixed overhead. The paper shows that the problem of finding the degree distribution of some LT code is a nonlinear programming problem.

The short LT codes play an important role in different telecommunication technologies upon implementation modern intelligent transportation systems. The proposed bio-inspired optimization method of the degree distributions of these LT codes allows improving their characteristics. The simulation showed that the use of the truncated degree distribution and cuckoo search procedure has reduced the computational complexity of optimization process. As a result of the research, optimized degree distributions for short LT codes with given number of source symbols were obtained. The experimental results showed that longer LT codes with optimized degree distributions has better the failure rate performance.

Thus, the approach to optimization rateless codes presented in the paper increases the efficiency and reliability of network-based intelligent transport systems.

References

1. Williams, B.: Intelligent Transport Systems Standards. Artech House Publishers, Norwood (2008)
2. Kythe, D.K., Kythe, P.K.: Algebraic and Stochastic Coding Theory. CRC Press, New York (2012)
3. Luby, M.: LT codes. In: The 43rd Annual IEEE Symposium on Foundations of Computer Science on Proceedings, Vancouver, BC, pp. 271–280. IEEE (2002)
4. Shokrollahi, A.: Raptor codes. IEEE Trans. Inf. Theory **52**(6), 2551–2567 (2006)
5. Chen, C.-M., Chen, Y.-P., Shen, T.-C., Zao, J.-K.: On the optimization of degree distributions in LT code with covariance matrix adaptation evolution strategy. In: IEEE Congress on Evolutionary Computation, Barcelona, pp. 1–8. IEEE (2010)
6. Zao, J.K., Hornansky, M., Diao, P.-L.: Design of optimal short-length LT codes using evolution strategies. In: 2012 IEEE Congress on Evolutionary Computation, Brisbane, QLD, pp. 1–9. IEEE (2012)
7. Yang, X.-S., Deb, S.: Cuckoo search via Lévy flights. In: 2009 World Congress on Nature & Biologically Inspired Computing (NaBIC), Coimbatore, pp. 210–214. IEEE (2009)

Improving Handover Mechanism in Vehicular WiFi Networks

Roman Yeryomin[1], Arnis Ancans[1(✉)], Ernests Petersons[1], and Anita Gerina-Ancane[2]

[1] Institute of Telecommunications, Riga Technical University, Azenes Street 12-201, Riga 1048, Latvia
roman.yeryomin@gmail.com, arnis.ancans@gmail.com, ernests.petersons@rtu.lv

[2] Institute of Mechanics and Mechanical Engineering, Riga Technical University, Kipsalas Street 6B, Riga 1048, Latvia
anita.gerina-ancane@rtu.lv

Abstract. Timely made handover decision between wireless access points (APs) in extended service set (ESS) WiFi network infrastructure with fast moving objects can reduce goodput degradation and interruptions of wirelessly transmitted data and consequently increase the average goodput and decrease the average response time of the network. Handover process has to occur on several layers (e.g. data link layer, network layer, application layer), therefore all these layers contribute to the common delay in the handover process. The 802.11k/r/v standards can be inspected together to effectively improve the handover performance. The handover issue in the data link layer can be mitigated using 802.11v standard to implement network assisted handover for directing poorly connected clients. This work presents recommendations and practical solution with improved handover procedure in WiFi based vehicular communication networks (VCNs).

Keywords: Vehicular communication network · WiFi handover · Standardisation

1 Introduction

In IEEE 802.11 networks with fast moving objects, where velocities are higher than 20 km/h, current handover mechanisms do not ensure the timely made handover decision. This issue becomes especially critical for WiFi based roadside vehicular communication networks (VCNs) where vehicles and their passengers use data transmission services. Such conclusion is made from experiments performed in field conditions [1, 2].

In the WiFi world handover, handoff and roaming, are all the same. Moving seamlessly from one BSS (Basic Service Set) to another is called handover or roaming. To switch BSSs seamlessly, the mobile client must recognize that it is nearing the cell boundary and that it needs to find other potential cells to move into before losing the signal completely. Handover is the process of moving an association from one AP to the next, so that the wireless connection is maintained as the mobile node moves.

E. Ginters et al. (Eds.): ICTE ToL 2019, LNITI, pp. 243–261, 2020.
https://doi.org/10.1007/978-3-030-39688-6_32

There are two types of handovers: homogeneous (horizontal) handovers and heterogeneous (vertical) handovers. Homogenous or horizontal handovers usually take place within a single access network. Mobility is localized and limited to the network range. Heterogeneous or vertical handovers happen across different network access technologies e.g. IEEE 802.11 to LTE [3]. In this article we consider the WiFi network and the related handover actions that take place within it.

IEEE 802.11 is a collection of physical layer (PHY) specifications and media access control (MAC) for implementing WLAN in the 2.4, 5 and 60 GHz frequency bands. They are maintained by the IEEE 802 LAN Standards Committee. The base version of wireless networking standard first was published in 1997. The standard and its amendments provide the fundamentals of WiFi technology. While each amendment is officially rescinded when it is incorporated in the latest version of the standard, the corporate world tends to market to the revisions because they concisely denote capabilities of their products. Consequently, each revision is preserved as a new standard in the networking community [4].

In practice, roadside vehicular communication infrastructure based on IEEE 802.11 will be integrated into two-rank network, such as, WiFi and LTE/5G or WiFi and fiber optical network, which will be integrated into the network of some Internet service provider (ISP). The WiFi network is access network, not counting specific point-to-point or point-to-multipoint WiFi solutions.

To successfully use the WiFi technology in VCNs improvements of existing handover algorithms are required or even more – completely novel design of handover solution.

The decision on the necessity to roam from the current WiFi coverage area to an adjacent area is made by the mobile client itself. There are special handover algorithms developed on client's side, which carry out all the necessary measurements and calculations for roaming. These algorithms are vendor proprietary solutions, not part of IEEE 802.11 standard and they are not publicly revealed. Thereby we can only determine the effectiveness of these algorithms by performing experimental measurements. However we cannot change or improve them. Usually there are only minimal customization possibilities and cannot be improved without deep collaboration with vendor which is typically impossible because of non technical reasons.

The handover process between access points (APs), in practice, observed as data traffic hollows, can be improved by using IEEE 802.11k [5], 802.11r [6] and 802.11v [7] amendments, now incorporated into the IEEE 802.11-2016 standard.

IEEE 802.11k/r/v standards are actively tested by many manufacturers of WiFi equipment, but they are not tested for VCNs. 802.11k/r/v standards are not mandatory still, therefore, the implementation of these standards into WiFi infrastructure hardware and customer devices is not common and are not being sufficiently encouraged.

Handover process has to occur on several layers of OSI (Open Systems Interconnection) model (e.g. data link layer, network layer, application layer). Thereby all these layers contribute to the common delay in the handover process. In order to ensure a seamless connectivity at application layer, different mechanisms can be used on different layers. IEEE 802.11 technology operates on physical and data link layers.

When roaming, mobile client in WiFi networks goes through several steps, and with each step it adds its contribution to the overall delay before a new link is established. Such steps as channel scanning, authentication, and association are executed before being attached to a new AP. Data link layer contributes the maximum to the overall delay and scanning operations contribute to most of the delay [8].

Handover algorithm can be improved by making mobile client roaming decision to a better AP timely. Thereby, it is possible to effectively minimize data traffic disruptions. Desirable improvements should be made by using the available IEEE 802.11 standard.

Reducing handover delay is important in vehicular context because it prevents the vehicle from spending a considerable portion of connection time, which already is small, in executing the handovers [9].

In this paper we will consider mechanisms used for improving the handover on the data link layer in relation with characteristics of roadside network. The goal of our work is to take a brief overview of IEEE 802.11k/r/v standards and enhance handover mechanism by preparing WiFi AP and mobile client to work with IEEE 802.11v. It is planned to perform experimental measurements and thereby to prove the benefit and effectiveness of 802.11v. Authors expect that it will increase common performance of WiFi network.

IEEE 802.11 based networks with improved handover mechanism can be used to cover specific areas ensuring better connectivity services for both automotive communication and vehicles' passengers' needs.

During the handover, both service continuity and quality of service (QoS) should be ensured to be seamless. In WiFi networks key performance metrics are throughput, packet loss, jitter, latency and response time. Authors will evaluate improvements in handover by conducting goodput and response time measurements in WiFi network.

The rest of this paper is organised as follows. Section 2 provides an overview of related works. Section 3 provides an overview of 802.11k/r/v standards. Section 4 describes proposal for handover algorithm improvement. Section 5 shows experimental setup and measurement scenarios. Measurement results are given in Sect. 6. Concluding remarks and future work are presented in Sect. 7.

2 Related Work

Several methods and practical solutions are offered for reducing problems related to handover in WiFi networks.

In paper [10] the authors design an efficient MAC protocol for drive-thru Internet specific to the special scenario in sparse highway environment and develop an accurate and scalable 2D Markov chain model to investigate the throughput performance under different driving speeds and number of vehicles. In turn experimental study [11] shows that people drive on familiar routes frequently, and thus the mobility and connectivity related information of their routes can be predicted with good accuracy by using historical information – such as GPS tracks with timestamps, radio frequency (RF) fingerprints, and link and network-layer addresses of visible APs. Authors use this information to develop new handoff and data transfer strategies. Experimental performance evaluation reveals that the predictability of mobility and connectivity is high enough to be useful in WiFi protocols. Article [12] is focused client mobility in Software-Defined Networking (SDN) for 802.11 standards. According to SDN principles, the authors proposed the enhanced architecture which resulted in merging multiple control channels into one for wired part of the network. The proposed approach significantly improves network's architecture in following points: network scalability, security and simultaneously supports seamless handover. Paper [3] takes a deeper look at predictive triggers at the link layer. With an appropriate implementation of predictive triggering algorithms, early handover initiation and preparation could be performed. A few statistical prediction methods have been studied thoroughly by using IMSL ARIMA (Autoregressive Integrated Moving Average) modules. They found that by applying ARIMA, they can predict future RSSI (Received Signal Strength Indicator) values with high accuracy, especially when only one or two steps ahead are predicted. In article [13] authors propose a Mobility and Signal Strength-Aware Handoff Decision (MSSHD) approach to predict the handover decision in WiFi networks and hence switching to the best preferable AP. The RSSI and the direction of mobile node parameters are considered as inputs to the Fuzzy logic system. Research [14] represents approach where new handover concepts for roadside networks using existing mechanisms of IEEE 802.11k/r/v are delivered in order to provide real-time capable handover mechanisms that can be used for VCNs' applications, especially in combination with value added services. Authors [15] are focused to theoretically evaluate 802.11k/r/v standards, analyse the functionality they provide, and assess their existing practical applications. Also the possible factors that influence standards adoption are evaluated. Researchers conclude that 802.11k/r/v support can increase overall performance of the wireless network.

Studying relevant articles, no experimental results of researches and implementation methods were found that shows expedience of IEEE 802.11v to solve the handover issues on data link layer, i.e. the timely made handover decision and increasing common performance of WiFi VCN networks. Therefore that is our aim and main motivation of our work.

3 Overview of 802.11k/r/v

IEEE 802.11k: Radio Resource Measurement of Wireless LANs allows 802.11k standard capable clients to request a neighbour report containing information about known neighbour APs that are candidates for roaming. The use of the 802.11k neighbour list can limit the need for active and passive scanning. The neighbour or candidate list includes information about BSS identifier (BSSID), channel, and operation details of the neighbouring radios. The neighbour list contains only neighbours in the same band with which the client is associated. Thereby, it reduces channel utilization. It reduces roam time and improves the decisions taken by the client. Additionally, it increases battery life of the device [16].

IEEE 802.11r: Fast BSS Transition addresses fast roaming. The initial handshake with the new AP occurs before client roams to the target AP, called as Fast Transition (FT). Initial handshake allows the client and APs to do Pairwise Master Key (PMK) calculation in advance. Once the client performs the re-association request or response exchange with the new AP, the PMK keys are applied to the client and AP. The FT key hierarchy allows clients to make fast BSS transitions between APs without the need for re-authentication at every AP. 802.11r eliminates the handshake overhead while roaming and thereby reduces the handover times between APs, which provides security and QoS [16, 17].

IEEE 802.11v: 802.11 Wireless Network Management provides information to discover the best available access point. Standard enables the WiFi network to send messages to associated clients, to offer better APs to associate with. This is useful for both load balancing and in directing poorly connected clients. 802.11v BSS Transition is applied to the following three scenarios: solicited request – client can send an 802.11v BSS Transition Management (TM) Query before roaming for a better option of AP to re-associate with a client; unsolicited load balancing request – if an AP is heavily loaded, it sends out an 802.11v BSS Transition Management Request (TMR) to an associated client; unsolicited optimized roaming request – if a client's RSSI and rate do not meet the requirement, AP sends out an 802.11v BSS Transition Management Request to this client. 802.11v BSS Transition Management Request is a suggestion given to client. Client can make its own decision whether to follow the suggestion or not. To force disassociating a client, it is possible to turn on the Disassociation Imminent function. This function will disassociate the client after a period of time if the client does not re-associate to another AP [16]. In this way 802.11v can reduce the handover times between APs.

4 Handover Mechanism Improvement

The decision to roam is driven by the wireless client NIC (Network Interface Card) driver – not by the AP. Wireless clients decide that it is time to roam based on a variety of conditions. The IEEE 802.11 standard does not address this at all; therefore roaming algorithms are vendor specific. Since the roaming algorithms are usually "secret recipes", the exact thresholds and conditions are hidden. Some of the ingredients in the roaming algorithm are the received RSSI, SNR (Signal to Noise Ratio), a count of missed AP beacons, errors due to collisions or interference, etc.

Some clients do not roam until connection is lost completely. Most of clients' NICs have advanced options. By selecting such parameters as *Roaming Aggressiveness* that impact RSSI threshold (TH) value and *Transmit Power* adaptation, is possible to optimize roaming procedure affecting how quickly the client will try to roam away from an AP association. Hence different clients use different thresholds; some will try to roam earlier than others at a given location within a cell. Some clients will tend to stay to an existing association until the AP can hardly be heard, whereas others will attempt to roam whenever a better AP is discovered [18]. However in VCNs these available options on client side do not solve handover delay problem.

One way to solve a discussed problem would be to introduce new additional handover metrics and other – enhance handover mechanism in conjunction with IEEE 802.11k/r/v.

In conventional handover RSSI and SNR are the major parameters used. As earlier experiments showed, on the client's side changing the RSSI threshold values using the *Roaming Aggressiveness* options, traffic disruptions practically remain the same. This method gave minor improvements at higher mobile client velocities.

4.1 Proposed Handover Mechanism

The goal is to create a mechanism which would allow WiFi network to give a command to the mobile client to disassociate from current AP and make handover to a "better" AP. The main questions are when to roam and where.

Proposed solution of handover mechanism improvement implies the use of IEEE 802.11v standard. 802.11k and 802.11r standards were not used in the implementation due to the high workload. Their practical implementation is the goal of further research.

Proposed solution is based on 802.11v unsolicited optimized roaming request scenario, that is assisted and directed handover which would allow the network to explicitly command a client to handover to a designated new AP.

A number of modified IEEE 802.11v handover procedures were implemented and tested for the effectiveness.

It should be noted that in the case of a standard handover, the decision of the moment of switching and choosing the new AP, is up to the client.

At first, a solution was developed where the handover decision is made on the network side. It was named *802.11v network directed (manual)* handover. This modification was implemented for manual execution. In addition, an automatic *802.11v network directed (auto)* handover solution was tested. However, this turned out to be quite unstable as this solution lacks information about received signal on client's side.

Eventually a solution was developed where the handover decision is made on the client side in combination with network support. The network indicates to the client which AP client must switch to. The solution was named *802.11v network assisted* handover. Here the client or station (STA) roams between APs automatically.

For comparison in Fig. 1 two previous mentioned generalized algorithms of 802.11v handover procedures are summarized. As well as the algorithms are supplemented with standard handover procedure. In our experiments these algorithms work separately.

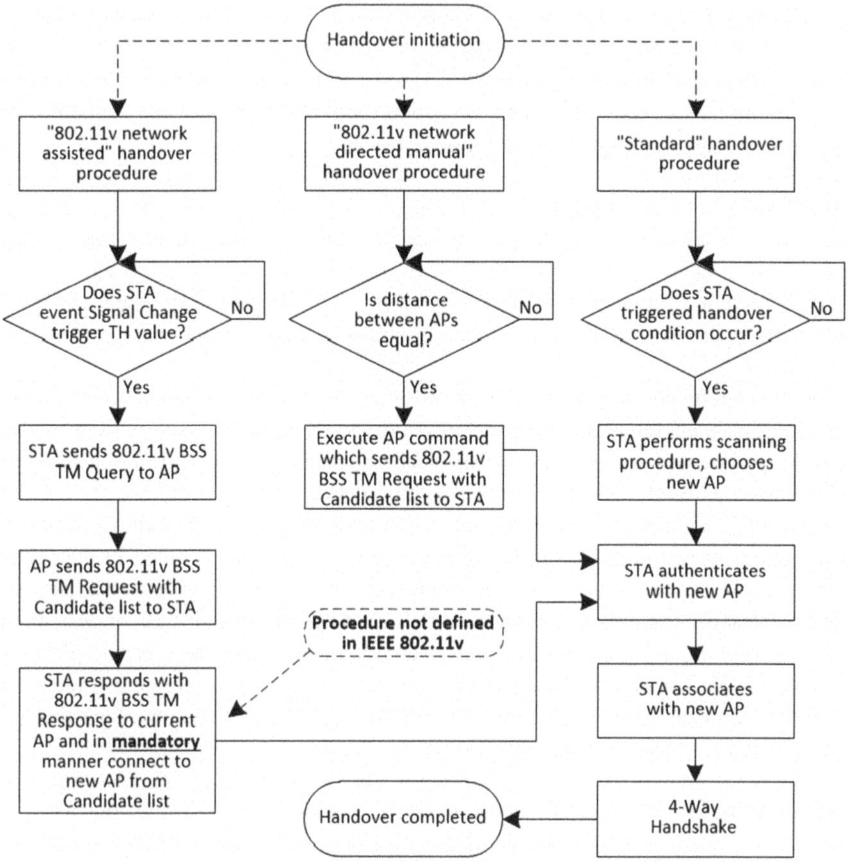

Fig. 1. Summarized algorithms of handover procedures.

The main difference and efficiency of the developed *802.11v network assisted* handover solution compared to the standard handover solution lies in two aspects: (1) the current AP sends to the client information about the next AP, so the client does not have to spend time on network scanning; (2) the current AP sends out a command to make a mandatory roam to the new AP.

4.2 Equipment Programming

Since WiFi network infrastructure's hardware and client's hardware by default does not support 802.11v standard, it had to be programmed.

The client and Wireless LAN ESS use TP-Link Archer C7 (AC1750) routers with OpenWrt OS [19]. There are four routers in total. In all measurements all routers were configured and enabled to use IEEE 802.11ac standard.

WiFi infrastructure hardware was prepared and a demo client created based on 802.11v standard, *hostapd* was configured on the AP routers and the *wpa_supplicant* was configured on the client router. 802.11v BSS Transition Management Request frame structure was repaired, which was incorrectly formed. 802.11v standard uses action frame, which is management frame subtype.

In the case of a *standard handover*, the customer can obtain information about the surrounding APs by scanning the wireless network in two ways. In the first case, the client simply listens for any management beacon frames (management frame subtype) broadcast from nearby APs (passive scan). In the second case, the client can send out the management probe request frames and the surrounding APs send out probe response frames (active scan). When receiving these messages, the client decides which AP to switch to.

For measurements with a standard handover client, only the standard parameters were configured: channel number, channel width, SSID name and password. The client handover mechanism was not modified. In all measurements with different handover mechanisms was used static IP address for the client.

In the *802.11v network directed (manual)* handover implementation, both the AP and the client were modified. On the client side the code was changed in way that it makes mandatory to execute handover when receiving a BSS Transition Management Request message containing candidate AP list. When the mobile client was in the middle between the APs, a command was executed manually on the current AP to send the client a BSS Transition Management Request message. The message contains the BSS Transition Candidate List or the next AP MAC address. Thus the client roams to the new AP when this message is received.

In the *802.11v network assisted* handover implementation, the client receives the broadcast Management Beacon Frames from the associated AP, sent by the AP every 100 ms. From these messages the RSSI level for the connection to the AP is obtained. On the client side, the RSSI threshold level of −62 dBm was configured. This is quite high as the distance between the APs was only 100 m. If client triggers a low RSSI signal level twice when moving within the AP coverage area, the mobile client processes and sends 802.11v BSS Transition Management Query message. Upon receipt of this message, the AP generates the 802.11v BSS Transition Management Request message. The BSS Transition Candidate list field in this message contains the next AP BSSID MAC address where the client will have to switch to. A database with the next AP's MAC address was placed on each AP. The AP Candidate list includes only the next AP for assuring a minimal handover decision delay. The client, upon receipt of such message from the current AP, responds with the 802.11v BSS Transition Management Response message and immediately, on a mandatory basis, switches to the next AP. This procedure is not defined in the IEEE 802.11v standard and was specially programmed.

5 Experimental Setup

Experimental goodput and response time measurements of vehicular communication roadside WiFi network were taken in real field conditions. Experiments were realized for three scenarios: using *standard* handover implementation (A); implementation with developed *802.11v network directed (manual)* handover (B); and implementation with developed *802.11v network assisted* handover method (C). For these experiments a vehicle moving at various velocities was used: 20, 50, 70 and 90 km/h. Vehicle moved along a straight 260 m long road section where experimental testbed was located. At least three measurements at each velocity were taken.

Goodput and response time measurements were performed and results were fixed using IxChariot software. In our case, the mobile client end device and the server were business-class laptops (i5 CPU, 8 GB RAM, NIC 802.11n/ac) which ran IxChariot software accordingly. Uninterruptible power supply (UPS) devices were used as power sources to ensure the operation of the testbed network for at least three hours.

The testbed was built from three attached APs, a server and a mobile client. Three TP-Link Archer C7 routers with pre installed OpenWrt software configured in AP mode served as the APs. Mobile client's set consisted of laptop and a TP-Link Archer C7 router with pre installed OpenWrt software which were configured in STA (station) mode.

In overall were used four UPS devices both for WiFi roadside infrastructure and also one UPS for mobile node, which was placed in the car. For the experiments a VW e-UP! electric car was used, which provided excellent acceleration to reach required velocity within relatively short distance.

5.1 Scenario A: Standard Handover

In this scenario the testbed which already is described in detail in the beginning of section five was used. Unmodified, standard configuration APs and STA were used.

In Fig. 2 the testbed scheme of first experiment for VCN based on WiFi technology is shown.

5.2 Scenario B: 802.11v Network Directed Handover

In this scenario the testbed which is already described in the beginning of section five was used. The WiFi network routers (AP) and mobile client router (STA) were modified by implementation of IEEE 802.11v standard and adding previously described *802.11v network directed (manual)* handover functionality. It should be noted that markers were placed in the testbed in the middle between the APs. When the mobile client was directly in front of the marker, i.e. in the middle, a command was executed manually to the current AP, which sent a BSS Transition Management Request message to the mobile client. In Fig. 3 the testbed scheme of second (B) experiment for VCN based on modified WiFi technology is shown.

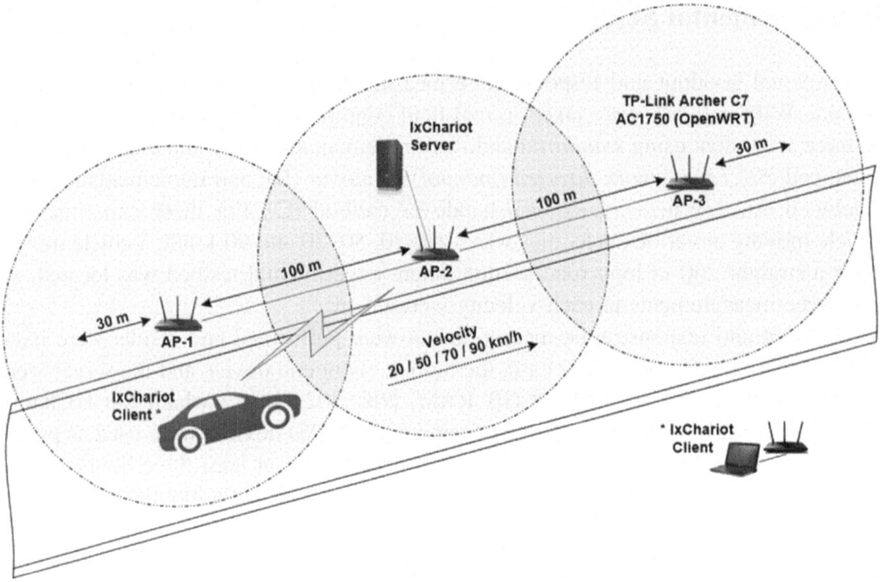

Fig. 2. Testbed topology for standard VCN based on 802.11ac.

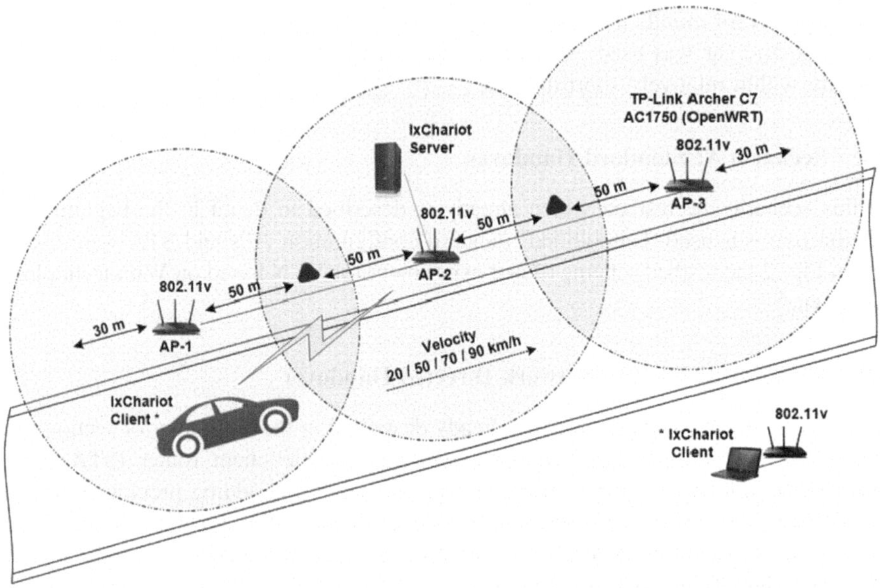

Fig. 3. Testbed topology for modified VCN based on 802.11ac and 802.11v.

5.3 Scenario C: 802.11v Network Assisted Handover

In this scenario the testbed which is already described in the beginning of section five was used and depicted in Fig. 3. Unlike scenario B, here the handover procedure takes place completely automatically and roadside markers are not required. The WiFi network routers (AP) and mobile client router were modified by implementation of IEEE 802.11v standard and adding previously described *802.11v network assisted* handover functionality.

6 Measurement Results

In this section experimentally obtained network performance parameters goodput, response time and handover delay are estimated.

6.1 Goodput

For comparison in Figs. 4, 5, 6 and 7 experimentally obtained goodput graphs for standard handover and *802.11v network assisted* handover mechanisms at different vehicle moving speeds over WiFi network are shown. In case with *802.11v network assisted handover*, the switching process between APs, which are marked with circles on horizontal axis, is performed pretty accurate.

In case with *802.11v network directed (manual)* roaming (not shown), the switching moments or roaming moments are perfect because they are controlled and directed by experiment performer and are executed almost exactly in the middle between APs ($\sim$80 m and $\sim$180 m). In *standard* roaming case the roaming is delayed or vague.

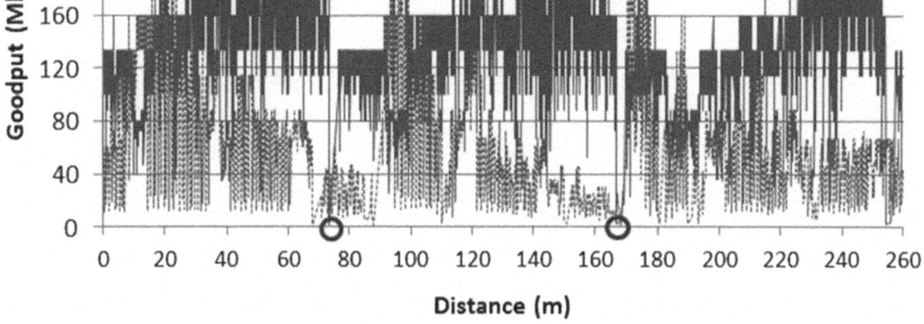

Fig. 4. Goodput at the velocity of 20 km/h.

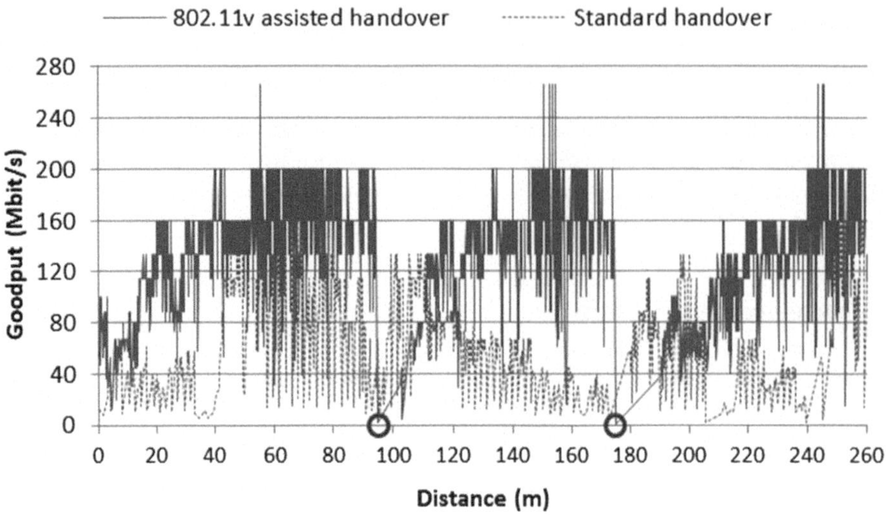

Fig. 5. Goodput at the velocity of 50 km/h.

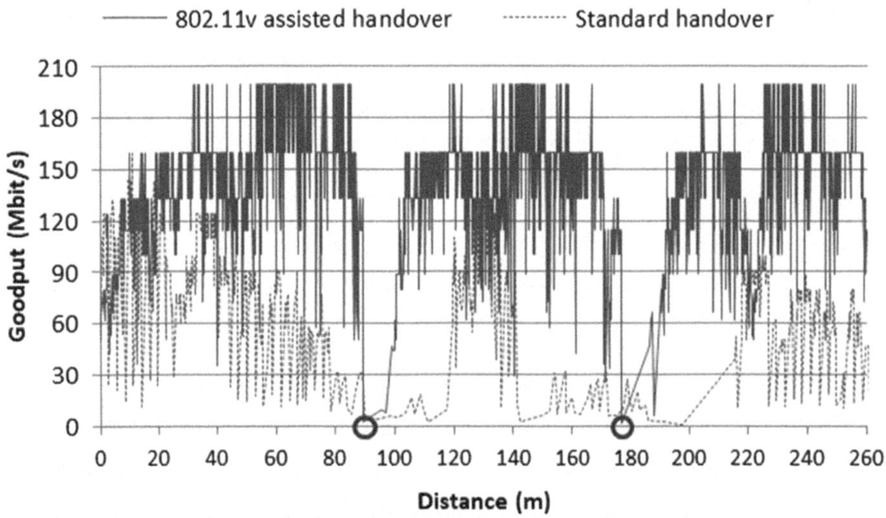

Fig. 6. Goodput at the velocity of 70 km/h.

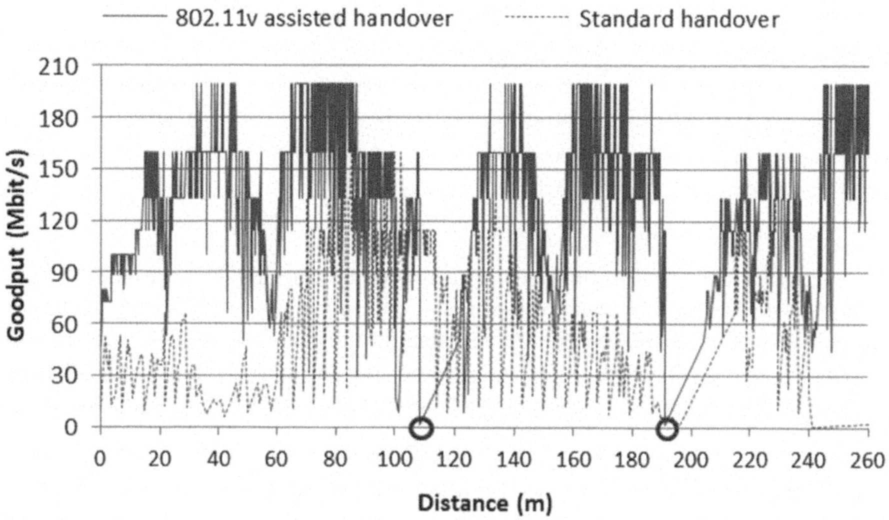

Fig. 7. Goodput at the velocity of 90 km/h.

In Fig. 8 goodput curves are given, which are obtained by calculating goodput average values at different moving velocities and by using different handover methods.

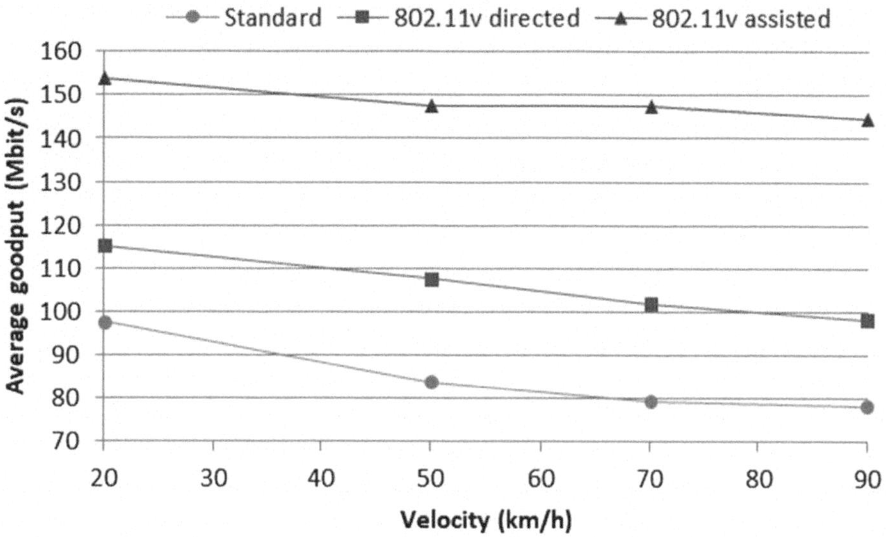

Fig. 8. Average values of goodput.

In Table 1 the evaluation of the nature of obtained goodput curves are given. Average values of goodput curves between all driving speeds were calculated and such distribution parameters as standard deviation and coefficient of variation were estimated.

Table 1. Average goodput and distribution parameters.

Handover method	Average goodput, Mbit/s	Standard deviation, σ	Coefficient of variation, %
Standard	84.80	7.70	9.09
802.11v directed (manual)	105.67	6.42	6.07
802.11v assisted	148.44	3.43	2.31

The results show that the highest average goodput value and the lowest distribution parameters are obtained by the developed *802.11v network assisted* handover solution. Particular attention should be paid to distribution indicators as they characterize the relationship of how goodput value varies with movement velocity. The smaller the parameters, the smaller the decrease in the goodput value at higher movement velocities.

6.2 Response Time

Network response time was estimated experimentally with the IxChariot software. Network response time is the total time taken from mobile client making a request till the time it receives a response. Response time can be affected by changes to the processing time of remote server and by changes in latency, which occurs due to changes in network hardware resources or utilization. Handover procedure and its delay must be taken into account, which increases the total network latency. Therefore response time consists of the sum of latency, processing time and handover delay.

In Fig. 9 response time curves are given, which are obtained by calculating average values of response time at different moving velocities and by using different handover methods.

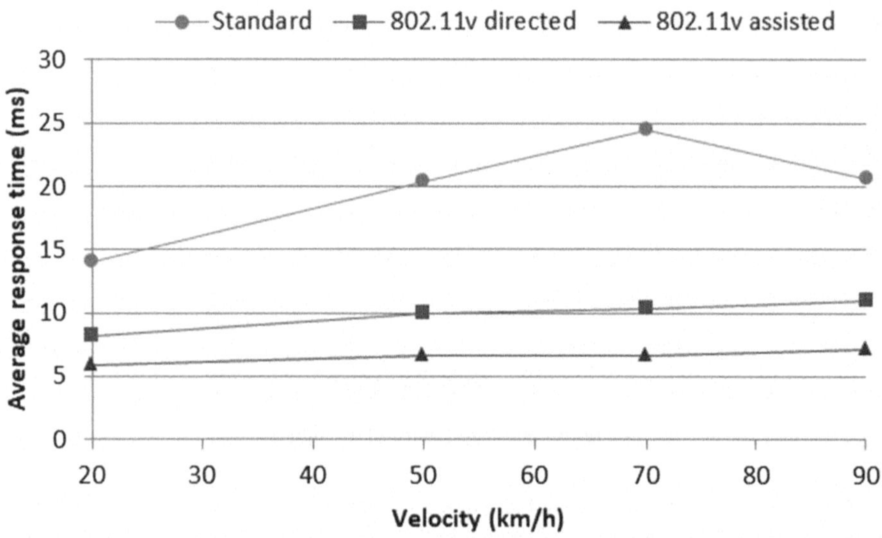

Fig. 9. Average values of response time.

As the mobile client speed increases, network response time also increases. The drop in the standard handover curve at 90 km/h can be explained by the fact that in this case the mobile client did not switch between all APs. Each handover procedure increases the overall response time value. This demonstrates the problem that, with the standard handover procedure, when driving at speeds above 90 km/h, there is a problem of switching correctly between APs.

In Table 2 the evaluation of the nature of the response time curves is given. Average values of response time curves between all driving speeds were calculated and such distribution parameters as standard deviation and coefficient of variation were evaluated.

Table 2. Average response time and distribution parameters.

Handover method	Average response time, ms	Standard deviation, σ	Coefficient of variation, %
Standard	19.87	3.75	18.89
802.11v directed (manual)	9.90	1.04	10.54
802.11v assisted	6.60	0.44	6.73

As we can see, the lowest average values of response time and the lowest dispersion indicators are obtained by the developed *802.11v network assisted* handover solution.

6.3 Handover Delay

Examining the response time graphs obtained experimentally with the IxChariot software, approximate handover procedure duration (handover delay) was determined. In the graphs, the handover delay reflects as the increase in the response time value (peak) and this moment coincide with the drops in traffic observed in the goodput graphs. As an example, in Fig. 10 the response time graph for 802.11v network assisted handover is given at moving velocity 70 km/h (see also Fig. 6).

In Fig. 11 experimentally estimated average handover delay between AP1 and AP2, as well as between AP2 and AP3 using different handover methods at different velocities is shown. Three measurements were taken at each driving speed.

In Table 3 the evaluation of the nature of obtained handover delay curves are given. Average values of handover delay curves between all driving speeds were calculated and distribution parameters like standard deviation and coefficient of variation were evaluated. From obtained results we can see, that the smallest average value of handover delay is using *802.11v network assisted* handover solution. Higher handover delays come from authentication, association and 4-way handshake operations that all refers to connection setup time. It has to be noted that the handover delay values were obtained from response time measurements, therefore, they also could be affected by both the network latency and processing time delays of client and server.

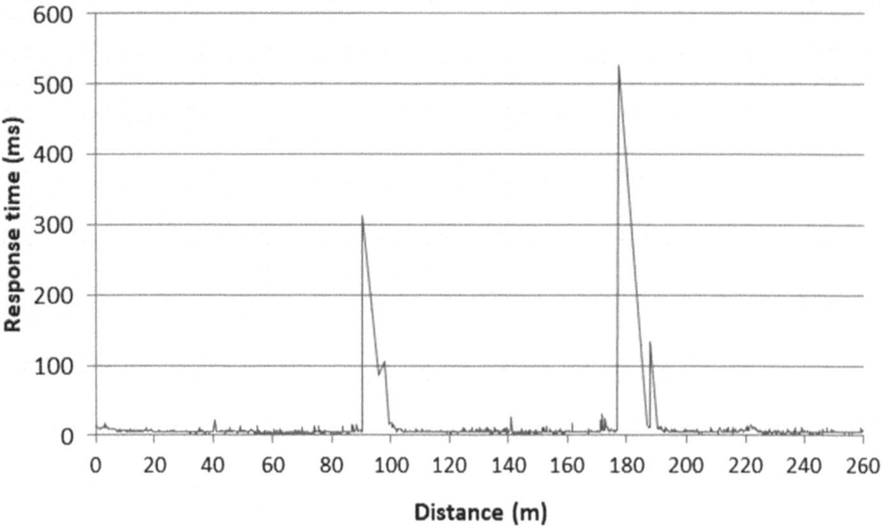

Fig. 10. Response time at the velocity of 70 km/h.

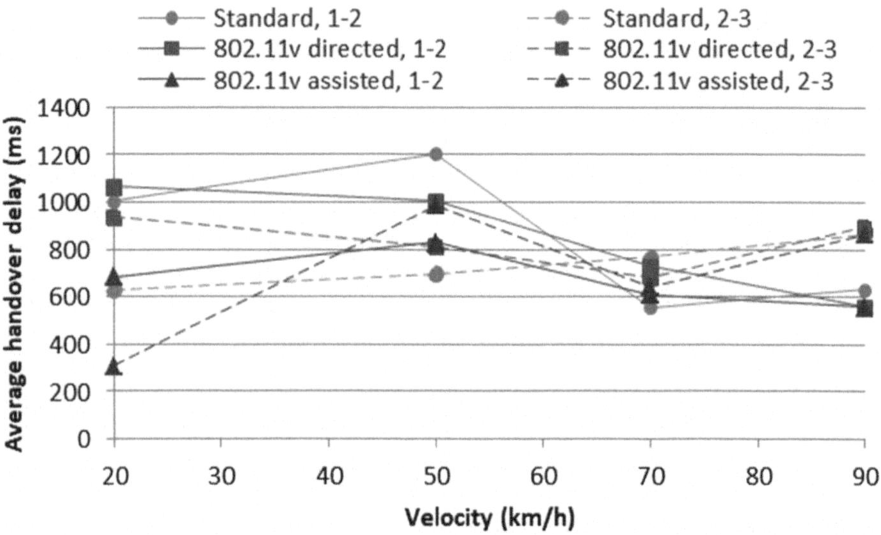

Fig. 11. Handover delay between APs using different handover methods.

Table 3. Average handover delay between APs and distribution parameters.

Handover method between APs	Average Handover delay, ms	Standard deviation, σ	Coefficient of variation, %
Standard, 1-2	848.42	266.61	31.42
Standard, 2-3	739.79	87.92	11.88
802.11v directed (manual), 1-2	838.67	205.20	24.47
802.11v directed (manual), 2-3	831.08	97.67	11.75
802.11v assisted, 1-2	669.25	102.00	15.24
802.11v assisted, 2-3	700.58	257.88	36.81

Several other methods can be used to determine the handover delay parameter with high accuracy. For example, the logging for client MAC address association/disassociation events with timestamp turned on must be enabled on APs or STA of client. 802.11 packet capturing also can be performed, and further analysis of the obtained data using the client association/disassociation action frame must be performed using Wireshark or a similar application.

7 Conclusion

Conducted experiments show that the standard WiFi technology must be improved in order to successfully use it for automotive communication network needs. Technology is cheap, widely available and it uses unlicensed frequency bands. It is a wireless network, which we can manage ourselves.

The handover decision must involve both parties the WiFi network and the mobile client. The network may provide an answer to which AP to switch and the client may provide a response at which moment to execute the handover procedure.

The incorporated former amendments IEEE 802.11k/r/v in the IEEE 802.11 standard are optional and thus do not oblige WiFi hardware manufacturers to fully incorporate the standard functionality into their products. That is the interest of the manufacturers themselves. The ESS Disassociation Imminent flag in the BSS Transition Management Request messages only recommends that the client switches to another AP. The client may also disregard this recommendation and not switch to the better (next) AP that is listed in the Candidate List field. It is therefore necessary to introduce a new flag in the BSS Transition Management Request messages, the existence of which would indicate that forced roaming is mandatory. This procedure is not defined in the IEEE 802.11v standard, and it is a new concept for the handover.

The best goodput and response time results with the lowest dispersion indicators at different vehicle moving velocities are obtained by the method with automatic *802.11v network assisted* handover solution. It was also shown that this mode can provide an accurate switching time (moment) between infrastructure APs. Consequently the handover decision becomes more efficient. The proposed solution will also be applicable to 802.11ax networks.

Because the BSS Transition capability and the improvement of 802.11v standard was proved to be beneficial and effective, which promotes the acceleration of handover process and the increase of average value of goodput, and the decrease of average value of response time, authors think that the proposed handover solution is perspective and our next work will be dedicated to its further development.

In future studies it is planned to test the functionality of neighbour report described in 802.11k standard and functionality of Fast BSS Transition described in 802.11r standard. Theoretical evaluation shows that the combination of these standards can be an effective solution for the handover process.

Main factors of handover are low latencies (delay) for handover and connection setup time. The highest increase of network response time parameter brings exactly the handover procedure. Handover delay in the vehicular IEEE 802.11 networks must be reduced for uninterrupted use of data transfer services from vehicles.

It is also necessary to find out the direction of motion and velocity, so that the Wi-Fi network could timely provide the list of candidate APs. Methods that can be used: online GPS data, GPS predictive tendency (inertial navigation), AP triangulation, etc. For known road sections, a precompiled AP database can be used, and skipping the scan operation can speed up client switching between APs.

In order to reduce the handover delay in vehicular WiFi networks by the methods described above, and to facilitate the maintenance of the wireless network, a centralised design model of network management should be implemented.

There are planned future study to verify the performance of improved WiFi VCN which will be integrated with LTE/5G network.

References

1. Ancans, A., Bogdanovs, N., Petersons, E., Ancans, G., Umanskis, A., Vishnevskiy, V.: Evaluation of Wi-Fi and LTE integrated channel performance with different hardware implementation for moving objects. Procedia Comput. Sci. **104**, 493–500 (2017)
2. Ancans, A., Bogdanovs, N., Petersons, E., Ipatovs, A.: Integrated wireless network performance estimation for moving vehicles. In: RTUWO 2017, Riga, pp. 203–207 (2017)
3. Chien, S.F., Liu, H., Low, A.L.Y., Maciocco, C., Ho, Y.L.: Smart predictive trigger for effective handover in wireless networks. In: IEEE International Conference on Communications 2008, Beijing, pp. 2175–2181 (2008)
4. Abdelgader, A.M.S., Lenan, W.: The physical layer of the IEEE 802.11p WAVE communication standard: the specifications and challenges. In: WCECS 2014, San Francisco, vol. 2, pp. 691–698 (2014)
5. IEEE Computer Society: IEEE Standard for Information technology – Telecommunications and information exchange between systems – Local and metropolitan area networks – Specific requirements, Part 11: Wireless LAN Medium Access Control (MAC) and Physical Layer (PHY) Specifications Amendment 1: Radio Resource Measurement of Wireless LANs, IEEE Std 802.11k-2008 (Amendment to IEEE Std 802.11-2007), 12 June 2008

6. IEEE Computer Society: IEEE Standard for Information technology – Telecommunications and information exchange between systems – Local and metropolitan area networks – Specific requirements, Part 11: Wireless LAN Medium Access Control (MAC) and Physical Layer (PHY) Specifications, Amendment 2: Fast Basic Service Set (BSS) Transition, IEEE Std 802.11r-2008 (Amendment to IEEE Std 802.11-2007 as amended by IEEE Std 802.11k-2008), 15 July 2008
7. IEEE Computer Society: IEEE Standard for Information Technology – Telecommunications and information exchange between systems Local and metropolitan area networks – Specific requirements, Part 11: Wireless LAN Medium Access Control (MAC) and Physical Layer (PHY) specifications, Amendment 8: IEEE 802.11 Wireless Network Management, IEEE Std 802.11v-2011 (Amendment to IEEE Std 802.11-2007), 9 February 2011
8. Dutta, A., Schulzrinne, H.: Mobility Protocols and Handover Optimization: Design, Evaluation and Application. Wiley-IEEE Press, Hoboken (2014)
9. Hasan, S.F., Siddique, N., Chakraborty, S.: Intelligent Transport Systems: 802.11-based Vehicular Communications, 2nd edn. Springer, Heidelberg (2018)
10. Li, B.Z., Luan, T.H., Hu, B., Chen, S.Z.: Efficient MAC protocol for drive-thru Internet in a sparse highway environment. IET Commun. **11**(3), 428–436 (2017)
11. Deshpande, P., Kashyap, A., Sung, C., Das, S.R.: Predictive methods for improved vehicular WiFi access. In: Proceedings of the 7th ACM International Conference on Mobile Systems, Applications, and Services, pp. 263–276 (2009)
12. Bencel, R., Kostal, K., Kotuliak, I., Ries, M.: Common SDN control channel for seamless handover in 802.11. In: 2018 Wireless Days (WD), pp. 34–36 (2018)
13. Sadiq, A.S., Abu Bakar, K., Ghafoor, K.Z., Gonzalez, A.J.: Mobility and signal strength-aware handover decision in mobile IPv6 based wireless LAN. In: International Multi Conference of Engineers and Computer Scientists, vol. 1, pp. 664–669 (2011)
14. Meschke, R., Krohn, M., Daher, R., Gladisch A., Tavangarian, D.: Novel handoff concepts for roadside networks using mechanisms of IEEE 802.11k & IEEE 802.11v. In: ICUMT 2010, pp. 1232–1238 (2010)
15. Sanchez, M.I., Boukerche, A.: On IEEE 802.11k/r/v amendments: do they have a real impact? In: IEEE Wireless Communications, vol. 23, no. 1, pp. 48–55 (2016)
16. Enterprise Mobility 8.1 Design Guide, 802.11r, 802.11k, 802.11v, 802.11w Fast Transition Roaming. https://www.cisco.com/c/en/us/td/docs/wireless/controller/8-1/Enterprise-Mobility-8-1-Design-Guide/Enterprise_Mobility_8-1_Deployment_Guide/Chapter-11.pdf. Accessed 21 Sept 2019
17. Dondurmacioglu, O.: Handover mechanics for Wi-Fi phones, Rev. 1. https://community.arubanetworks.com/t5/Community-Tribal-Knowledge-Base/Handover-mechanics-for-Wi-Fi-phones/ta-p/25696. Accessed 21 Sept 2019
18. Hucaby, D.: CCNA Wireless 200-355 Official Cert Guide. Pearson Education, Cisco Press (2016)
19. OpenWrt Wiki. https://wiki.openwrt.org/doc/howto/wireless.utilities. Accessed 21 Sept 2019

Ultra Wideband Communication Technology in the Transport and Logistics Systems

Aleksandr Serkov[1], Natalya Panchenko[2], Karina Trubchaninova[2(✉)], and Maxim Kurtsev[2]

[1] National Technical University "Kharkiv Polytechnic Institute", 2, Kyrpychova Street, Kharkiv 61002, Ukraine
aleksandr.serkov@hotmail.com

[2] Ukrainian State University of Railway Transport, Feierbakh Square 7, Kharkiv 61050, Ukraine
{panchenko074,kurtsev_m}@ukr.net, karyna.trubchaninova@kart.edu.ua

Abstract. The concept is based on the ultra-wideband technology, which consists in the transmission of low-power coded pulses in a very wide frequency band without carrier frequency. The task is to ensure sustainable and safe operation of mobile transport objects. The goal is to increase the capacity and interference immunity of control and communication systems under the action of natural and intentional interference in transport and logistics systems. The use of the Gaussian monocycle with information coding by means of temporary position-pulse modulation as a code signal ensures high volumes and speeds of information transmission with high noise immunity of the communication channel and its protection from interception is justified. It is shown that for the organization of independent channels in one frequency band it is advisable to use a system of orthogonal codes. A technical solution for the design of an ultra-wideband receiving – transmitting antenna system is proposed. Work with low radiated power allows for stable communication in transport and logistics systems.

Keywords: Transport communications · Position-pulse modulation · Ultra-wideband technology

1 Introduction

At the present stage, to include transport in the logistics system, informatization of transport and provision of telecommunication technologies are required. Analysis of existing technologies using wireless communications in logistics shows that the most promising range for use is the frequency range from 0.5 to 11 GHz. The use of this range is due to the low attenuation of signals in the propagation medium. The use of precisely the range of ultra-wideband communications ensures stable and safe operation of mobile transport facilities, as a whole leads to an increase in the bandwidth capability and noise immunity of control and communication systems under the influence of natural and deliberate interference in transport and logistics systems (TLS). Thus, the task of ensuring the electromagnetic compatibility (EMC) requirements of

E. Ginters et al. (Eds.): ICTE ToL 2019, LNITI, pp. 262–270, 2020.
https://doi.org/10.1007/978-3-030-39688-6_33

mobile wireless devices becomes urgent, due to the requirements for a high density of communication channels per square meter of the working area, their bandwidth and noise immunity under unintentional and deliberate interference. The peculiarity of ensuring the requirements of EMC in wireless communication systems is due to the need to organize radio access of many users to a limited resource - transmission medium. There are a number of methods based on the distribution of parameters such as frequency, time, code and space between individual mobile communications with a minimum of mutual interference and maximum use of the characteristics of the transmission medium [1]. However, the main method of providing EMC in wireless communication systems is to reduce the level of electromagnetic fields [2, 3]. Thus, the aim of the work is to develop ultra-wideband communication technology to ensure the electromagnetic compatibility requirements of wireless mobile systems.

2 Problem Analysis and Statement

Mobile radio communication between mobile objects is carried out when one or both mobile devices move, and occupy a random position relative to each other. In this case, the parameters of the communication channel change in time, since moving in space leads to a change in the propagation conditions of the signal. In most cases, the mobile communication channel is characterized by the lack of direct visibility between the transmitter and receiver, especially in dense urban areas. According to the results of reflection, scattering and diffraction of electromagnetic waves in the interaction with various objects in space, *multipath propagation* occurs. The superposition of these waves leads to a change in the amplitude and phase of the signal and creates a complex electromagnetic environment at the receiving site.

The movement of the subscriber in space also causes a *Doppler shift*, which leads to a change in time of the amplitude and phase of the received signal. The magnitude of the Doppler shift is proportional to the transmission frequency and speed. Even movement over short distances, which are comparable to the wavelength of the radiation signal, can cause significant changes in its parameters.

Objects that arise along the signal propagation path and limit the direct visibility between the transmitter and receiver cause *eclipse and path loss*, changing the signal parameters over time.

Signal distortion caused by dimming and path loss is usually compensated by signal power control systems, increasing it. However, this significantly complicates the electromagnetic environment at the receiving site, it does not allow to meet the requirements of electromagnetic compatibility of mobile systems. Compensation of signal distortions caused by Doppler shift and multipath propagation of radio waves is a more difficult task, which requires complex signal processing at both the receiving and transmitting sides.

The consequence of multipath propagation of radio waves is the distortion of the shape of the received signal. It is characteristic of any type of signal, but becomes especially dangerous in the case of the use of broadband signals. This is due to the fact that due to interference, the individual frequencies are in phase, which leads to an increase in the signal level, and vice versa, the out-of-phase composition leads to its weakening.

In this case, two extreme cases are distinguished. In the first, the maximum delay between different signals does not exceed the duration of one character. Thus, interference occurs within a single character. Otherwise, the maximum delay between different signals exceeds the duration of one symbol, and based on the interference results, signals representing different symbols are added together, causing Intersymbol interference. It most significantly affects the distortion of the signal, since the amplitude and phase of the signal change for different symbols, which makes it impossible to reproduce the initial state.

Thus, the main task in meeting the electromagnetic compatibility requirements of wireless mobile systems is to reduce the level of electromagnetic interference and compensate for signal distortions caused by intersymbol interference and multipath propagation of radio waves.

3 Ultra Wideband Communication Technology

The use of ultra-wideband communication technology, in which the ratio between the frequency bandwidth of the signal and its center frequency is greater than unity, allows to reduce the level of electromagnetic interference environment [4, 5]. In this case, the information signal is distributed evenly in a continuous frequency-time space, and the essence of the technology is the transmission of low-power encoded pulses in a very wide frequency band without a carrier frequency. The most attractive for wireless communications, due to minimal loss in free space, is the frequency band of 2–10 GHz. The use of a Gaussian Monocycle with subsequent encoding of information by means of temporal position-pulse modulation (PPM) is proposed as a code signal. Thus, each information bit is encoded by a temporary shift of the pulse train (chips) relative to the reference one (see Fig. 1), depending on whether it is encoded—zero or one [6]. The offset value is usually a quarter of the time duration of the encoding chip.

Moreover, each information bit is encoded by several hundred ultrashort pulses - chips that arrive with a specific sequence and are emitted directly into free space. The encoding of an information bit by a chip sequence is limited by the following requirements. On the one hand, in order to avoid the effect of intersymbol interference, the arrival period of the T_b chips should be large enough so that the energy of the received chip has time to completely dissipate before the arrival of the next chip.

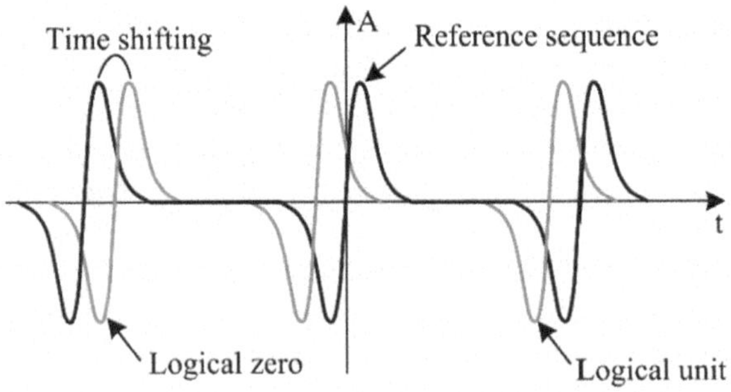

Fig. 1. Illustration of the PPM encoding method.

The period of arrival of T_b chips strives to reach the level of information bit duration, which in modern error-correcting codes (see Fig. 2) is within 1–4 s [6]. At the same time, the number of coding chips in the information bit decreases, correspondingly reducing its energy.

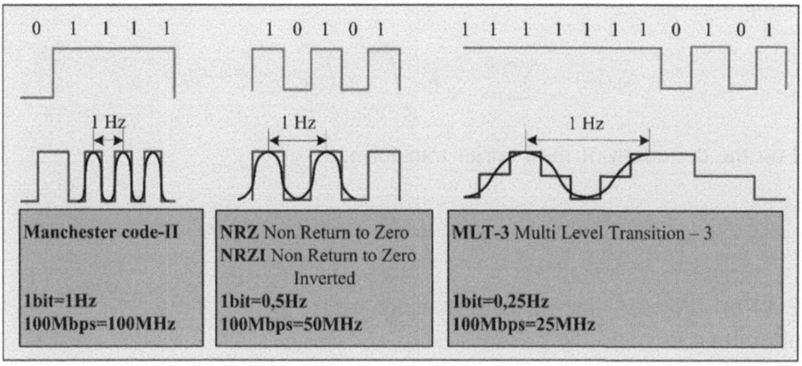

Fig. 2. Illustration of the dependences of the period of receipt of information bits on the encoding method.

On the other hand, in order to restore an information bit, the period of arrival of coding chips should be significantly less than the period of arrival of information bits.

Theorem. An information signal $x_a(t)$ with a limited frequency band is transmitted with a period T_a and can be restored directly from the received signal $x_b(t)$, which is a sequence of chips, with an arrival period T_b then, and only if $T_a \geq 2T_b$.

Evidence. The information bit of the signal $x_a(t)$ with a limited frequency band from 0 to ω_a and the period of their arrival T_a is encoded by a sequence of chips with a period of arrival T_b. As a result, a sample signal $k(t)$ is obtained, which is a sequence of a certain number of chips, with an arrival period T_b. Thus, the obtained discrete signal $x_b(t)$ will have the following form: $x_b(t) = x_a(t)k(t)$.

Since the signal $k(t)$ makes up a uniform sequence of pulses, it is a periodic signal and can be represented as a Fourier series.

$$k(t) = \sum\nolimits_{n=-\infty}^{\infty} K_n \, e^{in\omega_b t}. \quad (1)$$

Therefore:

$$x_b(t) = x_a(t)k(t) = \sum\nolimits_{n=-\infty}^{\infty} K_n \, x_a(t)e^{in\omega_b t}. \quad (2)$$

Fourier - the image of the function $x_b(t)$ will be:

$$X_b(\omega) = \int_{-\infty}^{\infty} x_b(t)e^{-i\omega t}dt. \quad (3)$$

Substituting the values of $x_b(t)$:

$$X_b(\omega) = \int_{-\infty}^{\infty} \sum\nolimits_{n=-\infty}^{\infty} K_n\, x_a(t) e^{in\omega_b t} e^{-i\omega t} dt. \tag{4}$$

After simplification:

$$X_b(\omega) = \sum\nolimits_{n=-\infty}^{\infty} K_n \int_{-\infty}^{\infty} x_a(t) e^{-i(\omega - n\omega_b)t} dt. \tag{5}$$

Based on the definition of the Fourier transform:

$$X_b(\omega - n\omega_b) = \int_{-\infty}^{\infty} x_a(t) e^{-i(\omega - n\omega_b)t} dt. \tag{6}$$

Where $X_a(\omega)$- Fourier - the image of function $x_a(t)$.

Substituting this into the previous ratio:

$$X_b(\omega) = \sum\nolimits_{n=-\infty}^{\infty} K_n X_a(\omega - n\omega_b). \tag{7}$$

Thus, the spectrum of the signal $x_b(t)$ is the spectrum of the signal $x_a(t)$ and the spectrum that arises from the spectrum of the signal $x_a(t)$ when each harmonic is transferred to the carrier frequency. Each of the transferred spectra is multiplied by the corresponding coefficient of the Fourier series for $k(t)$. And when $T_a \geq 2T_b$, then these transferred spectra do not overlap, and the spectrum of the signal $x_a(t)$ is multiplied by the coefficient K_0, which is included in $X_b(\omega)$.

$$X_b(\omega) = K_0 X_a(\omega), \frac{-\omega_b}{2} \leq \omega_a \leq \frac{\omega_b}{2}. \tag{8}$$

The restoration of the spectrum of the signal $x_a(t)$ will take place when the signal $X_b(\omega)$ is transmitted through a band-pass filter.

Thus, the fulfillment of the condition $T_a \geq 2T_b$ for restoring the information signal limits the minimum number of coding pulses to three chips. The maximum number of coding pulses depends on their temporal characteristics. The current technical state of the element base allows you to encode one information bit with several hundred chips, the time duration of which is in the range of 0.2–2.0 nS with a arrival period of 10–100 nS. Typically, such encoding chips are in the form of idealized Gaussian monocycles, the main part of the spectrum of which is in the range from 1 to 10 GHz [6].

To organize independent channels in one frequency band, it is advisable to additionally apply a system of orthogonal codes. These are sets of sequences for which all pairwise values of the cross-correlation function are equal to zero. At the same time, all channels simultaneously use the same frequency. In such a system, each mobile user is assigned their own unique code sequence, which allows only the proper signal to be allocated from the air, and the cross-correlation between all users is zero. Information channels are separated by shifting each bit of information encoded by the chips by a time proportional to the current value of a pseudorandom sequence. Moreover, the time range of the shift is one to two orders of magnitude higher than the offset value in PPM modulation. And noise immunity reception of an informational message occurs only when the receiver and transmitter use the same channel code. This significantly increases the noise immunity of the signals, the level of EMC in mobile wireless communication systems.

The signals correspond most fully to the conditions of orthogonality; they are the terms of the Hadamard matrix, which are more often called Walsh functions. So a set of Walsh codes of length n includes $\boldsymbol{n}$ the term of the Walsh matrix $\boldsymbol{n} \times \boldsymbol{n}$. Thus, the expansion of the spectrum of the digital information signal is carried out using a special code sequence. As a result, the spectrum of the digital information signal is expanded by direct multiplication by a pseudo-random sequence, as a result of which it becomes noise-like, and the signal power is distributed in a wide frequency band. The width of the spectrum of the information signal is determined by the spectrum of the pseudo-random sequence and is expanded 10^1–10^4 times. The amplitude of the pulsed electromagnetic field is also reduced, which ensures the requirements of EMC.

The extraction of the useful signal against the background of noise is carried out by correlating the received and reference signals. The correlator convolves the received signal with the reference. It is an ideal detector for determining the time shifts of received pulses relative to the reference ones. So when receiving a unit, the correlation function is +1, and when receiving 0, it takes the value −1. In any other cases, the correlation function is 0. But in the case of representing an information bit, for example, 200 ultra-short pulses with a chip, if the code matches, they accumulate in the receiver integrator. The information bit will be detected even when, for example, 99 of the 200 chips are distorted. Several hundred ultrashort pulses encoding each of the information bits accumulated in the correlator of the receiver make it possible to significantly increase the signal-to-noise ratio, providing the possibility of transmitting information in a wide frequency range well below the white noise level, while ensuring EMC requirements for mobile devices. Due to the various paths of propagation of radio waves (multipath propagation), signal interference occurs and a complex electromagnetic environment is created at the signal receiving site. The digital encoded signal comes in the form of several time-shifted copies. However, when the offset difference is longer than the duration of one ultrashort coding pulse, the receiver is synchronized with the most powerful component of the received signal, while others are rejected. In addition, the use of ultrashort coding pulses prevents intersymbol distortions. This is due to the fact that the energy of the received pulse signal almost always has time to dissipate before receiving its subsequent copy. Thus, the ability to work with a small signal power and its high insight through obstacles can effectively transmit information indoors and objects with a complex architecture and provide EMC requirements in multipath propagation of signals. The organization of communication in a system of mobile devices imposes additional restrictions due to the uncertainty in time and space of the polarization vector of the information signal relative to the receiving antenna. So, in the case of their orthogonal arrangement, the level of the received signal will be zero. Therefore, it is proposed to introduce a flickering polarization, according to which each of the ultrashort pulses encoding the information bit is alternately supplied to one or the other antenna, which are located orthogonally in the antenna unit.

Due to the fact that coded impulse signals in a very wide frequency band are subject to radiation, the requirements for the wide-range of the transceiver antenna are basic. According to technical characteristics, the antenna element is most suitable, which is an antenna with an expanding slot TSA (Tapered Slot Antenna) [7–11]. The shape of the open slit determines the frequency band, and its energy orientation is characterized by a narrow main beam and the practical absence of side lobes in the frequency range 2–6.5 GHz. Due to such technical characteristics, it is used as the main component of each of the two antenna units, which in turn create an antenna system (see Fig. 3). In this system, the axis of symmetry of the blocks are arranged orthogonally, which allows for arbitrary flickering polarization. The arrival of ultrashort pulses encoding the information bit to another antenna unit is carried out similarly to the first. This arrangement of both antenna units provides reception of electromagnetic radiation of arbitrary polarization, which is typical for mobile devices.

Fig. 3. Antenna system with flickering polarization.

Thus, the developed technology allows to reduce the level of interference electromagnetic environment and to provide EMC requirements for mobile wireless communication systems.

4 Analysis

Depending on the relative position of the polarization vector of the electromagnetic signal and the receiving antenna of the mobile device, the level of the induced signal changes in proportion to the directivity function:

- for vertical polarization of the information signal

$$D_v(\psi, \varphi) = sin\psi cos\varphi/(1 - cos\psi cos\varphi); \tag{9}$$

- for horizontal polarization of the information signal

$$D_v(\psi, \varphi) = sin\varphi/(1 - cos\psi cos\varphi), \tag{10}$$

where: ψ – lifting angle; φ – azimuth angle.

An analysis of the above relations shows that, depending on the type of linear polarization of the information signal, the function $D_v(\psi, \varphi)$ varies from 0 to 2.4, which excludes signal reception for some combinations of the mutual orientation of the vectors. At the same time, the use of flickering polarization allows the use of both types of polarization, excluding combinations in which the signal will be absent. In this case, the directivity function $D_v(\psi, \varphi)$ will be in the range from 1 to 2.4, which allows more than double the radius of action of electromagnetic radiation.

5 Conclusion

The use of ultra-wideband signals, flickering polarization and antenna systems in communication systems of mobile devices allows increasing the signal-to-noise ratio at the receiver input, which allows decreasing the level of electromagnetic radiation, thus ensuring the electromagnetic compatibility requirements of mobile wireless communication systems. The effectiveness of the application of the developed technology is due to the possibility of realizing a potentially high density of communication channels per square meter of the working area. Due to the decrease in the level of the information signal with respect to white noise in the operating frequency range, both traditional narrow-band communication systems and communication systems using ultra-wideband signals are simultaneously operated independently. The introduction of the technology for ensuring the electromagnetic compatibility requirements of mobile wireless communication systems makes it possible to ensure their implementation at all stages of the development and implementation of mobile wireless communication systems in transportation.

Acknowledgements. This work was partially funded by the European Union in the context of the project "dComFra - Digital competence framework for Ukrainian teachers and other citizens" (Project Number: 598236-EPP-1-2018-1-LT-EPPKA2-CBHE-SP) under the ERASMUS + program. The European Commission's support for the creation of this work does not mean full approval of its content, which reflects only the views of the authors. The Commission cannot be held responsible for any use of the information contained in this work.

References

1. Klymash, M., Pelishok, V., Mikhailenich, P.: Mobile Technologies. NVED WAD, Lviv (2007)
2. Serkov, A., Breslavets, V., Tolkachov, M., Churyumov, G., Saad, I.: Noise-like signals in wireless information transmission systems. Adv. Inf. Syst. **1**(2), 33–39 (2017)
3. Serkov, A., Lazurenko, B.: Prospects for development of wireless communication systems. In: Problems of Informatization, Abstracts of the VI International Technical Scientific Conference, 14–16 November 2018, Cherkasy-Baku-Belsky-Byala-Kharkiv, p. 22. National Technical University "Kharkiv Polytechnic Institute", Kharkiv (2018)

4. Serkov, O., Churyumov, G.: On the issue of solving the problem of electromagnetic compatibility of the wireless telecommunication systems. Appl. Radio Electron. **16**(3–4), 117–121 (2019)
5. Harmuth, H.: Non-sinusoidal Waves for Radar and Radio Communication. Academic Press, New York (1981)
6. Serkov, A., Breslavets, V., Tolkachov, M., Kravets, V.: Method of coding information distributed by wireless communication lines under conditions of interference. Adv. Inf. Syst. **2**(2), 145–148 (2018)
7. Serkov, O., Churyumov, G., Breslavets, V., Tolkachov, M.: Model TSA. In: Abstracts of the XVII International Scientific Conference "PIM-2017", p. 76. National Technical University "Kharkiv Polytechnic Institute", Kharkiv (2017)
8. Serkov, O., Breslavets, V., Tolkachov, M., Churyumov, G.: The wideband pulsed antenna and its application. In: 9th International Conference on Ultrawideband and Ultrashort Impulse Signals (UWBUSIS-2018), Odessa, Ukraine, 4–7 September 2018, pp. 340–343 (2018)
9. Serkov, A., Kravets, V., Kasilov, O., Lazurenko, B., Mickus, A.: The concept of information security in the IoT system. Adv. Inf. Syst. **3**(1), 136–139 (2019)
10. Serkov, A., Pustovoitov, P., Yakovenko, I., Lazurenko, B., Churyumov, G., Tokariev, V., Nannan, W.: Ultra wideband technologies in mobile object management systems. Adv. Inf. Syst. **3**(1), 109–115 (2019)
11. Serkov, A., Kravets, V., Yakovenko, I., Churyumov, G., Tokariev, V., Nannan, W.: Ultra wideband signals in control systems of unmanned aerial vehicles. In: 10th International Conference on Dependable Systems, Services and Technologies (DESSERT), Leeds, UK, pp. 25–28 (2019)

The Assessment of the Possibility of Using Open Systems of Radio Transmission for the Purposes of Railway Transport

Marcin Chrzan(✉)

Kazimierz Pulaski University of Technology and Humanities,
26-600 Radom, Poland
m.chrzan@uthrad.pl

Abstract. The main aim of the article is to investigate the influence of the method of railway telegram encryption on the delay of transmission in open radio systems. It is intended for persons actively involved in the security of radio transmission for railway tasks and persons involved in the legal legislation of standards. The development of the systems of wireless communication and their availability causes that various branches of transport imply wireless technologies for their needs. Although every radio transmission may be systematically protected without unauthorized access, additional methods of encryption improving safety according to the designers of such systems are applied in specific applications. The norm EN 50159:2011 was analyzed in this article. The main assumptions contained in the standard are presented in this article and it is addressed to persons dealing with railway safety. The results presented in this article can be used to verify the usefulness of the application of encryption methods described in the standard in terms of parameter - delayed transmission and changes in legal regulations in this area.

Keywords: Railway traffic control systems · Delay of transmission · Safety functions

1 Introduction

Due to degradation of wired infrastructure connecting element of systems of railway control, related above all to their devastation, the medium of radio transmission for independent management of railway traffic control devices is applied more and more often [5, 6]. The problem connected with implying such solutions is safety of transmission in such systems [10].

It must be emphasized that safety at specific level of transmission is currently offered by the very systems of radio transmission, which was also used in GSM-R (Global System for Mobile Communications – Railway) standard, which is a subsystem of ERTMS (European Railway Traffic Management System). Creating separate system for the purposes of railway industry requires huge expenditure on designing, research, certification and building and implementation of such technology. Therefore, it is possible to use public open radio networks for the purposes of railway industry, which considerably reduces costs because such system is based on existing infrastructure of

E. Ginters et al. (Eds.): ICTE ToL 2019, LNITI, pp. 271–278, 2020.
https://doi.org/10.1007/978-3-030-39688-6_34

local operators. It is therefore necessary to develop a way of transmitting information that meets the requirements of secure radio transmission. Open radio systems are currently considered around the world as one of the elements of the system connected with safety that uses radio transmission for exchange of data between specific elements of railway infrastructure. This issue is regulated by the norm EN 50159:2011 [7]. Therefore, there is a need to meet safety requirements defined by the norm mentioned above in order to achieve assumed level of SIL (Safety Integrity Level) by the whole system using open systems of wireless communication. The basic threats and methods of preventing these threats were also presented in this article. The criteria that must be taken into account while designing open systems and results of the research conducted based on the author's program on selected railway route in Poland were presented.

2 The Safety Requirements for Transmission in Open Radio Network

Based on the norm EN 50159:2011, Table 1 shows the list of threats and methods of preventing threats (safety function). Within the scope of encryption algorithms in the norm EN 50159:2011, the application of reliable encryption algorithms such as DES (Data Encryption Standard) is recommended. An official successor of DES is AES (Advanced Encryption Standard) algorithm. The level of safety for AES higher due to longer encryption key in comparison with DES. When it comes to selection of encryption algorithms norm EN 50159:2011, it is not recommended to apply the mode of block encryption ECB (Electronic Codebook) when the size of the blocks of input data (before encryption) is larger than after encryption.

2.1 Threats and Safety Functions Based on the Norm EN 50159:2011

Open transmission system requires to include the following actions for the safety of exchange of information in every system participating in transmission [7]: treating transmission system as a system not fully trusted, regardless of applied internal security protection: the use of safe transmission function, the use of safe access methods.

The basic threat for the system, resulting from untrusted transmission system, is failure to receive by the recipient important and authentic telegram. It may happen in the cases described in Table 1. The threats mentioned in Table 1 are the result of, among others, unknown number of users who would want to have access to network and unknown number and type of equipment that can be included in the network. It creates potential threat to safety of the system, mainly due to the possibility of occurring data in unknown formats and amounts, as well as hacker attacks. The structure of the system related to safety that uses connection to open transmission system is presented in Fig. 1. The environment of the systems of radio transmission, from the point of view of safety, should be treated as untrusted. In order to limit the threats from open transmission system, the following conditions can be taken into consideration: authenticity of telegrams, integrity of telegrams, duration of sending telegrams, defined sequence of telegrams.

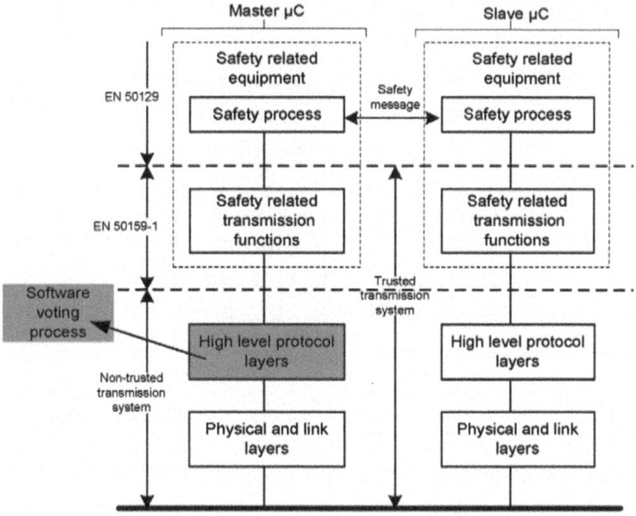

Fig. 1. Structure of a safety system that uses a non-trusted transmission system [11].

The systems executing safety functions using system of radio transmission or connected to public network, may be exposed to malicious attacks from other network users. In order to protect data against illegal access and modification, that is, confirming authenticity of telegrams, it is recommended to apply cryptographic techniques. The degree of effectiveness of this method depends on the power of the algorithms and discretion of the keys. Therefore, proper algorithm and length of a key must be selected and confidentiality through proper management of the keys must be ensured.

Table 1. A list of threats and defense methods [6] (A - numbering of telegrams; B - application of time stamps in the telegrams; C - definition of maximum time of awaiting response; D - identification of the participants of transmission; E - application of return messages; F - the use of procedures authorization; G - application of safety codes; H - data encryption).

	A	B	C	D	E	F	G	H
Repetition	X	X						
Deletion	X							
Insertion	X			X	X	X		
Resequence	X	X						
Corruption							X	X
Delay		X	X					
Masquerade					X	X		X

EN 50159:2011 also gives recommendations within the scope of selection of algorithms depending on the types of the telegrams (see Table 2).

Table 2. Safety codes for various types of telegrams [6].

Type	Type of a telegram model for transmission safety system			
	A0	A1	B0	B1
CRC	R	US	–	R
MAC	R	HR	R	R
Hash code	R	US	HR	HR
Digital signature	R	R	R	R

Symbols: "HR" - means that this type of safety code is highly recommended; "R" - means that this type of safety code is recommended; "-" - means that this type of safety code has no recommendations for or against using it; "US" - means that this type of safety code is not recommended.

The analysis showed that with the application of appropriate techniques of securing transmission, they can be broadcasted in open channel for the purposes of controlling railway devices.

2.2 Methodology for Measuring Transmission Delay

For modelling of a system of open radio transmission of the telegrams between railway traffic control devices and local control station, the following assumptions were made [3, 4, 7–9, 12–14]: railway traffic control devices of specific functionality and characteristics are the part of the system; railway traffic control devices have operational permissions and safety certificates on the specific level of SIL; radio transmission changes transmission medium from wired to wireless, which causes that behaviour of these devices should not change – their parameters are determined; radio transmission uses LoS (Line-of-Sight) and NLoS (Non-Line-of-Sight) [1, 2], radio transmission in open systems, constant velocity of a train - 160 km/h; network load is natural load, no impact on network traffic generated by measuring system – real conditions, in which system can work; the parameters of GSM-R transmission with the assumption that fulfilment of these conditions meets safety requirements of transmission are applied for modelling. In order to take the measurements in physical environment, a model of transmission for LTE open system presented in Fig. 2 was built. A system, in which transmission takes place at the first stage of the research between virtually programmed LCS (Local Control Station) and virtual railway traffic control device, on the route of a train will be applied during the research. The systems not being real-time systems from the environment of operating systems - Microsoft Windows and Apple Mac OS X were applied for the measurements. The research were conducted on the railway line Cracow – Warsaw line No. 4, which is a part of line E65.

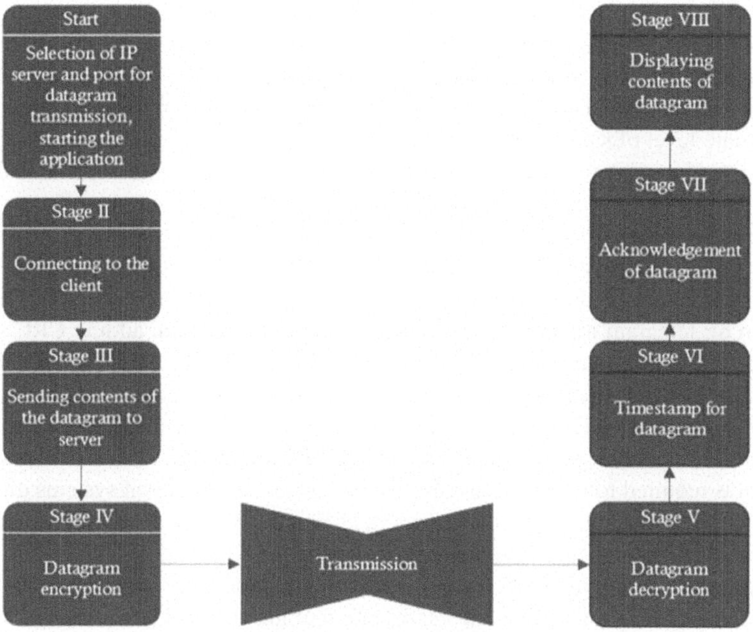

Fig. 2. The block diagram for the algorithms used to transmit datagrams [5].

A computer with Mac OS X 10.12.6 and LTE MOXA OnCell G3470ALTE was applied during the research, block diagram of a measuring system is shown in Fig. 2.

During the measurements, encrypted message was sent from a mobile station placed on the train to the local control station located at the Faculty of Transport and Electrical Engineering of the University of Technology and Humanities in Radom.

The random messages were sent (16B, 32B, 64B, 128B, 256B, 512B, 1024B, 2048B, 4096B, 6144B, 8192B and 10240B). The measurements referred to delay of transmission of a message in time interval of 15 min. For encryption of random railway messages of various sizes, the following algorithms were used: AES CBC No Padding, AES CBC PKCS5 Padding, AES ECB No Padding, AES ECB PKCS5 Padding, DES CBC No Padding, DES CBC PKCS5 Padding, AES ECB No Padding, AES ECB PKCS5 Padding, RSA with 1024 and 2048 bit key. The measuring program consists of two modules: Server (LCS simulator) and Client (transmission receiver). Server Program allows to connect with Client Program after determination of common parameters of transmission such as: TCP/IP (address and port), determination of a method of signal encoding. In the event of Algorytm Rivesta-Shamira-Adlemana (RSA) transmission with the use of a key – server program allows to generate a key based on inserted password. Server Program receives a telegram, decrypts it, checks integrity and sends back acknowledgment of receipt to the Client. Client Program initiates exchange of transmission with specific data encryption. In addition, it allows to, send a permanent message (telegram) of specific cyclical length, in specific time, send a variable message, generated separately for each connection from a server, of constant length, generated cyclically, in specific time, send one-time message of specific length. Telegram sent to a client from a server is constructed in such a way to meet safety requirements for open and closed transmission systems contained in EN50159:2011 norm, which is connected with the use of equipment and software working on it. Appropriate level of safety transmission of a telegram should be ensured in a way that allows detection of errors of signal transmission from a sender to a recipient, and in the event of break in transmission above specified time, the system should cause transition into the state ensuring safety.

The program consists of two modules. Server Module (a simulator of Local Control Station) and Client Module (transmission receiver). Server Program sends a telegram in the form of: telegram number, time stamp, content of a telegram, adding CRC (Cyclic Redundancy Code), encrypted AES (key 128, 192 or 256 bits, ECB or CBC (Cipher Block Chaining) mode can be selected), the key is generated automatically based on the password. Client Program receives a telegram, decrypts it, checks its integrity and sends back acknowledgment of receipt to the Server Program. Such structure of a telegram is required to meet safety requirements for closed and open systems described in EN 50159:2011. It is particularly important because safety of transmission is connected with the use of this equipment and its software [5].

Appropriate safety of signal transmission from a sender to a recipient of a telegram should be ensured in a way that allows: detection of the errors of transmission, break in transmission above specified time should result in transition of the system to the state ensuring safety. The list of the results of the measurements are shown in Table 3 and Fig. 3.

Table 3. The comparison of delays of transmission of 16 B–10 kB messages with various methods of encoding.

Average delay $\bar{x}$ [ms]								
Length of a message [B]	AES CBC		AES ECB		DES CBC		DES ECB	
	$\bar{x}$	σ	$\bar{x}$	σ	$\bar{x}$	σ	$\bar{x}$	σ
16	24,011	5,096	24,195	5,201	24,012	5,120	24,189	5,346
32	26,063	4,609	24,104	5,332	26,172	4,514	26,246	4,529
64	26,954	4,887	26,350	4,558	26,787	4,930	27,072	5,072
128	27,691	5,168	27,932	5,243	27,695	5,494	27,839	5,285
256	29,978	6,160	29,910	6,099	29,966	6,232	29,994	6,302
512	32,712	6,846	33,168	6,746	32,559	7,115	32,968	7,051
1024	39,608	8,038	39,481	7,986	39,485	8,480	39,730	8,474
2048	49,966	9,807	49,478	9,937	49,884	10,020	49,459	10,277
4096	74,977	11,723	73,984	10,502	74,049	10,841	73,761	12,692
6144	107,533	12,958	107,186	10,761	111,451	15,415	108,728	13,585
8192	154,259	14,582	150,367	13,973	152,047	13,265	150,715	15,626
10240	212,673	14,689	206,338	11,823	209,598	15,073	207,205	14,070

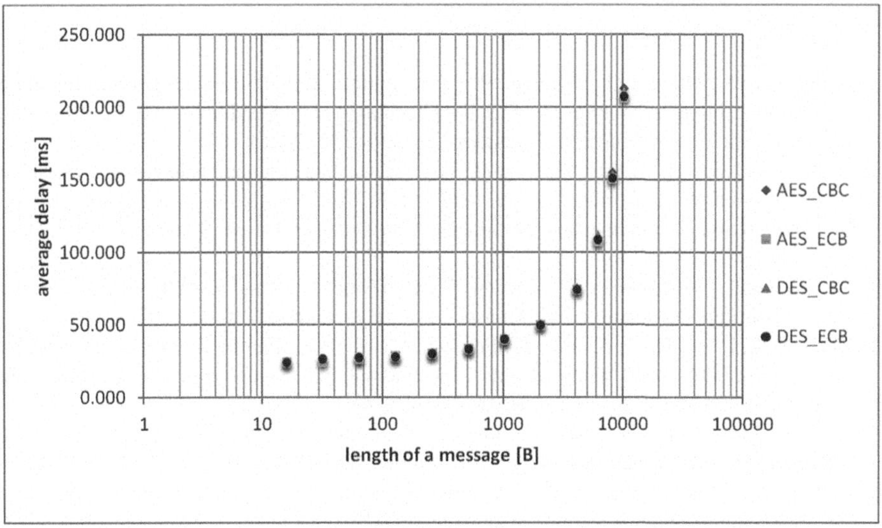

Fig. 3. The comparison of average values of delays of transmission in various methods of encoding 16 B–10 kB messages (logarithmic scale).

The most powerful RSA algorithm is omitted from the figure because it introduced long delays for small data packets.

3 Conclusion

So far, few studies were carried out on the use of open radio systems for railway traffic. And even more so, for the public Long Term Evolution (LTE) standard. Research results showed that for 16 byte messages the same LTE network parameters have the greatest influence on delay, while for 10 kB messages the most important influence on transmission delay has encoding and decoding time of the messages. However, the encoding method does not affect the encoding and sending times (see Fig. 3). The conducted research proves that only the AES algorithm should be used in open railway telecommunication networks. Its structure and security level guarantee confidentiality and reliability of information at the appropriate security level without affecting the transmission delay in case of small packets. Therefore, encryption systems based on older techniques reducing transmission security should be eliminated from the standards. In the future will be examined, the resistance of the Virtual Private Network (VPN) channel transmitting railway messages in the LTE open system algorithm RSA to attacks of unauthorized persons. The use of the 5G network for the transmission of railway telegrams will also be considered.

References

1. Barazzetta, M., Micheli, D., Diamanti, R., et al.: Optimization of 4G wireless access network features by using reverberation chambers: application to high-speed train LTE. In: IEEE 46th European Microwave Conference, London, pp. 719–722 (2016)
2. Baek, J.H., Choi, H.Y., Kim, G.Y., et al.: A LTE wireless communication interface test for on-board oriented train control system field test. In: IEEE 5th International Conference on Information and Communication Technology Convergence, Busan, South Korea, pp. 690–694 (2014)
3. Ciszewski, T., Nowakowski, W., Chrzan, M.: RailTopoModel and RailML – data exchange standards in railway sector. Arch. Transp. Syst. Telemat. **10**(4), 10–15 (2017)
4. Chou, S.-J., Chou, H.-T., Kuo, L.-R.: Potential causes of PIM problems in the LTE outdoor base station multi-band antennas. In: 21st International Symposium on Antennas and Propagation, Okinawa, Japan, pp. 1080–1081 (2016)
5. Chrzan, M.: Bezpieczeństwo kolejowych systemów telekomunikacyjnych, 1st edn. Kazimierz Pułaski Technical University of Radom, Radom (2018)
6. Chrzan, M.: Nowoczesne systemy łączności dla potrzeb PKP. Prace Naukowe Elektryka No. 2 (8), Politechnika Radomska, Zakopane (2004)
7. Chrzan, M.: Metoda bezpiecznej transmisji danych pomiędzy urządzeniami sterowania ruchem kolejowym z wykorzystaniem systemów radiowych czwartej generacji, 1st edn. UTH im. Kazimierza Pułaskiego w Radomiu, Radom (2013)
8. Dahlman, E., Parkvall, S., Sköld, J., Beming, P.: 3G Evolution – HSPA and LTE for Mobile Broadband, 2nd edn. Academic Press, Cambridge (2008)
9. Kim, J., Choi, S.W., Song, Y.S., Yoon, Y.K., Kim, Y.K.: Automatic train control over LTE: design and performance evaluation. IEEE Commun. Mag. **53**, 102–109 (2015)
10. Kornaszewski, M., Chrzan, M., Olczykowski, Z.: Implementation of new solutions of intelligent transport systems in railway transport in Poland. In: Mikulski, J. (ed.) Smart Solutions in Today's Transport. Communications in Computer and Information Science, vol. 715, pp. 282–290. Springer, Heidelberg (2017)
11. Idirin, M., Aizpurua, X., Villaro, A., Legarda, J., Melendez, J.: Implementation details and safety analysis of a microcontroller-based SIL-4 software voter. IEEE Trans. Ind. Electron. **58**(3), 822–829 (2011)
12. Nguyen, K., Beugin, J., Berbineau, M., et al.: Analytical approach for evaluating LTE communication errors in train control application. In: IEEE International Conference on Communication Workshop, London, UK, pp. 2369–2374 (2015)
13. Parichehreh, A., Savazzi, S., Goratti, L., Spagnolini, U.: Seamless LTE connectivity in high-speed trains. Wirel. Commun. Mob. Comput. **16**(12), 1478–1494 (2016)
14. Puttonen, J., Kurjenniemi, J., Alanen, O.: Radio problem detection assisted rescue handover for LTE. In: IEEE 21st Annual IEEE International Symposium on Personal, Indoor and Mobile Radio Communications, Istanbul, Turkey, pp. 1752–1757 (2010)

Electromagnetic Compatibility of WiFi Networks on Rolling Stocks in the Railway Transport

Valentin Popov and Alexey Shevchenko(✉)

Riga Technical University, Azenes Street 12a, Riga 1048, Latvia
popovs@latnet.lv, alexey@cryptolab.net

Abstract. This work considers the main problems associated with electromagnetic compatibility (EMC) in WiFi networks, widely used in modern rolling stock vehicles of railway transport. Typical internal electromagnetic (EM) interference is caused by the high density of equipment layout in the rolling stock, and external EM interference is created by public cellular networks of mobile communication and other radio engineering systems. The main parameters are estimated of EMC in WiFi networks, which allows to increase the EMC at the time of configuring these networks. The results of experiments are shown to determine the parameters of WiFi networks within the selected railway wagon, which show a high level of fluctuations in the useful signal along the wagon and a sufficiently high level of EM interference signals, which ultimately affects the EMC of WiFi network.

Keywords: Electromagnetic compatibility (EMC) · WiFi networks · Railway transport

1 Introduction

Modern digital wireless mobile communication networks such as WiFi (802.11 standard), which in recent years have become widely used in passenger trains of railway transport, impose increased requirements on the parameters of their electromagnetic compatibility (EMC), as the rolling stock uses an extensive network architecture, consisting of the Wi-Fi devices themselves, various radio stations, sensors, microprocessor control devices (equipped with high density), conductive elements, pantograph, and other components, which distorts the normalized characteristics of the operating radio receivers and radio transmitting devices of the WiFi network. The development of high-speed European railways, the introduction of European Railway Traffic Management System (ERTMS) and centralized management of automation, telemechanics and communication systems, leads to the fact that the communication of both GSM-R and WiFi networks, is becoming increasingly vulnerable to internal and external electromagnetic interference (EMI). Therefore, when designing such systems, it is necessary to conduct studies of compliance with EMC conditions, in accordance with accepted standards IEEE 802.11 (also Vieira, ANSI, ETSI). It should be noted that in the countries of Eastern Europe, whose railways are gradually installing the GSM-R

E. Ginters et al. (Eds.): ICTE ToL 2019, LNITI, pp. 279–286, 2020.
https://doi.org/10.1007/978-3-030-39688-6_35

standard, analogue radio stations in the ranges of 150, 300 and 450 MHz are still in operation, and in case of their use, they will also affect the interference of radio communication channels of WiFi networks. In addition, the WiFi network zones fall within the radio coverage clusters of public cellular networks of the corresponding generations of 2G, 3G, 4G mobile communication. Providing reliable and widespread uniform radio coverage of the WiFi network on the rolling stock can be quite a challenge. Therefore, a family of channel-forming devices is required, which would provide both the transmission of Wi-Fi signals to all related customers, and the choice of the best in its characteristics of the communication path and provide a constant reduction of EMI levels in order to implement the best possible Wi-Fi radio coverage and ensure a high level of EMC. Thus, the problem of electromagnetic compatibility of WiFi networks on the railway transport nowadays is quite sharp and requires minimizing the effects of EMIs on their functioning [1–13].

2 Potential Configuration of WiFi Networks on Rolling Stocks

To improve customer service, railway operators are installing WiFi networks on their trains to provide customers with access to Internet services: WiFi router in train - WiFi connection in stations – access to Internet. As an example, one of the options for placing routers in the passenger wagon, which allows to create within the wagon a uniform radio coverage of the network, and gives to passengers the ability to connect quite effectively to access points. [3] WiFi coverage area provides control of subscribers' authorization and providing them with connection to the Internet network as on board of the train, as well as when passengers are at the station. Data transmission along the train can be provided by WiFi network between wagons using 802.11 technology. Data is transmitted from one wagon to another with the help of bridge antennas located at each end of the wagon. [5] It would seem that creating such a simple configuration of the WiFi network at small distances between access points and subscriber terminals should provide high-speed transmission of voice and multimedia information, but the work of the WiFi network can be regularly interrupted due to internal and external EM interference, the main of which are:

- Internal EM interferences of WiFi network: the effect of signals from neighboring cells, which are sources of EM interference for the useful channel, leads to a collision effect (the so-called hidden node problem); mutual influence of mobile stations and access points working on overlapping frequency channels; As the number of subscribers grows, the mutual EM interference increases and the number of collisions and retransmission packets increases;
- External EM interferences of WiFi network: Effects of public (e.g. urban) mobile networks, as well as Wi-Fi networks that may be present in other trains when two trains are moving on counter-courses (co-channel EM interference comes mainly from other Wi-Fi networks that are in close proximity to the train), causing blocking, intermodulation and cross-distortion, jitter and other negative processes; Environmental impact on radio wave propagation conditions, i.e. objects that cause

attenuation, reflection, refraction or scattering of 2.4 GHz 802.11bgn and 5 GHz 802.11an radio waves; Influence of the number of users who have mobile devices on and around the receptor; Weak protection from artificially created interference, the possibility of invisible control over the technological process of the WiFi network by hostile persons.

3 Distinctive Features Between Access Points and Receptor Related with EM Interference

If to consider connection of the mobile terminal R_1 (receptor - mobile phone, smartphone, tablet, etc.) to the access point TR_1 (AP, 802.11), as shown in Fig. 1, the aggregate amount of interference will be determined by the form:

$$\Omega_{int\,j} = \sum\nolimits_{j=1}^{j=n} \Omega_j,\ \Omega_{int\,i} = \sum\nolimits_{i=1}^{i=m} \Omega_i \qquad (1)$$

where summation is implemented from $j = 1$ to $j = n$, and from $i = 1$ to $i = m$.

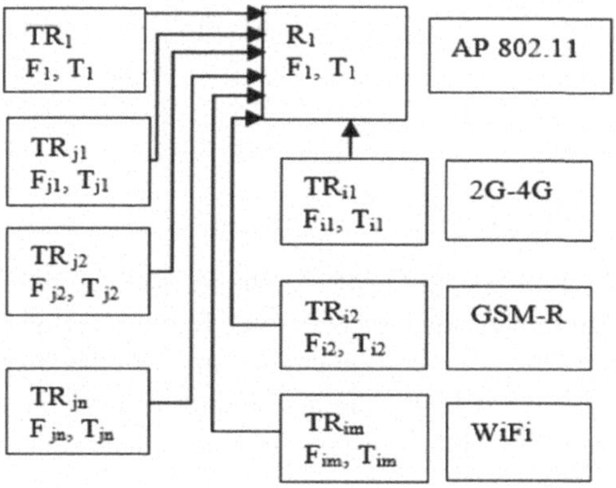

Fig. 1. Connection of the mobile terminal *R* to the access point *TR*.

Mobile terminal R_1 - as a receptor, is an object of external influence and should have a certain spatial, frequency and time selectivity, as well as selectivity in signal level and polarization plane. In turn, this can be interpreted as the existence of some area Ω_1^1 in the space $V_1^1(x, y, z) \cdot V_{1s}^1$, within which this receiver is able to respond to external electromagnetic influences. Thus, region Ω_1^1 represents a "band of transparency" of the receptor as a filter with a certain selectivity by spatial coordinates (antenna pattern and EMW polarization) and by signal volume. If *n* and *m* - sources of EMI at the receiver input creates interference signals, it is equivalent to the intersection of subsets of the useful signal and subsets of EMI:

$$\Omega_{1^1} \cap \Omega_{int} \tag{2}$$

However, if below condition is fulfilled, then:

$$\Omega_{1^1} \cap \Omega_{int} = \emptyset \tag{3}$$

where $\emptyset$ is an empty space sign, it indicates that there is no EM interference with the receptor and electromagnetic compatibility conditions in the TR1 radio channel $\leftrightarrow$ R1 are complied with, and the EM interference level can be assumed to be corresponding to the level of natural electromagnetic background. But if next below condition is fulfilled, then:

$$\Omega_{1^1} \cap \Omega_{int} \neq \emptyset \tag{4}$$

If to imagine a set of radio communication facilities in the form of radio channels of transmission and reception in spatial, frequency and time coordinates (taking into account energy and polarization parameters), it is possible to use the concept of Radio Resources (RR) as a set of possibilities of placement in spatial, in frequency and in time areas of non-crossing subsets and realizing their electromagnetic compatibility. Consequently, EMC of WiFi network can be provided [9] by:

- Radio frequency resource (according to IEEE 802.11);
- Energy resource;
- Spatial resource;
- Polarization resource;
- Time resource;
- Combination of complexes of the above resources.

Therefore, in order to assess the EMC, strategies and algorithms are developed to manage the above parameters such as operating frequencies of channels and transmission power, the formation of directional diagrams with the appropriate polarization of the electromagnetic wave (EMW), the distribution of users in space, the choice of data transmission speeds, the choice of error coding schemes, etc.

4 Requirements for the Structure and Equipment of the WiFi Network on Rolling Stock

Let's consider the basic requirements for the equipment of WiFi network for railway rolling stock:

- The WiFi network should support mechanisms to prioritize the transmitted data and ensure the quality of service (QoS);
- WiFi network equipment should support the mechanism of administrative access control, which guarantees access to active network equipment with administrator's rights only in case of authentication process, which allows to enter different access levels with the help of authorization system, as well as to keep records of changes made;

- When power is supplied to the WiFi network equipment, the operating mode of the equipment should be set automatically;
- WiFi network should be built on the principle of hierarchical structure and have the following elements: - Access level, - Aggregation level, - Core level, - Channel level.

The access level of subscribers in the rolling stock should be organized on the basis of access points and optionally wire-less network control units (wireless network controllers).

5 Methods for Evaluating Electromagnetic Compatibility of WiFi Networks

In general case when analyzing EMC of radio stations there are used a number of approximate estimates: pair, group and complex. In general - time, frequency, energy, polarization and spatial parameters characterizing EMC can be as different as desired in the context of both deterministic and probabilistic approaches. Estimation of EMC is usually realized on several groups of sources of EM interference. So, for EM interference from internal sources, which are fixed in space and time it is expedient to use a deterministic approach, because the power levels of EM interference sources, their location, orientation and polarization of their antennas are strictly fixed. Classical criterion of EMC in this case is a condition at which the level of unintentional EM interference at the receptor input does not exceed the permissible level:

$$P_{ij} \leq P_{per} \tag{5}$$

where *Pij* is the level of EM interference in *dBm*, *Pper* is the permissible level of unintentional interference in *dBm*.

In general, it is necessary to evaluate the cumulative effect of multiple independent noise signals on the receptor Eq. (1), characterized by different random power levels, frequencies, directional properties and location of antennas, etc. Thus, for example, EM interference signals from public base stations at the receptor location in the wagon are generally random values. Therefore, it is most expedient to carry out a statistical evaluation of EMC taking into account a variety of random parameters of radio networks surrounding the railway rolling stock. For statistical evaluation of the mutual influence of mobile devices of users inside the wagon it is possible to use the probability of deterioration of communication (malfunction in the form of errors in the transmitted data), which is described by the following dependence [8]:

$$P_c = P_1 \times P_2 + (1 - P_1) \times P_3 \tag{6}$$

where P_c is the full probability of communication degradation, P_1 is the probability of frequency coincidence of different transmitters, when interference occurs (conflict probability); P_2 is the probability of communication deterioration when frequencies coincide (socket EM interference); P_3 is the probability of communication deterioration when frequencies do not coincide when the signal/noise ratio in the radio channel is low (e.g. poor quality radio coverage of the WiFi network, etc.), when interference from a neighboring frequency channel occurs.

However, with a large number and different types of interference sources, mathematical models of EMC may not adequately reflect the real situation, and therefore the most appropriate are measurements and monitoring of WiFi networks, which should include following information: the number of EMI sources, their spatial location; EMI power levels; EMI frequency and spectral characteristics; statistics of transmitted packet data, etc.

There are exist software complexes [14], most of which operate either Windows, or MacOS, or Android, and which provide basic information about the existing WiFi signals in the coverage area: signal levels, used channels, SSIDs, MAC-addresses and types of protection of a particular network. Some of them can detect hidden SSIDs, determine noise levels, or provide statistics about successfully and unsuccessfully sent and received packets of wireless connection.

6 Experimental Model Studies of Electromagnetic Compatibility of WiFi Networks on Railway Transport

An experimental studies [3] of the change of the useful signal level at two carrier frequencies of 2.4 GHz and 5 GHz along the wagon space were carried out. The useful signal level was measured by the receiver-receptor when the access point (AP) was in the center of the wagon, and there were significant fluctuations in the signal level due to reflections of EMWs from the walls and ceiling (three cases), as well as the fading effect at 2.4 GHz frequency.

The authors conducted the following experiments in rolling stocks by using software system complexes, this work presents the results of experiments that can be useful in organizing a WiFi network on rolling stock in the railway transport. The bridge antenna TRW327 is located in the center between railway rolling stocks and connected with several access points within the wagons, see Fig. 2. The results of measurement of WiFi network in the railway transport, dependence of the radio signal power level for the standard 802.11 are shown in Fig. 3, part of modeling experiments [15].

The first wagon has access points: 213-TR218 and 217-TR211, the second wagon: 502-TR151 and 503-TR152. As can be seen from the data obtained: 213-TR218 access point uses channel 1 with a frequency of 2412 MHz and a channel bandwidth of 40 MHz; 217-TR211 access point uses channel 3 with a frequency of 2422 MHz and a channel bandwidth of 40 MHz; 502-TR151 access point uses channel 7 with a frequency of 2442 MHz and a channel bandwidth of 40 MHz; 503-TR152 access point uses channel 6 with a frequency of 2437 MHz and a channel bandwidth of 40 MHz. This WiFi network operates in 2.4 GHz band, where 11(13) channels are used, it is worth noting that this configuration creates a high level of interference and channel overlapping, causing a degradation of WiFi network performance. To improve the performance of this WiFi network, it is necessary to reduce interference and channel overlapping, for this would be recommended to lower the bandwidth frequency of access points from 40 MHz to 20 MHz and switch to non-overlapping channels of 2.4 GHz band (1, 6, 11), as example (see Fig. 2):

- AP 213-TR218 channel 1 with a frequency of 2412 MHz, bandwidth of 20 MHz;
- AP 217-TR211 switch to channel 6 with a frequency of 2437 MHz, bandwidth of 20 MHz;

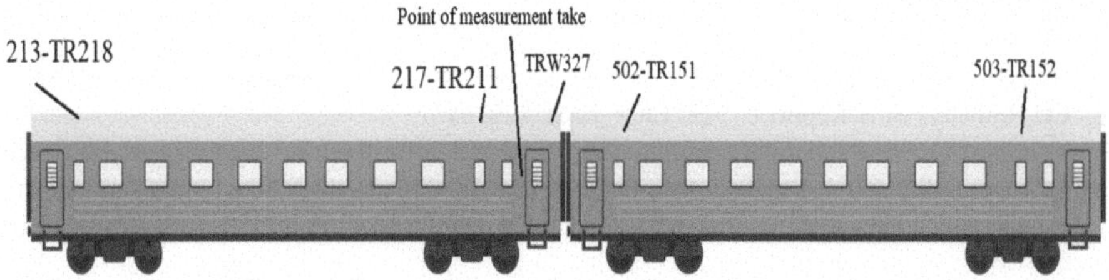

Fig. 2. Position of installed bridge antenna, access points of WiFi network on rolling stocks.

- AP 502-TR151 switch to channel 11 with a frequency of 2462 MHz, bandwidth of 20 MHz;
- AP 503-TR152 is recommended to choose channel 2, with a frequency of 2417 MHz, bandwidth of 20 MHz, as it will not have overlapping with the nearest 502-TR151 access point, and the rest of the access points will be located quite distant.

7 Conclusion

Sources of unintentional external and internal EMI on railway rolling stock for WiFi networks from power supply systems, automatics and telemechanics, from analogue radio stations, from GSM-R train radio communication system, from public cellular mobile communication networks 2G, 3G, 4G, WiMAX networks, from VHF public radio communication, radio relay and radar stations, etc. lead to the fact that the problem of electromagnetic compatibility of WiFi networks on railway rolling stock is currently quite sharp and requires the modeling of the basic parameters of cellular WiFi networks, careful measurement and monitoring of these parameters (including the use of software systems), compliance with the requirements of minimizing external interference with the functioning of the WiFi network, in accordance with standards EN 50121 (EN 50121-3-2 Railway applications - EMC - Rolling Stock, EN 301 489-7 for GSM and DCS mobile equipment) for EMC. Experiments have shown that within the wagons the signal levels and the rate of transmission of packet information depend not only on external EM interference, but also on the conditions of radio waves propagation: reflections, scattering on inhomogeneities, absorption, etc. The current weaknesses of WiFi networks in railway transport compared to ordinary WiFi network, such as the range, quality of service, mobility and (especially) interference forcing some railway operators and equipment suppliers to look for new solution and alternatives.

In the near future, new standards and protocols such as 5G and 802.11ax, 802.11ay will introduce significant changes in data transfer rates, which can create new challenges for electromagnetic compatibility and losses of performance/quality in case of high levels of interference.

References

1. Humbert, H.J.: EMC rolling stock, vol. 11/12, pp. 399 – 410. Elektrische Bahnen (2000)
2. EU Railways Stds Report 67575_ERA_EMC (2010)
3. Verginie, D.: EMC between rolling stock and rail infrastructure encouraging European interoperability. Analysis of the attenuation of the WiFi signals inside and outside a railway vehicle. HAL Id: hal-00544468. https://hal.archives-ouvertes.fr/hal-00544468. Accessed 10 Oct 2019
4. Rioult, J., Dudoyer, S.: Measurement reports – study of the signal attenuation inside and outside trains at 2.4 GHz and 5 GHz. Railcom Project (2009)
5. Ogunsola, A., Mariscotti, A.: Electromagnetic Compatibility in Railway. Analysis and Management. Lectures. Springer (2013)
6. Popovs, V., Golovins, J., Toršins, A.: 802.11 standarta bezvadu lokalie tikli (WLAN). Instalacijas (802.11b WLAN) un eksperimentalie petijumi. (Latvian). RTU, Riga (2006)
7. Popov, V.: Electromagnetic compatibility of radio systems on transport. Lectures. Riga-Almaty, RTU/KazAT (2012)
8. Popov, V., Shevchenko, A.: Problems of electromagnetic compatibility in GSM-R. In: Proceedings of 23rd International Scientific Conference. Transport Means 2019, vol. 3, pp. 1090–1096 (2019)
9. Popov, V.I., Skudnov, V.A.: Fundamentals of cellular networks of mobile communication design. Gorjachaja-Linija-TELECOM, Moscow (2017)
10. Railway Standards En 50121, 17 November 2007
11. Grandy, S.: Vietnam – New EMC Requirement for 2.4 GHz and 5 GHz WiFi Devices (2014)
12. Lecoq, X., Rousseau, D.: EMI/EMC and co-existence simulation methodology for high performance digital, mixed signal and RF wireless products. France (2017)
13. Marshman, C.A.: EMC in Railways. Professional Development Course on Electric Traction Systems, IET 13th, pp. 283–290 (2014)
14. https://play.google.com/store/apps/details?id=com.textron.greenlee.airscout2&hl=en_US. Accessed 10 Oct 2019
15. Shevchenko, A.: Modeling experimental investigations of electromagnetic compatibility of WiFi networks, widely used on modern railway vehicles. Report. RTU, Riga (2019)

Improving and Interference Immunity of Railway Transport Control Systems

Olha Ananieva, Mykhailo Babaiev, Vasyl Blyndiuk, and Mykhailo Davidenko(✉)

Ukrainian State University of Railway Transport, Feierbakh sq., 7, Kharkiv 61050, Ukraine
romashka13052015@gmail.com, mmbxiit@gmail.com, akademyzd_bvs@ukr.net, davdenk@kart.edu.ua

Abstract. The modern vector of the development of railway transportation logistics puts as the main goals forward uninterrupted and timely delivery of freights to destinations, reducing the time of freight cars turnover, optimizing the process of passengers conveyance. An important role in the implementation of this strategy in the railway sector is assigned to centralized traffic control systems and supervisory control systems. These systems regulates the movement of trains, increase train-handling capacity, and ensure safe and uninterrupted operation of railways. A transmission and reception of information in such systems occurs under simultaneous impact of several interference. Some of the interference are random, and some of them are described by deterministic functions of time. The effect of interference on the receiving device leads to errors in discrimination of information signals and in interpreting of information messages. The average probability P_{EM} of discrimination error is an objective indicator of the interference immunity of the reception. In this paper, we consider the case of discrimination of three information signals. The interference is the sum of the deterministic sinusoidal and impulse components, as well as Gaussian noise with a uniform spectrum. The discrimination is made by joint parameters estimation of information signals and deterministic interference based on the minimum mean square error criterion. The authors evaluated the potential lower limit for P_{EM}. It is shown that in a wide range of signal and interference parameters, the potential interference immunity is several orders of magnitude better than the boundary value, which is determined by the requirements for the functional safety of railway control systems.

Keywords: Interference immunity · Signal discrimination · Railway transport control systems

1 Introduction

The electromagnetic environment in the communication channels of modern systems of centralized traffic control and supervisory control of railway transport is characterized by a large number and variety of interference. The sources of this interference are both technogenic and natural by origin. Interference factors can be both external and internal. The influence of interference on information signals leads to an increase in the

E. Ginters et al. (Eds.): ICTE ToL 2019, LNITI, pp. 287–294, 2020.
https://doi.org/10.1007/978-3-030-39688-6_36

relative number of incorrect decisions made during the recognition of these signals. As a result, there is an incorrect interpretation of information messages. An excessive relative number of such errors is unacceptable primarily where communications are part of the technological process safety system (for example, in railway transport). The number of such systems is growing steadily, therefore, determination of numerical characteristics of the interference immunity of the means of communication integrated into them is increasingly important.

2 Related Work

The problem of evaluation of the ability of newly developed or already existing receiving communication and radar devices to correctly discriminate information signals from mixture with interference is constantly in the field of vision of specialists. In [1], the question of the reliability of detecting a point object in a noisy radio image was considered. The signal-interference environment is characteristic for a relatively narrow range of problems, therefore the application area of the proposed method is narrow. The use of the classical theory of detection of a single radio pulse as applied to air targets was considered in [2]. The results were obtained for a single signal and are intended for reliable tracking of the target. Therefore, their application area is limited. In [3], interference immunity is understood in the sense of the variance of the estimate of the angular position of the target. It does not directly characterize the errors of discriminated signals (targets). In [4], the results of constructing a set of transmitter signals providing a given probability of errors at the receiving end are described. However, these results cannot be applied in the opposite situation, when the set of signals is rigidly defined, and these signals are non-orthogonal and have different energies. As noted in the fundamental monograph [5], methods for solving the problem of interference immunity evaluation for such a situation have not been developed. The situation has changed little since the time it was published. Among the sources available to the authors, only work [6] attracts attention. There developed an approximate method for calculating the interference immunity of a receiver of a system of non-orthogonal signals. However, the assumption made on the Gaussian form of the multidimensional probability density of the joint distribution of the likelihood ratio functionals (LRF) of all signals limits the application area of the results obtained. This assumption for a set of Gaussian quantities is a priori valid only when these quantities are independent [7]. With nonorthogonal signals, the LRF values are not independent.

3 Formulation of the Problem

No results have been revealed that would accurately evaluate the interference immunity of the device for a discrimination of several given non-orthogonal signals received against a background of Gaussian interference (such a device, for example, is described in [8]). Therefore, it is of interest to develop a method for obtaining an approximate estimate of interference immunity in the situation described above, which is typical for signal transmission subsystems of railway automation.

The purpose of the research was to obtain a numerical estimate of the mean error of non-orthogonal signals discrimination when they are observed against the background of additive three-component interference in communication channels of a centralized traffic control systems of a railway transport. The first two components of this interference are a pulse and a sinusoidal oscillation, described by deterministic functions of time. The third component is Gaussian noise with a uniform spectrum.

To achieve the chosen purpose, the following tasks were solved:

- Identification of conditions under which the average discrimination error can be approximately analytically described for given signals;
- Signal transform in accordance with the identified conditions;
- Quantitative estimation of the average error of given signals discrimination.

4 Estimation of the Measurements Error

4.1 Conditions for Approximate Analytical Calculation

In [8], the problem of synthesizing an optimal signal reception device against the background of three-component interference was solved. However, the interference immunity indicators are not given in that work. The value of the mean error of information signals discrimination is accepted as such an indicator in this work. The synthesis is based on a "signal + interference" mixture model $u(t)$ introduced in the work [9] in the form:

$$u(t) = s_i(t) + v(t) + n(t), \quad i = 1, 2, 3, \tag{1}$$

Here, t is a time, $s_i(t)$ is the i-th signal (a centralized traffic control system with three information signals is considered), $v(t)$ is the deterministic interference and $n(t)$ is the noise. The device developed in [8] performs indirect discrimination of information signals: it compares the output effects of channels corresponding to deterministic functions of time $z_i(t)$ (the first two terms of mixture (1)):

$$z_i(t) = s_i(t) + v(t). \tag{2}$$

This problem is equivalent to the problem of signals discrimination against a background of random noise [5, 10]. The probability of such a discrimination error is denoted as P_{Ei}. With three information signals, the mean probability of error P_{EM} is :

$$P_{EM} = \sum_{i=1}^{3} P_i \cdot P_{Ei}, \tag{3}$$

where P_i is a priori probability of the presence of the i-th signal at the input of the receiver.

It was shown in [10] that the probability of an error when receiving the i-th signal (i = 1, 2, 3) depends on the value of the joint probability density of the output effects of the discrimination channels. As noted in the monograph [5], in the general case, methods for determining this joint density do not exist, even if the density of each of the signals is Gaussian. Only for a specific case in which the discriminated signals are orthogonal and their energies are equal, a known analytical solution exists [10]. For three a priori equally probable signals, this solution is expressed in terms of the error function as:

$$P_{EM} = 1 - erf\left(\sqrt{0,5 \cdot E/N_0}\right), \tag{4}$$

where E is the energy of each of the signals; N_0 is the doubled power spectral density of the noise, uniform in the frequency band of the signals.

This solution was the basis for calculating the minimum achievable P_{EM} value for the device described in [8]. Namely: the P_{EM} value was calculated, which can be provided after orthogonalization of real signals and alignment of their energies.

4.2 Signals Transform

For the most favorable discrimination of signals $z_i(t)$, they are orthogonalized using the Gram-Schmidt procedure [11]. Orthogonalized signals are denoted as $z_{0i}(t)$. The transformation rules are:

$$z_{01}(t) = z_1(t), \tag{5}$$

$$z_{02}(t) = z_2(t) - \lambda_{21} \cdot z_1(t), \tag{6}$$

$$z_{03}(t) = z_3(t) - \lambda_{31} \cdot z_{01}(t) - \lambda_{32} \cdot z_{02}(t). \tag{7}$$

Here orthogonalization factors are:

$$\lambda_{ij} = \frac{\int_{T_1}^{T_2} z_i(t) \cdot z_{0j}(t)dt}{\int_{T_1}^{T_2} z_{0j}^2(t)dt} \tag{8}$$

where T_1 and T_2 are consequently initial and final time of the signal processing interval.

When using the descriptions of signals and interference accepted in [9], discriminated signals are described by expressions of the form:

$$z_i(t) = s_i(t) + v_p(t) + v_E(t), \tag{9}$$

here

$$s_i(t) = U_{ms} \cdot f_i(t - \tau_s) \cdot \sin(\omega_s t + \varphi_s) \tag{10}$$

- is the i-th information signal with a binary code envelope $f_i(t-\tau_s)$, time delay τ_s relative to the moment T_1, amplitude U_{ms}, angular frequency ω_s and initial phase φ_s;

$$v_p(t) = U_{mp} \cdot e^{-\alpha(t-\tau_p)^2} \cdot \sin[\beta(t-\tau_p)+\gamma\pi] \quad (11)$$

- pulse component of the interference with amplitude U_{mp}, pulse width regulator α, pulse oscillations frequency β, initial phase regulator γ and with a maximum offset τ_p relative to the moment T_1;

$$v_E(t) = U_{mE} \cdot \sin(\omega_E t + \phi_E) \quad (12)$$

- sinusoidal component of the interference with an amplitude U_{mE}, angular frequency ω_E and initial phase ϕ_E.

To solve the problem of potential interference immunity evaluation, all signal and interference parameters were assumed to be known. The orthogonalization coefficients were calculated in accordance with (8)–(12). In this case, the values of the mutual correlations of the signal $s_i(t)$ with the interference $v_p(t)$ and $v_E(t)$, as well as the values of the correlations of the interference with each other, were assumed to be negligible. When deriving the calculation relations, the integration interval $[T_1, T_2]$ in formula (8) is taken equal to 1.6 s (information signal $s_i(t)$ repetition period generated by the track code transmitter [8]). The pulses and pauses of the binary code envelope $f_i(t-\tau_s)$ are also corresponding to this transmitter.

As follows from the technical data [8] and also noted in [12], the signal $s_i(t)$ of the track code transmitter have different energy, respectively, the signal $z_i(t)$ have different energy too. It is easy to see that the orthogonalization process makes additional changes to its value. Taking into consideration approximate nature of the calculations, it was assumed that all discriminated signal $z_{0i}(t)$ have the same energy. Its value is equal to the post-orthogonalization energy of the signal whose energy was greatest before orthogonalization. Such a signal is the "Green" signal of the track code transmitter [8]. It is denoted in this paper by $s_2(t)$ and, respectively, $z_{i2}(t)$ and $z_{02}(t)$. The energy is

$$\begin{aligned} E = {} & U_{ms}^2(0{,}395 - 0{,}39\lambda_{21} + 0{,}23\lambda_{21}^2)0{,}8(1-\lambda_{21})U_{mE}^2 \\ & + 0{,}25(1-\lambda_{21})^2 U_{mp}^2\left\{erf\left[\sqrt{2\alpha}\cdot(1{,}6-\tau_p)\right] + erf\left(\sqrt{2\alpha}\cdot\tau_p\right)\right\}. \end{aligned} \quad (13)$$

The fact of the orthogonality of discriminated signals is implicitly taken into account by the subsequent use of relation (5).

4.3 Quantitative Evaluation of the Interference Immunity

Formulas (4) and (13) were used to calculate estimates of the mean probability P_{EM} of the discrimination error. The parameters of the pulse interference were $\alpha = 10^{-3}\text{s}^{-2}$, $\tau_p = 0.77$ s. It corresponds to locomotive speed 75 km/h at the track switch and to the location of the interference pulse in the middle of the information signal. The dependences of P_{EM} on U_{mp} for a fixed U_{mE} and P_{EM} on U_{mE} for a fixed U_{mp} are calculated. Examples of graphical dependencies showing the results of this calculation with a signal amplitude of $U_{ms} = 0.05$ V and $N_0 = 10^{-5}$ W/Hz are shown in Figs. 1 and 2. These magnitudes of amplitude and spectral density are characteristic of automatic cab signaling circuits. From the calculation results it follows that the maximum approximate value of P_{EM} is of the order of $3{\cdot}10^{-14}$. The decrease in P_{EM} observed in the graphs with an increase in the amplitudes of the determined interference components is an artifact generated by the process of orthogonalization and energy equalization of all three signals.

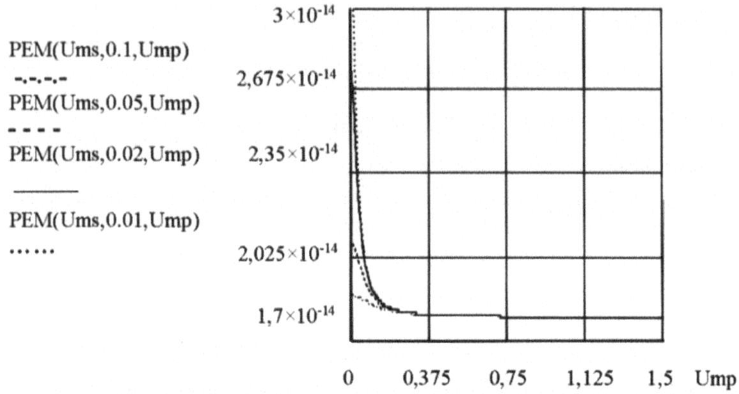

Fig. 1. The dependence of the mean probability of discrimination error on the amplitude of the impulse interference at $U_{ms} = 0.05$ V and $N_0 = 10^{-5}$ W/Hz.

It is easy to verify that post-orthogonalization energy equalization reduces the amplitudes of deterministic interference and increases the amplitude of the information signal.

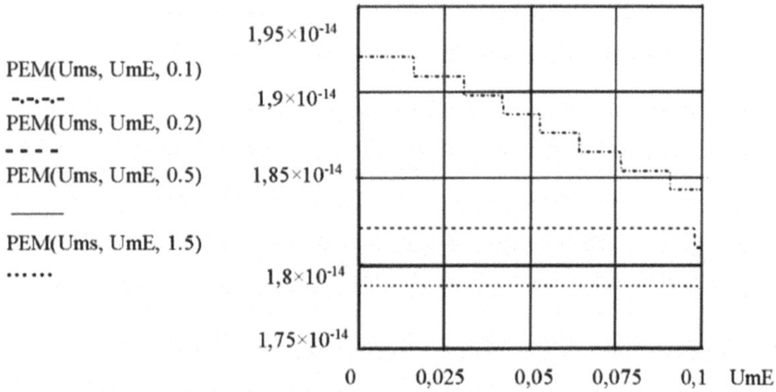

Fig. 2. The dependence of the mean probability of discrimination error on the amplitude of the sinusoidal interference at $U_{ms} = 0.05$ V and $N_0 = 10^{-5}$ W/Hz.

The analyzed receiving device is intended for use in a traffic control system. Therefore, its interference immunity must comply with the requirements of the

Ukrainian standard DSTU 4178-2003 (its European counterpart is CENELEC EN 50126: Railway Application – The Specification and Demonstration of Reliability, Availability, Maintainability and Safety), namely: it must enforce the probability of failures leading to dangerous failures not more than $0.14 \cdot 10^{-10}$ per one hour. An algorithm has been adopted in the automatic cab signaling system: in order to maintain the state of the correspondence relay, it is necessary that at least one correct signal is present among three consecutive information signals. The number of such triples per hour will be 750. Then the probability that at least one error will occur during an one hour will be $1 - (1 - P_{EM}^3)^{750} = 2{,}025 \cdot 10^{-48}$. This is much less than the probability of failures allowed by regulatory documents.

5 Conclusion

An introduction of interference-resistant methods and means for transmission, reception and processing of information signals in the communication channels of centralized traffic control systems and supervisory control will improve the management of railway rolling stock, reduce losses from train delays and reduce transportation costs due to failures of railway automation devices that lead to violations of a train schedule. The reliability of the reception and discrimination of information signals directly affects all these indicators. In the study, a method for the approximate calculation of the non-orthogonal signals discrimination error is proposed. It consists in preliminary orthogonalization of the discriminated signals. Brief summary of the results is following: It is shown that the mean error of discrimination for given non-orthogonal signals can be approximately described analytically if these signals are pre-orthogonalized; Discriminated signals are orthogonalized using the Gram-Schmidt procedure; The numerical values of the mean discrimination error are calculated as a function of the amplitudes of the determined interference. Given the algorithm for changing the indications of a locomotive light signal, the value of this error is more than six orders of magnitude less than the numerical value that is admitted by the current requirements for the functional safety of train control systems.

This work acquaints the reader with the reliability characteristics of perspective centralized traffic control systems of a railway transport. It also provides the tools for analyzing the reliability of such systems. The approximate nature of the estimates obtained is a disadvantage of the proposed method. A quantitative evaluation of approximation and the development of mathematical ways to improve accuracy are seen through further research.

References

1. Vio, R., et al.: The correct estimate of the probability of false detection of the matched filter in weak-signal detection problems. Astron. Astrophys. **604** (2017)
2. Yue, K., Chen, S., et al.: Calculathion of aircraft target's single-pulse detection probability. J. Aerosp. Technol. Manag. **7**(3), 314–322 (2015)

3. Cziporenko, V.V.: Analiz pomekhoustojchivosti bespoiskovogo metoda korrelyaczionno-interferometricheskogo pelengovaniya s rekonstruirovaniem prostranstvennogo analiticheskogo signala. Vostochno-Evropejskij zhurnal peredovykh tekhnologij **2/9**(86), 45–52 (2017)
4. Narueporn, N., Ahmad, S., Tunietti, D., Devroye, N.: Communications system performance and design in the presence of radar interference. IEEE Trans. Commun. **66**(9), 4170–4185 (2018)
5. Trifonov, A.P., Shinakov, Y.S.: Sovmestnoe razlichenie signalov i oczenka ikh parametrov na fone pomekh. Radio i svyaz', Moskva (1986)
6. Radchenko, Y.S.: Statisticheskie metody' szhatiya, vosstanovleniya i obrabotki signalov v informaczionny'kh sistemakh. Avtoref. dis. dokt. fiz.-mat. nauk 01.04:03. Voronezh. https://fizmathim.cjm/read/151961/a?#?page=1. Accesses 30 Oct 2019
7. Tikhonov, V.I.: Statisticheskaya radiotekhnika. Radio i svyaz', Moskva (1982)
8. Ananieva, O., Babaiev, M., Blyndiuk, V., Davidenko, M.: Development of a device for the optimal reception of signals against the background of an additive three-component interference. East.-Eur. J. Enterpr. Technol. **2/9**(98), 6–13 (2019)
9. Ananieva, O.M.: Matematicheskaya model` smesi signala i mnogokomponentnoj pomekhi na vkhode putevy`kh ustrojstv zheleznodorozhnoj avtomatiki. Informatsiino-keruiuchi systemy na zaliznychnomu transporti **16**, 16–19 (2017)
10. Falkovich, S.E., Khomyakov, E.N.: Statisticheskaya teoriya izmeritel'ny'kh radiosistem. Radio i svyaz', Moskva (1981)
11. Sibert, U.M.: Czepi, signaly', sistemy'. Ch. 2. Mir, Moskva (1988)
12. Goncharov, K.V.: Sintez czifrovogo lokomotivnogo priemnika avtomaticheskoj lokomotivnoj signalizaczii. Visnyk Dnipropetrovskoho natsionalnoho universytetu zaliznychnoho transportu **1**(43), 30–38 (2013)

Aspects of Remote Monitoring of the Transport Vessel Under Operating Conditions

Andrii Golovan[1(✉)], Igor Gritsuk[2], Maksym Kurtsev[3], Oksana Ischuka[4], and Roman Vrublevskyi[2]

[1] Odessa National Maritime University, Odessa 65029, Ukraine
andrew.golovan@gmail.com
[2] Kherson State Maritime Academy, Kherson 73009, Ukraine
{gritsuk_iv, amor-vr}@ukr.net
[3] Ukrainian State University of Railway Transport, Kharkiv 61001, Ukraine
kurtsev_m@ukr.net
[4] Riga Technical University, Azenes Str. 12, Riga 1048, Latvia
oksieca@inbox.lv

Abstract. Remote monitoring is one of the most important steps in operating of a modern transport vessel. Monitoring of technical condition includes the use of vehicle-to-infrastructure (V2I) systems for the formation and application of separate systems of the ship power plant maintenance and repair. The information V2I model developed by the authors is characterized by the vibrational field of the transport vessel, means of monitoring of the technical condition parameters and infrastructure components for monitoring of any transport vessel. The article describes the features of the information V2I system, the process of preparation, monitoring and assessment of the technical condition of the transport vessel power plant under operating conditions with the possibility of predicting its technical condition. The main idea is to implement the information V2I system for monitoring of the technical condition of the transport vessel power plant in the practice of transport operation.

Keywords: Maritime transport · Vehicle-to-infrastructure · Power plant monitoring

1 Introduction

In order to improve the efficiency of technical use of a transport seagoing vessels (TSV) under operating conditions, it is required to regularly carry out its constructive modernization and use contemporary technological and informational methods and tactics when choosing a maintenance and repair strategy (MRS) of the TSV [1–3]. Constructive upgrading of the TSV is based on the development of new and the use of existing microprocessor and embedded systems that control the various processes going on in the mechanisms in the performance of the tasks assigned to the TSV. Microprocessor systems provide automation of these processes, optimize and plan the work of controlled mechanisms and make conditions for monitoring the parameters of

E. Ginters et al. (Eds.): ICTE ToL 2019, LNITI, pp. 295–301, 2020.
https://doi.org/10.1007/978-3-030-39688-6_37

the technical state in the V2I system [4–6]. Simultaneously, an important task is to enhance and verify a set of technical actions aimed at improving MRS and increasing the coefficient of technical willingness of the TSV in time-varying operating conditions (OC) [1–6].

The use of the TSV in non-stationary operating conditions involves constant monitoring of the parameters of the actual state of its power plant, carrying out the necessary technical arrangements for maintenance and ensuring uninterrupted operation. Such an interpretation of the conditions of operation of the TSV is possible only by monitoring the parameters of the technical state of the power plant, based on the processing of information about the operating modes of the power plant, continuous diagnostics and prediction of technical parameters. Incidentally, the researchers involved in the operation of the TSV face the problem of harmonizing the operating processes of the TSV and the parameters of operating conditions, generalizing and improving methods for predicting the technical state of the TSV. A lot of consideration is paid to accounting and monitoring the technical state parameters of the TSV under operating conditions, so the issue of building monitoring systems based on V2I systems for marine operating conditions of the TSV is relevant. A significant task is also to create an information space to monitor the efficiency of the operation of the TSV, taking into account the various operating conditions. Monitoring of the TSV power plant under operating conditions is one of the management functions of the state and operating conditions of the TSV, based on collecting information about the object of research and is aimed at observing the operating modes, set values of use and performance of planned activities, as well as prediction and control, specified by the design particularity and operating conditions of the TSV and the tasks assigned to it. The technical operation processes of modern TSVs should be considered as a complex dynamic system, functioning of which happens under the action of various random factors from both internal processes in their components and systems and under the influence of environmental conditions. That is, they are completely applied to structured systems such as V2I [7–13].

The proposed approach contains the system information interaction of the monitoring components: a transport vessel with an operator and an onboard information system; operating conditions of the TSVs (atmospheric and climatic conditions, water environment conditions, transport conditions and operating culture); transport and maritime infrastructure [9–13]. The structure of the model for providing information by the vehicle technical condition monitoring system under operating conditions has been developed. The aim of the study was to develop an information monitoring system designed to ensure the efficient operation of a transport vessel in non-stationary operating conditions coupled with the absence of high-quality data transmission networks except for expensive satellite data transmission systems. Database monitoring system models were created [9, 14]. Algorithms for identification, monitoring and diagnostics have also been developed with the possibility of prediction of the transport vessel power plant technical state parameters.

2 Models of Assessing the Technical State

The effectiveness of the TSV, as a complicated technical system, depends on its technical state. Hence the problem of managing a vehicle based on data obtained in the monitoring process and in predicting its main parameters attends. It is exposed that the external environment can bring in uncertainty and randomness of the raw data and situations, and randomly change the nature of the interaction between TSV mechanisms and systems. In such dynamic computing systems, random disturbances may occur, being measurement errors of diagnostic parameters. The impact of various kinds of nosiness that can cause an increase in the error in the conversion of diagnostic information is associated with the presence of unrecorded but objectively acting external causes. The record of the opportunity of moving over long distances, the complexity and variety of operating conditions of the TSV and the characteristics of the transport infrastructure can be considered as the object of automation of modern information systems such as V2I under operating conditions [7–13].

The theory of the TSV operation is the basis for the development of modern intellectual systems of regulating and planning in transport, built with the use of modern solutions and methods of obtaining information. Most of the tasks in the process of improving operational control methods for the performance of the TSV, being solved by the technical services of the TSV operation, have an information element of the evaluation: marine conditions of the TSV operation, monitoring of external conditions; prediction of possible emergencies, transport conditions in terms of saturation and intensity of the TSV movement, features of the transported cargo, mode and speed of movement; atmospheric and climatic conditions, culture of operation of the TSV etc. [7–13]. The listed and similar tasks are largely solved by obsolete methods that no longer provide the required quality and effectiveness.

The system of remote monitoring of the transport vessel technical state under operating conditions is a complex dynamic system with a well-ordered structure, an extensive network of relationships between its elements, which is developed in space and time. This system consists of many elements, which are characterized by the relevant initial conditions, are at different hierarchical levels and have their own characteristics of functioning. The activity of a complex manufacture system is determined by technological processes implementing its subjects, in accordance with the target orientation within the subject area. Each subject of monitoring has its inherent only functioning algorithms, calculating parameters, the results of which are used in the algorithms of subjects at various levels. After completion of control procedures and improvement of relevant technological processes, a change in the level of performance evaluation of the object of research takes place [7–9]. In general, the model of the efficiency of the subjects of monitoring can be represented as a combination of the following components: models of objects involved in structural relationships; models of parameters determining the performance results of the subjects; modeling algorithms establishing rules for ensuring the performance of objects and changes in the values of their parameters, etc. In the article the development of the monitoring model of the technical state parameters of the transport vessel power plant is based on the general approach to the research of the system “Transport vessel - Operator (Person) - Operating conditions - Transport vessel operation infrastructure", which contains the system interaction of the monitoring components: a transport vessel with an operator and an onboard information system, the operating conditions of the transport vessel (water environment conditions, atmospheric and climatic conditions, transport conditions and operating culture); transport infrastructure and maritime infrastructure (see Fig. 1) [7, 9, 13].

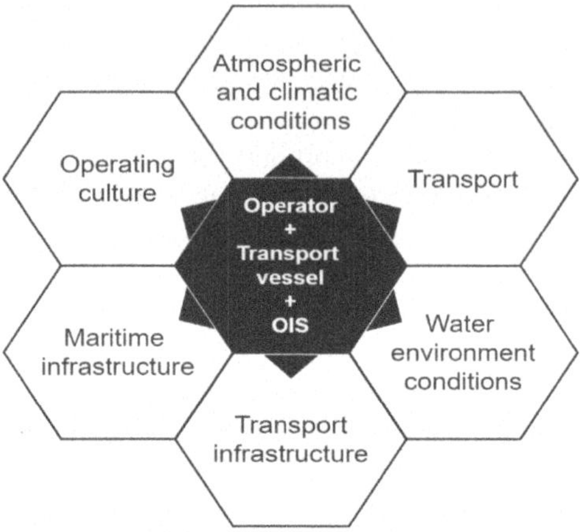

Fig. 1. The general scheme of the system interaction "Transport Vessel - Operator (Person) - Operating Conditions - Transport vessel operation infrastructure" [7, 9, 13].

In general, the standard form of the TSV technical state monitoring process under operating conditions is a formal process of changing and transforming information about the TSV technical state and processes of functioning and its operating conditions. Formally, the TSV technical state monitoring process under operating conditions is shown in Fig. 2 [7–9, 15–19, 21].

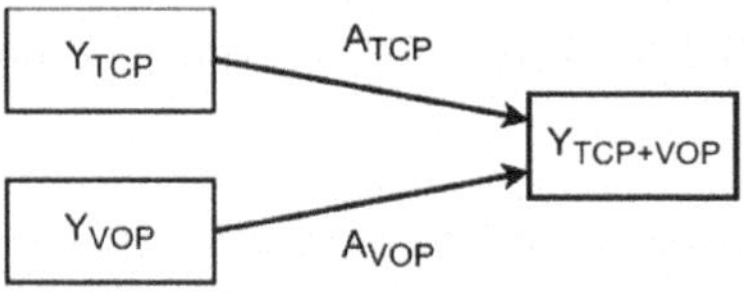

Fig. 2. TSV technical state monitoring process.

where Y_{TCP} - a set of aggregate models of the TSV technical state parameters; Y_{VOP} - a set of combined models of the parameters of the TSV operating conditions; $Y_{TCP\ VOP}$ - a set of combined models of the TSV technical state parameters in the relevant operating conditions; functional display of combined models of the TSV A_{TCP} - technical state parameters, but A_{VOP} - operating conditions.

3 Results of the Experimental Implementation

In the process of developing the information support for the processes of monitoring the parameters of the TSV technical state, taking into account the operating conditions, existing information sources were collected in terms of TSV coordinates on the territory in real time, models of maritime infrastructure objects, territorial natural and man-made systems and results of TSV tracking.

Fig. 3. An example of installing a vibroacoustic sensor and a diagnostic module on the EGT annular filter.

Fig. 4. An example of installing a current transformer on the power cable of an auxiliary diesel generator.

For a more convenient visualization of the analysis results, raster fragments of the investigated sea areas are added to the system. Sources of information to ensure the functioning of the information system V2I of monitoring the TSV technical state taking into account the operating conditions were the following in terms of: parameters of the state and position of the TSV on the map: https://www.shipmonitoring.org/; TSV transport operating conditions and geolocation: https://maps.google.com; TSV atmospheric and climatic operating conditions: http://ready.arl.noaa.gov/READYcmet.php. As a result of the experimental application of the system, reliable fixation of changes in the parameters of the technical state of the vessel was obtained, taking into account variations in operating conditions in combination with the transport and maritime infrastructure on an example of a separate TSV. The basic tasks of the system include monitoring the GPS parameters of the vessel's movement, operating modes of the main objects of the TSV power plant: main engines - crankshaft rotation speed; gas turbine rotation frequency [22] (sensors installation location is shown in Fig. 3), auxiliary diesel generators - electrical load (sensor installation location is shown in Fig. 4) and a ship boiler (operating time). Based on the data obtained, the calculations of power of the main engines and fuel consumption are made. At the stage of development the resolution of data recording and processing was 1 s, however in the final version the discreteness was equal to 1 min. The results of changing the parameters of the transport vessel technical state, taking into account changes in operating conditions in conjunction with the transport and maritime infrastructure, make it possible to establish and investigate on the example of a separate transport vessel, the patterns of changes in monitored values, determine characteristics of operating conditions or changes of the TSV parameters depending on operating conditions. The direct tasks of the system include monitoring the GPS parameters of the vessel's movement, operating modes of the main objects of the TSV power plant: main engines - crankshaft rotation speed is measured using inductive sensors; gas turbine rotation frequency is measured using a vibroacoustic sensor [20], auxiliary diesel generators - electrical load and the ship boiler operating time is measured using current transformers.

4 Conclusion

Conducted experiments show that the developed V2I model of remote monitoring under operating conditions is applicable to transport seagoing vessels. The use of remote monitoring of the technical condition parameters and infrastructure components for monitoring helps to improve the efficiency of technical use of a transport seagoing vessel. Using a vibrational field to determine engine power is cheaper than using torque meters, indicator card measurements, digital pressure monitoring and intelligent combustion monitoring. In future studies is planned to test the applicability of the V2I model on transport inland vessels, passenger seagoing and inland vessel as well as on stationary power plants equipped with diesel generators.

References

1. Grynkiv, A.V.: The use of forecasting methods in managing the technical state of aggregates and vehicle systems. Agric. Eng. Ind. Eng. Autom. **29**, 25–32 (2016)
2. Troitskiy-Markov, T.Y., Sennovskiy, D.V.: Principles for energy efficiency monitoring system. Monit. Sci. Saf. **4**, 34–39 (2011)
3. Davydenko, L.V., Davydenko, V.A., Komenda, N.V., Yarmolska, N.V.: Energy monitoring functions of compound production systems and their objectives for increasing energy efficiency. The Herald of KhNTUA. Technical sciences, Issue 153, Problems of energy supply and energy saving in agricultural sector of Ukraine, pp. 125–127 (2014)
4. Alekseyev, V.V., Kurakina, N.I., Orlova, N.V., Minina, A.A.: GIS monitoring of transport networks. Data +. Geo information systems for business and society, no. 2(69) (2014). https://www.dataplus.ru/news/arcreview/detail.php?ID=17802&SECTION_ID=1058. Accessed 21 July 2019
5. Anfilatov, V.S., Emelyanov, A.A., Kukushkin, A.A.: Systems Analysis in Management. Finance and Statistics, Moscow (2002

6. Borukayev, Z.H., Ostapchenko, K.B., Ericyuk, L.I.: Computer model of energy efficiency monitoring: aspects of information modeling. Power Eng. Electrif. **1**, 3–7 (2007)
7. Gritsuk, I.V., Volkov, V., Mateichyk, V., Grytsuk, Y., et al.: Information model of V2I system of the vehicle technical condition remote monitoring and control in operation conditions. SAE Technical Paper (2018). https://doi.org/10.4271/2018-01-0024
8. Gritsuk, I., Gutarevych, Y., et al.: Improving the processes of preheating and heating after the vehicular engine start by using heating system with phase-transitional thermal accumulator. SAE Technical Paper (2016) https://doi.org/10.4271/2016-01-0204
9. Gritsuk, I., Volkov, V., Gutarevych, Y., Mateichyk, V., et al.: Improving engine pre-start and after-start heating by using the combined heating system. SAE Technical Paper (2016). https://doi.org/10.4271/2016-01-8071
10. Atroshchenko, V.A., Shevtsov, Yu.D., et al.: Technical possibilities for increasing the capacity of autonomous power plants of power systems. Monograph. Publishing House-South (2010)
11. Vlasov, V.M., Nikolaev, A.B., et al.: Information technologies for vehicles. Vehicles Efficiency Monitoring. MAHI State Technical University. Nauka, Moscow (2006)
12. Akhmedov, T.N., Zhankasiev, S.V., Finkel, A.E.: Basics of vehicle condition monitoring system in the process of carriage. Scientific aspects of the development of transport and telematics systems, pp. 138–164 (2010)
13. Haydamakin, N.A.: Automated Information Systems, Bases and Databases. Introductory Course. Study Allowance. Gelios ARV, Moscow (2002)
14. Volkov, V.P., Mateichyk, V.P., Nikonov, P.B., et al.: Integration of Vehicle Technical Maintenance in the Structure and Progress of the Intelligent Transport System. Monograph. Publishing House Noudlijj, Donetsk (2013)
15. Kuleshov, A.P.: Information model as a basis for designing the corporate automated information systems. Inf. Technol. **3**, 26–30 (2006)
16. Makhammad, M.D.: The development of an information system for diesel power stations with the possibility of prediction of their technical condition. The author's abstract of Ph.D. thesis, Kuban State Technological University, Krasnodar (2009)
17. Govorushchenko N.Ya.: System engineering of motor transport (calculation methods of research). Monograph. KhNAHU, Kharkov (2011)
18. Govorushchenko, N.Ya., Turenko, A.N.: System engineering of transport (the example of road transport). RIA KhAHI, Kharkiv (1999)
19. Varbanets, R., Karianskyi, S., Rudenko, S., Gritsuk, I., et al.: Improvement of diagnosing methods of the diesel engine functioning under operating conditions. SAE Technical Paper (2017). https://doi.org/10.4271/2017-01-2218
20. Golovan, A., Rudenko, S., Gritsuk, I., Shakhov, A., et al.: Improving the process of vehicle units diagnosis by applying harmonic analysis to the processing of discrete signals. SAE Technical Paper (2018). https://doi.org/10.4271/2018-01-1774
21. Gritsuk, I., Zenkin, E.Y., Bulgakov, N., Golovan, A., et al.: The complex application of monitoring and express diagnosing for searching failures on common rail system units. SAE Technical Paper (2018). https://doi.org/10.4271/2018-01-1773
22. Golovan, A., Gritsuk, I., Popeliuk, V., et al.: Features of mathematical modeling in the problems of determining the power of a turbocharged engine according to the characteristics of the turbocharger. SAE Int. J. Engines (2020). https://doi.org/10.4271/03–13-01-0001

Autonomous Vehicles and Mechatronics Education

Gilberto Marzano[1,2](✉), Andris Martinovs[1], and Svetlana Usca[1]

[1] Rezekne Academy of Tecnologies, Rezekne, Latvia
gilmarzano@gmail.com, {gilberto.marzano,andris.martinovs,
svetlana.usca}@rta.lv
[2] Spoleczna Akademia Nauk, Łódź, Poland

Abstract. Automotive mechatronics involves the multidisciplinary integration of automotive mechanical and electronic systems. As such, automotive mechatronics education integrates knowledge and competences across automotive-specific mechanics, electronics, communication, advanced control, and modeling. Over the last few years, the subject of autonomous vehicles has been attracting significant attention from researchers, as well as interest from the media and the public at large. It encompasses a myriad of new applications, ranging from driverless cars to airborne surveillance of sensitive areas. Designing solutions for autonomous vehicles requires the integration of hardware and software components, as well as interaction with cloud-based servers. This paper focuses on autonomous/unmanned ground vehicles within the scope of a mechatronics curriculum. It has been proposes an educational model for autonomous vehicles for inclusion in the new mechatronics curriculum of NewMetro, an EU funded project that aims to develop an innovative European framework of competences for mechatronics education.

Keywords: Automotive mechatronics · Autonomous vehicles · Education

1 Introduction

Today, automobile manufacturing arguably represents one of the most important industries in the world [1], and cars, trucks, and buses play a fundamental role in our everyday lives. Over the last century, the automobile industry has thoroughly changed people's habits and attitudes. In modern times, autonomous vehicles will thoroughly change our lifestyles once again. They will impact the way people commute to work, and transform the delivery of goods and packages. In short, they will enormously affect the automobile industry system [2], and, by increasing the safety of travel in general, will have a profound effect on the motor insurance system as well.

Autonomous driving encompasses three classes: semi-autonomous, sufficient-autonomous, and fully-autonomous driving. Semi-autonomous driving combines three main capabilities: adaptive cruise control, lane-keeping, and automatic stopping. Increasing these capabilities contributes to increasing the comfort of driving a vehicle, and make driving safer. Semi-autonomous driving comprises such aspects as automatic parking, retrieving a car from the garage, avoiding obstacles and pedestrians on the

E. Ginters et al. (Eds.): ICTE ToL 2019, LNITI, pp. 302–310, 2020.
https://doi.org/10.1007/978-3-030-39688-6_38

road, and managing alternative routes according to traffic and weather conditions. Self-sufficient-autonomous driving allows a vehicle to drive in a well-predetermined context. For instance, a self-autonomous vehicle is able to drive from point A to point B on a predetermined road. This type of vehicle can stop at traffic lights, avoid obstacles, park itself, etc.

Fully-autonomous driving can accomplish all the requirements of normal driving.

According to SAE (Society of Automotive Engineers) International's J3016 standard, there are six levels in on-road motor driving [3]:

- Level 0 – No Automation;
- Level 1 – Driver Assistance;
- Level 2 – Partial Automation;
- Level 3 – Conditional Automation;
- Level 4 – High Automation;
- Level 5 – Full Automation.

In the first three levels, a human has primary responsibility. In the last three, the computer is responsible for driving.

In the last few years, research on autonomous vehicles has grown exponentially, whilst also attracting attention from the media [3]. It encompasses myriads of new applications, including airborne surveillance of sensitive areas.

Designing solutions for autonomous driving requires the integration of mechanical, electronic, and software components as well as interaction with cloud-based servers. Consequently, the automotive mechatronics curriculum should include not only elements of Industry 4.0 but also notions of autonomous driving.

This paper focuses on autonomous driving within the scope of the new mechatronics curriculum that is currently developing within the NewMetro project.

2 The NewMetro Project

Mechatronics has been described as "an intelligent and unifying paradigm, that offers an area of interdisciplinary knowledge and interactions regarding ways of working and thinking, practical experiences, as well as theoretical knowledge" [4]. Mechatronics is generally considered to represent the merger of traditional Mechanical and Electrical/Electronic Engineering with Computing elements [5]. However, over the last decade, the spread of the Internet of Things (IoT) has been giving rise to significant changes in the way mechatronics and related systems are designed and configured [6].

Nowadays, the growing importance of mechatronics is due largely to a burgeoning consumer demand resulting from the extensive use of WI-FI and cloud computing. Over the last few years, there has been an increase in the need to provide users with a broader range of functions based on smart devices capable of being modified and integrated within conventional as well as advanced mechatronic components. For instance, new vehicles contain many mechatronic sub-systems used, for example, in drive-train management, smart lights, auto-parking systems, driver alertness, and pre-collision detection. Similarly, new healthcare devices include pulse, blood oxygen, temperature, and respiration sensors, and provide a recording hub to track and store a patient's vital signs.

As a result, worldwide, many universities, as well as vocational and engineering schools, have introduced Mechatronics Curricula to offer courses that meet the market demand and the future trends of industry 4.0. Nevertheless, as digital technology continues to evolve and bring significant changes, new mechatronics education programs are needed. Meanwhile, it is essential to create new teaching-learning models that are in line with the technological advances.

The NewMetro project has been designed to address the impact that the IoT has on mechatronics education. It is an EU funded, 36-month project that aims to develop an innovative European framework of competences for mechatronics education. The project involves 11 partners from 7 European countries (Italy, Latvia, Austria, Germany, Poland, Greece, and Spain). It was launched in December 2018 and, in May 2019, a research on the key factors that can influence the development of a new mechatronics curriculum was carried out. The research was articulated in three primary activities: (1) reviewing relevant surveys, articles, and projects connected with mechatronics competences; (2) identifying the current skills gaps in order to define the framework of competence for an international curriculum; (3) surveying the views of current students, young employees, and employers. From the research it has emerged that, in the mechatronics industry, competition is particularly strong, despite the highly customized nature of mechatronic products.

Table 1. The most important mechatronics competencies.

Technological area	All (N = 212)		Students/young employees (N = 155)		Employers (N = 58)		Differences between (p)	
	Mean	Number of cases	Mean	Number of cases	Mean	Number of cases	Employers students/employees	Counties
System design and integration of electronic and mechanical components (assembling mechatronic systems, setting up machine controls, etc.)	67	142	.71	99	.72	42	.270	.534
Robotics (setting up automotive robots, implementing autonomous driving devices)	.43	90	.43	66	.19	24	.999	.079

(*continued*)

Table 1. *(continued)*

Technological area	All (N = 212)		Students/young employees (N = 155)		Employers (N = 58)		Differences between (p)	
	Mean	Number of cases	Mean	Number of cases	Mean	Number of cases	Employers students/employees	Counties
Design virtual testing and validation using modelling and simulation tools (simulating mechatronic design concepts, using CAM software)	.41	86	.37	57	.07	28	.110	.254
PLC programming (programming a CNC controller)	.40	85	.37	56	.50	29	.049	.240

The results of the survey reveal that the most important mechatronics competencies for students, young employees, and employers are related to four primary technological areas (see Table 1). The survey results were processed using SPSS 25.0 software. Cronbach's alpha test has been used to determine the internal consistency of the survey, whilst the Kolmogorov–Smirnov test has been used to test the goodness of fit for the data distribution. In the evaluation of competencies for PLC programming, employers considered them more important than students/young employees do. In other cases, no statistically significant differences have been found.

Following the administration of the survey, a focus group of employers was organized. This group indicated the strategic importance of the technological area of Robotics, and also emphasized the importance of the sector of autonomous driving.

Accordingly, it has been decided to focus on how to modernize the current curriculum on automotive mechatronics in the light of the autonomous driving perspective.

This paper illustrates the proposal for a new mechatronics educational curriculum that includes a module in autonomous driving technologies.

3 Autonomous Driving Technologies

Autonomous driving encompasses several distinct technologies. It is a complex system that consists of three main subsystems: sensing, perception, and decision making [7]. An autonomous driving system can be viewed as a complex set of interacting intelligent sensors and software that are integrated with the vehicle driving components [8].

3.1 A Subsection Sample

The primary task of the sensing subsystem is that of extracting information from the sensors' raw data, recording and analyzing them to understand the context and so make decisions about the driving.

Usually, an autonomous vehicle utilizes many types of sensor:

- GPS and IMU (Global Positioning System and Inertial Measurement Unit), which allow the vehicle to localize itself, providing accurate and real-time updates of its localization;
- LiDAR (Light Detection and Ranging or Laser Imaging Detection and Ranging), which is used for mapping, localization, and the avoidance of obstacles.
- Cameras, which are used for several functions, notably lane detection, object recognition, traffic light detection, pedestrian detection, etc.
- Radar and Sonar, which are used to calculate the distance from an object as well as the velocity approaching it, in order to reduce the danger of collisions.

3.2 Perception Sub-system

Data coming from sensors are processed by the perception subsystem. It performs three main tasks:

- Localization, combining information from GPS and IMU and correcting errors;
- Object detection, using different filters to analyze the input image;
- Object tracking, whose goal is to automatically track the trajectory of an object, e.g., another vehicle or a person crossing in front.

Recently, Deep Learning Technology has been seeing significant advances in object detection and tracking. Convolution Neural Network (CNN) is a type of Deep Neural Network that is being successfully experimented. Nowadays, there are Deep Learning open platforms such as Spark and Paddle that are available in an open source network through Baidu.

Perception is a very big obstacle to face in implementing autonomous driving systems. The problem of visual inferences from images is the central issue of computer vision, which is a branch of artificial intelligence.

To support image recognition and classification, several datasets have been realized through different methods, e.g., crowdsourcing [9] and technical approaches [10–12].

3.3 Decision Sub-system

The aim of the decision sub-system is to oversee the other sub-systems, and to use the information they provide to direct and manage the vehicle [13]. Its role is to make decisions regarding the reactions the system should take, for example, following the rules of the road, or taking other actions such as applying the brakes in an emergency or turning the steering wheel [14].

The decision sub-system makes effective and safe decisions by leveraging information regarding:

- Vehicle position;
- Routing outputs;
- Previous behavioral decisions;
- Obstacle information;
- Traffic and map information;
- Local traffic rules.

Usually, a Markov Decision Process (MPD) is used for identifying and choosing the course of action from among the best alternatives. The solution for an MDP is a policy which describes the best action for each state in the MDP. The optimal policy can be obtained through a variety of methods, such as linear programming or dynamic programming.

3.4 Autonomous Driving Model

To represent the various components of an autonomous driving system, a model has been defined. It is based on the three sub-systems described above (namely, sensing, perception, and decision), an algorithm library, and a dataset of driver behavior including, for example, driver psychological uncertainties and inconsistent activities (see Fig. 1).

Analyzing and representing driver behavior is crucial if a Bayesian network is used to find the likely impact of traffic situations, in which participants are represented as nodes connected with conditional distribution functions.

4 Autonomous Vehicles in a Mechatronics Curriculum

It has been defined the "Autonomous Unmanned Vehicles (AUV)" module (that represents 15 credits in the European Credit Transfer and Accumulation System, ECTS), based on the general model for autonomous driving resulting from the literature analysis and the comparison of mechatronics curricula in the countries participating in the NewMetro project. Within the classical curricula, machine learning, and convolutional neural networks are not included [16] and courses at universities often lack integration of the topics included in the proposed autonomous driving model.

The AUV module foresees preliminary knowledge in a range of areas, namely: Technical Mechanics; Materials Science; Metrology; Thermodynamics and Heat Engineering; Technical Graphics; Basics of Design; Precision and Standardization of Parts; Metalworking Technologies and Technological Equipment; Hydraulic and Pneumatic Drives; Electrical Machines and Electric drives; Programming Basics; Electrical Engineering; Electronics; Power Electronics; Automatic Control and Regulation; Microprocessor Techniques; Programming for Microprocessors; and Computer Control Systems.

The AUV module includes the specific knowledge that are part of the autonomous driving model:

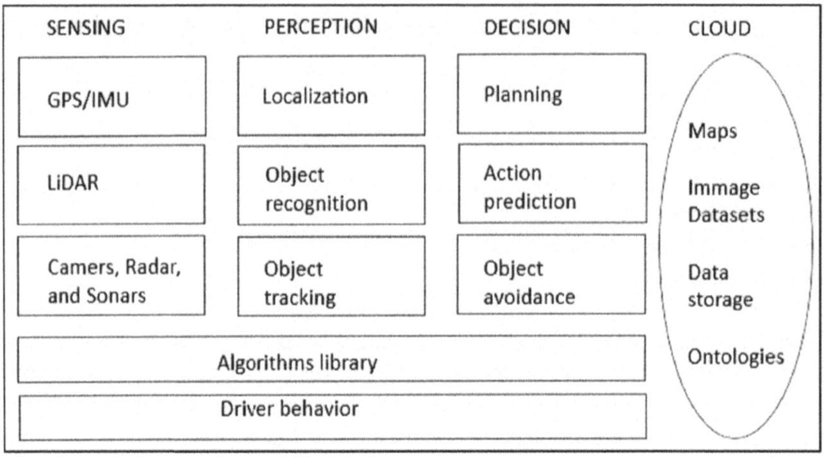

Fig. 1. Autonomous driving model overview.

- AUV Sensing (interfacing GPS/IMU, LiDAR, Cameras, Radar, Sonars, and intelligent devices);
- AUV Mechanics (interacting with driving devices);
- AUV Electronics (electrical supply, circuitry, driving controllers);
- AUV Software (metaheuristic methods, computer vision, machine learning, convolutional neural networks, image datasets);
- AUV behavioral driving (driver behavior, controlling, classification, prediction, planning, decision-making, etc.);
- AUV Design (design, modeling, simulation);
- AUV Prototypes implementation (at the end of the module, students, working in groups (of 3–4 students) present their prototype/model.

Following this module, students will achieve:

- An overview of autonomous driving technologies;
- Knowledge of Deep learning in autonomous driving;
- Knowledge of Behavioral decision in autonomous driving; and
- Competence in sensor interfacing and control.

It is expect that, at the end of the module, students will be able to:

- Design an AUV prototype using CAD/CAE technologies;
- Integrate mechanical and electronic parts of an AUV prototypes;
- Manage/simulate AUV communication with the control center; and
- Implement and use algorithms for autonomous driving.

5 Conclusion

The potential of autonomous driving is significant. Technologies enabling autonomous vehicles can open new opportunities in many fields, with positive repercussions on everyone's daily lives. In this sense, then, autonomous driving represents an extremely disruptive innovation [17].

Introducing AUV in a mechatronics curriculum will reinforce the integration of competence that has been presented is a preliminary attempt to introduce, into current mechatronics curricula, some notions, such as machine learning and deep machine learning, which are essential in robotics and behavioral robotics. Indeed, they closely interact with control theory.

The further step of the NewMetro project is to define the general framework for new mechatronics curricula. The idea is the creation of specific modules that integrates basic mechatronics knowledge. The AUV module has been conceived in this perspective. The research will continue from here, in order to realize an experimental educational module that will be tested within the NewMetro project.

Acknowledgements. This paper has been supported by the EU funded NewMetro project No. 600984-EPP-1-2018-1-IT-EPPKA2-SSA). The views expressed reflect those of the authors alone.

References

1. Papatheodorou, Y., Harris, M.: The Automotive Industry: Economic. Impact And Location Issues, IndustryWeek. https://www.industryweek.com/global-economy/automotive-industry-economic-impact-and-location-issues. Accessed 19 Oct 2019
2. McGrath, M.E.: Autonomous Vehicles: Opportunities, Strategies, and Disruptions (2018). Independently published
3. SAE, SAE. J3016 Standard: taxonomy and definitions for terms related to on-Road motor vehicle automated driving systems (2014)
4. Bagloee, S.A., Tavana, M., Asadi, M., Oliver, T.: Autonomous vehicles: challenges, opportunities, and future implications for transportation policies. J. Modern Transp. **24**(4), 284–303 (2016)
5. Habib, M.K., Davim, J.P.: Interdisciplinary Mechatronics: Engineering Science and Research Development. Wiley, Hoboken (2013)
6. Cetinkunt, S.: Mechatronics with Experiments. Wiley, Hoboken (2015)
7. Bradley, D., Russell, D., Ferguson, I., et al.: The Internet of Things-the future or the end of mechatronics. Mechatronics **27**, 57–74 (2015)
8. Liu, S., Li, L., Tang, J., Wu, S., Gaudiot, J.L.: Creating autonomous vehicle systems. Synth. Lect. Comput. Sci. **6**(1), i–186 (2017)
9. Behere, S., Torngren, M.: A functional architecture for autonomous driving. In: 2015 First International Workshop on Automotive Software Architecture (WASA), pp. 3–10 (2015)
10. Fei-Fei, L., Deng, J., Li, K.: ImageNet: constructing a large-scale image database. J. Vision, **9**(8), 1037. https://jov.arvojournals.org/article.aspx?articleid=2135307. Accessed 20 Oct 2019
11. Baker, S., Scharstein, D., Lewis, J.P., et al.: A database and evaluation methodology for optical flow. Int. J. Comput. Vision **92**(1), 1–31 (2011)

12. Scharstein, D., Hirschmüller, H., Kitajima, Y. et al.: High-resolution stereo datasets with subpixel-accurate ground truth. In: German conference on pattern recognition, pp. 31–42. Springer, Cham (2014)
13. Yang, G., Song, X., Huang, C., Deng, Z., Shi, J., Zhou, B: DrivingStereo: a large-scale dataset for stereo matching in autonomous driving scenarios. In: Proceedings of the IEEE Conference on Computer Vision and Pattern Recognition, pp. 899–908 (2019)
14. Hubmann, C., Becker, M., Althoff, D., Lenz, D., Stiller, C.: Decision making for autonomous driving considering interaction and uncertain prediction of surrounding vehicles. In: 2017 IEEE Intelligent Vehicles Symposium (IV), pp. 1671–1678 (2017)
15. Czubenko, M., Kowalczuk, Z., Ordys, A.: Autonomous driver based on an intelligent system of decision-making. Cogn. Comput. **7**(5), 569–581 (2015)
16. Frochte, J., Lemmen, M., Schmidt, M.: Seamless integration of machine learning contents in mechatronics curricula. In: 19th International Conference on Research and Education in Mechatronics. IEEE (2018)
17. Gao, P., Kaas, H.W., Mohr, D., Wee, D.: Disruptive Trends that will Transform the Auto Industry, vol. 1, pp. 1–9. McKinsey & Company, New York (2016)

Developing Teachers Digital Competence in Transport and Logistics Through Blended Learning Tools

Vasyl Cherniavskyi[1], Sergei Voloshynov[1](✉), Olena Volska[1], Natalya Panchenko[2], Aleksejs Vasiljevs[3], and Tatiana Bezverkhnuik[4]

[1] Kherson State Maritime Academy, Kherson 73000, Ukraine
ch_vv@i.ua, s_voloshinov@ukr.net, vol.3.77@mail.ru
[2] Ukrainian State University of Railway Transport, Kharkiv 61001, Ukraine
panchenkonatali69@gmail.com
[3] Riga Technical University, Azenes Str. 12, Riga 1048, Latvia
aequee@gmail.com
[4] Odessa Regional Institute for Public Administration of the National Academy for Public Administration under the President of Ukraine, Odessa 65000, Ukraine
tatiana.bezverkhniuk@gmail.com

Abstract. The e-course for teachers in the field or transport and logistics forms the motivation for professional communication, skills in using digital resources, reducing uncertainty over new technologies, recognition by colleagues, increasing self-esteem and striving for continuous improvement and development. The LMS Moodle learning environment is used for organizing the communicative interaction of all stakeholders in order to form professional competencies. The system tracks the dynamics, progress of training, and clearly shows which elements of the course have already been passed, and which are just to be completed. Teachers who study on the course have the opportunity to feel like students. The development of digital competence of teachers in the field or transport and logistics is a priority that can be solved with the help of a blended learning system.

Keywords: Transportation and logistics · E-learning · Digital competence

1 Introduction

The efficiency of the system of future specialists' professional training in the marine industry at the current stage of education development, taking into account the European integration processes, is determined by an adequate choice of goals and objectives, organizational forms, methods and means of training in their rational combination. Modern training in maritime affairs should be carried out under the conditions of widespread use of information and communication technologies, application of the module-rating system of educational process organization, which focuses on independent training and provides the possibility of remote access to various sources; the use of virtual computer models (simulators), multimedia tutorials and classic training tools.

E. Ginters et al. (Eds.): ICTE ToL 2019, LNITI, pp. 311–318, 2020.
https://doi.org/10.1007/978-3-030-39688-6_39

It is known that a significant improvement of professional training for the future specialists of the marine industry is possible if the development of modern theoretical and methodological principles of training based on a competent approach and the introduction in the educational process of maritime educational institutions methodical system based on the principle of integration humanitarian and professional development a complete competence model of a marine professional are provided. In Ukraine, the transition to digital society has been the boost for destroying traditional pedagogy foundations and has led to irreversible changes in organization of the educational process. As it turns out, creating computer classes, accumulating interactive multiboards, providing Internet and high-speed WiFi access cannot solve the most important problem – forming the teacher's and future professional's digital literacy.

The Horizon 2018 Annual Report defines that digital literacy goes beyond the acquisition of discrete technological skills, transforming into a deeper awareness of the digital environment that allows to adapt to new conditions. And these are higher education institutions that are entrusted with the responsibility to develop students' digital literacy and contribute to responsible and proper use of online technologies [1].

2 Competency Framework

The issue of digitization (converting information into a digital format) was recognized as a priority at the Conference of Education Ministers of the European Higher Education Area (The Fifth Bologna Policy Forum) (May 25, 2018). The call was proclaimed to support teachers in the creative digital environment, develop their digital skills and competences with a view of better use of digital and blended learning [2]. Thus, the description of the teacher's digital competence is continuing in Ukraine at the level of the Ministry of Education and Science according to the plan for implementation of the Concept of teacher education, which is an important step in contributing to the quality of educational services, integration of Ukraine into the European and world educational environment and increase of the graduates' competitiveness. The consideration of the existing experience in forming teachers' digital competence (DC) will allow to make this process more effective, real and not book-learned. In M. Shermar's apt words, over the past five years, humanity has slowly but irreversibly transformed from the top-down to bottom-up system - for one simple reason: both information and people seek to be free [3].

Digital literacy is recognized as the top priority for higher education, but despite the increasing importance of DC, defining its components remains a complicated subject [4]. The creation and appliance of information computer technologies (ICT) in pedagogical activity were investigated by the following scientists: V. Bykov, L. Petukhova, I. Robert, O. Spivakovskyi, O. Kukharenko, V. Yatsyshyn. In the literature there are many synonymous terms to the definition of "digital competence", namely: information competence, digital literacy (competence in the field of ICT (ICT competence)). According to researchers Alma Zh. Murzalinova and Natal'ya S. Kol'eva, interpretation of the concept of information competence reflects the prompt response of the conceptual framework of pedagogical science to the information society demands, which are gradually becoming more complicated, but, unfortunately, the learning results in building competence are limited to the ability to work with information.

DC consists of:

- Technical skills of using digital technologies;
- The ability to reasonably use digital technologies in work, study and everyday life in general;
- The ability to critically evaluate digital technologies;
- The motivation to participate in digital culture.

Digital competence is regarded as a core competence in policy papers; in research, however, it is not yet a standardized concept. Several policy- or practice-related projects are currently working for finding a common and acceptable definition [5]. Information competency is one of the essential qualities of a person living in the information age. It includes skills in information management in both educational domains and the outside world, as well as readability and capability to use modern information and communication technologies [6].

The components of DC also include additional knowledge, skills, aptitudes and attitudes, among which are ICT technical skills, the ability to use the mentioned resources in the educational process, the ability to plan, analyze and manage the educational process by using them [7]. Popova H. and Yurzhenko A. used blended learning to introduce competency approach in professional maritime education. They have proved that competency framework implemented in Kherson State Maritime Academy (KSMA) increases the effectiveness of the learning process management and implements a competency approach in the process of professional training [8].

3 Blended Learning

3.1 E-learning Environment

Learning management system is a learning platform designed to bring educators, administrators and students together into one robust, secure and integrated system in order to create a personalized learning environment. Teachers develop packages of educational and methodological online materials with interactive tasks of their disciplines, providing free access to them for students, which are located at LMS of an educational institution. All packages of educational and methodological online materials located at LMS must be fully compliant with the curricula of the academic disciplines. The advantages of using this platform are as follows: availability of e-learning content anywhere and anytime; integration with other platforms and tools; fast downloading of any document format; simple interface; free-of-charge basis; constant updating; multiple-language support, etc. The structure of electronic website of KSMA consists of following components: ship engineering faculty, ship navigation faculty, maritime college and training center.

3.2 E-courses

E-course (electronic course) is a part of blended learning, resource which supports educational process in the institution as well as self-education within curricula.

E-course on LMS is represented as separate educational electronic environment help by a teacher. E-course of the certain discipline is connected to the Standards of training, certification and watch keeping for seafarers (STCW) Convention and other disciplines through repository of competences. The example of e-courses on LMS Moodle is Navigational Safety at Sea, which includes 780 registered participants. The beginning of the course contains the following structural elements: competence-based requirements for professionals' skills; STCW-78/2010 Page; connection with other professional training disciplines; place of a discipline in the general structure of the training according to STCW; forum; information about the teacher and author of the discipline; examples of testing activities; a working training program; a course glossary.

3.3 E-course Structure

The compulsory structure of the course includes the following elements: course abstract, course performance map, syllabus, interactive content (gamified exercises), feedback on the course, work evaluation elements, e-gradebook. Feedbacks are generated through forums, private messaging and tasks with the built-in option of posting comments. Most course elements (assignments, forums) contain assessment elements (can be evaluated by the teacher).

4 Formation of Teachers' Digital Competence on LMS Moodle

The NMC Horizon-2017 Report states that blended learning is a short-term trend to be implemented that contributes to the introduction of educational technologies in higher education in the next one or two years. It has also been stated that blended learning has drawn together the best practices of full-time study and online learning, while the low cost of blended learning, its flexibility and availability, as well as the possibility of using sophisticated technologies in the learning process have made this alternative much more attractive [1].

The effectiveness of building professionals' competencies depends on the level of their preparation and on how relevant and up-to-date their knowledge is. Thus, one of the biggest obstacles to the introduction of e-learning and blended learning at Ukrainian higher education institutions is teachers' low digital literacy. Electronic LMS, in the framework of which blended learning is implemented, are widely used in practice by many higher education institutions in Ukraine. So for the purpose of staff development open distance-learning courses are createdю They allow teachers to learn how to use digital technologies in the educational process independently and at the time of their convenience.

In spite of the above-mentioned, distance-learning courses are not popular in practice, precisely because of the lack of involved teachers' skills to learn online. In order to support KSMA professionals and build teachers' digital competence, Innovative Technology Laboratory was established in 2017 to develop blended learning.

The laboratory provides practical assistance to professionals in mastering the latest ICT, creating e-courses, applying Web 2.0 and Web 3.0 technologies, implements e-learning and m-learning, which are effectively used for different modes of study, develops various e-learning software packages, including systems related to content delivery, organization and learning management - LMS, which comprise tools on administration, communication, knowledge assessment and development of training courses. The current direction of the laboratory work is the introduction of e-learning on the LMS Moodle [4]. The specialists of the laboratory provide assistance in choosing the structure of an e-learning course, the process of its creation etc. The teachers carry out tasks in their personal courses [8]. The peculiarity of the e-course "Moodle learning" is its interactive content (video, interactive pictures, screencasts and gamified exercises), communication tools (forums, chats), interactive lectures and exercises. All topics are formed with consideration to Bloom's taxonomy-based thinking and exercises are matched according to Web 2.0. services [9]. An example of effective use of digital competence is training of teachers in the course "Organization of technical operation of ship power plants" with use of the LMS Moodle system. The essence of the course is emphasized by the important role of the technical operation of ship power plants in solving the issue of energy security of vehicles. Increasing the efficiency of engines implies the improvement of their main indicators, which include, first of all, power performance - both economic and environmental [10–12].

5 Experimental Study of E-learning Implementation

At the beginning of the experiment at KSMA (e-learning implementation), 09/01/15, the quality of knowledge of future sea transport specialists as a whole at the Academy was 70.42%. Blended learning was not applied.

The first year of study (2015–2016) takes place using blended learning and a small part of simulators training (see Fig. 1). Teaching time is allocated to classroom work, 73%, e-learning 19%, and simulators training 8%. After the first year of blended learning, the quality of knowledge increased by 1.70% to 72.12%.

The second year of study (2016–2017) comes with the use of blended learning and a small fraction of training. Teaching time is allocated to classroom work 68%, e-learning 17% and simulators training 15%. After the second year of blended learning, the quality of knowledge increased by 1.24% to 73.40%.

In the third year of study (2017–2018) occurs using e-learning and practical training. Teaching time is allocated to e-learning 14%, to classroom training 25% and practical training 61%. After the third year of blended learning, the quality of knowledge decreased by 3.23% to 70.13%. This is due to the fact that future experts in sea transport go to practice on ships. They serve as cadets or ratings and fulfill the functions stipulated by the STCW. In this regard, more time is devoted to work rather than to learning.

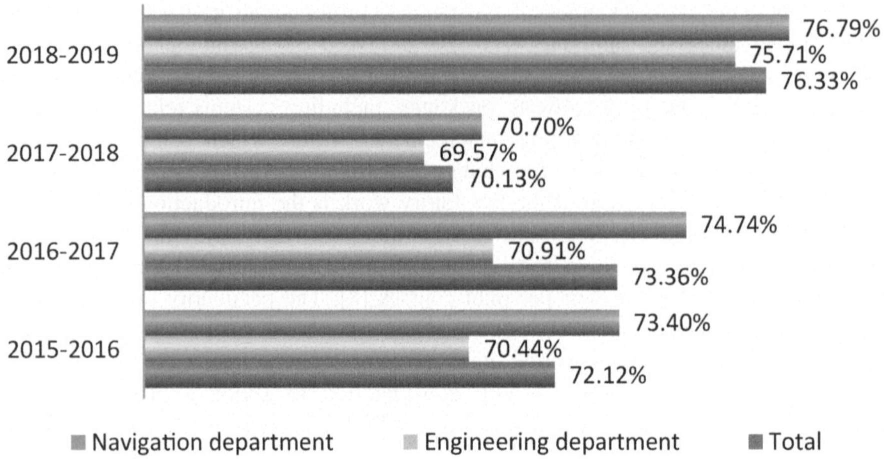

Fig. 1. Qualitative assessment (in percent) of the use of blended learning by sea transport professionals.

The fourth year of study (2018–2019 again) comes with the use of blended learning and a large proportion of training. Teaching time is allocated to 38% of the classroom work, 26% to e-learning and 36% to simulators training. After the fourth year of blended learning, the quality of knowledge increased by 6.30% to 76.33%. As a whole, the project implementation period is 5.91%. In general, the distribution of training load is presented in the diagram below (see Fig. 2).

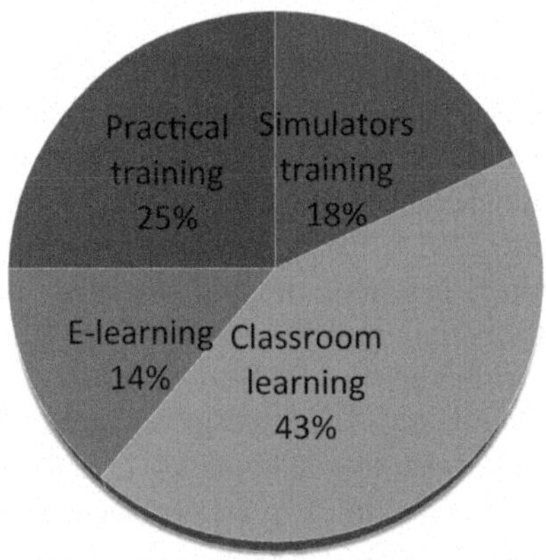

Fig. 2. Distribution of training time for future sea transport specialists.

As can be seen in Fig. 2 total time of sea transport specialists training is divided into four parts: practical training and classroom training (68%) goes as a traditional training; e-learning (global trend with the use of electronic means) and simulators training (use of virtual-real ship of KSMA).

Thanks to the introduction of e-learning and optimization of simulators training and practical training in the system of sea transport specialists education, the quality of knowledge was increased by 5.91%. Which in turn made it possible to provide for the employment of sea transport specialists in the first year 39% of graduates, 62% in the second year and 84% in the third year.

6 Conclusion

The efficiency of the system of professional training of future specialists in the marine industry at the current stage of development of education, taking into account the European integration processes, is determined by an adequate choice of goals and objectives, organizational forms, methods and means of training in their rational combination. The e-course for teachers in the field or transport and logistics forms the motivation for professional communication, skills in using digital resources, reducing uncertainty over new technologies, recognition by colleagues, increasing self-esteem and striving for continuous improvement and development.

The introduction of a competency-based approach with a focus on innovative learning technologies leads to significant changes in the content and procedural components of the traditional training system, the development and implementation of a new methodological system based on the following provisions that constitute theoretical and methodological principles of learning: systematic and continuity, computerization, approach and humanization.

It should be mentioned that sea transport specialists every five years have to undergo their certification and additional training, the long life learning is also present. Modern graduates should have high level of digital competence to be competitive at the labor market. Thus, teachers also should have such a level to teach in electronic virtual educational environment and organize study there. E-learning environment has a great potential in organizing the communicative interaction of all participants of the educational process in order to build professional competencies. The use of LMS Moodle will be also appropriate for training professionals of other transport specialties.

References

1. NMC Horizon Report. https://library.educause.edu/resources/2018/8/2018-nmc-horizon-report. Accessed 18 Aug 2018
2. Paris communiqué (2018). https://mon.gov.ua/storage/app/media/news/%D0%9D%D0%BE%D0%B2%D0%B8%D0%BD%D0%B8/2018/06/06/12/paris–communiqueenua2018.pdf. Accessed 22 Oct 2019
3. Brockman, J.: This will make you smarter. New Scientific Concepts to Improve Your Thinking. Edge Foundation 2012. https://www.rulit.me/books/eta-kniga-sdelaet-vas-umnee-novye-nauchnye-koncepcii-effektivnosti-myshleniya-read-424702-1.html. Accessed 22 Oct 2019
4. The Digital Competence Framework 2.0. https://ec.europa.eu/jrc/en/digcomp/digital-competence-framework. Accessed 22 Oct 2019

5. Ilomaki, L., Kantosalo, A., Lakkala, M.: What is digital competence? In Linked portal. Brussels, European Schoolnet (2011). http://linked.eun.org/web/guest/in-depth3. Accessed 20 Jan 2017
6. Bidaybekov, Y., Kamalova, G., Bostanov, B., Salgozha, I.: Development of information competency in students during training in Al-Farabi's geometric heritage within the framework of supplementary school education. Eur. J. Contemp. Educ. **6**, 479–496 (2017). https://www.researchgate.net/publication/320136915_Development_of_Information_Competency_in_Students_during_Training_in_Al-Farabi's_Geometric_Heritage_within_the_Framework_of_Supplementary_School_Education. Accessed 01 Sept 2017
7. Prohorova, S.: The concept of digital competence of a foreign language teacher in the world educational space Bulletin of the Ivan Franko Zhytomyr State University. Pedagogical Sci. **4**, 113–116 (2015). http://nbuv.gov.ua/UJRN/VZhDUP_2015_4_24. Accessed 22 Oct 2019
8. Popova, H., Yurzhenko, A.: Competency framework as an instrument to assess professional competency of future seafarers, In: Proceedings of 15th International Conference ICTERI 2019, vol. I, pp. 409–413. Kherson, Ukraine (2019). http://ceur-ws.org/Vol-2387/20190409.pdf. Accessed 22 Oct 2019
9. Voloshynov, S., Popova, H., Yurzhenko, A.: The application of augmented reality to the formation of English-speaking competence of future maritime professionals. In: Mathematics. Information Technology. Education, VIII International Sciences. Practice Conference pp. 73–75. Lutsk, Ukraine (2019)
10. Gritsuk, I., Gutarevych, Y., Mateichyk, V., Volkov, V.: Improving the Processes of Preheating and Heating after the Vehicular Engine Start by Using Heating System with Phase-Transitional Thermal Accumulator, SAE Technical Paper 2016–01-0204 (2016). https://saemobilus.sae.org/content/2016-01-0204/. Accessed 22 Oct 2019
11. Gritsuk, I., Volkov, V., Mateichyk, V., Gutarevych, Y.: The evaluation of vehicle fuel consumption and harmful emission using the heating system in a driving cycle. SAE Int. J. Fuels Lubr. **10**(1), 236–248 (2017). https://saemobilus.sae.org/content/2017-26-0364/. Accessed 22 Oct 2019
12. Gritsuk, I., Volkov, V., Mateichyk, V., Grytsuk, Y.: Information Model of V2I System of the Vehicle Technical Condition Remote Monitoring and Control in Operation Conditions, SAE Technical Paper 2018-01-0024 (2018). https://saemobilus.sae.org/content/2018-01-0024/. Accessed 22 Oct 2019

Process Support and Visual Adaptation to Assist Visual Trend Analytics in Managing Transportation Innovations

Dirk Burkhardt[1(✉)], Kawa Nazemi[1], and Egils Ginters[2]

[1] Human-Computer Interaction and Visual Analytics Group, Darmstadt University of Applied Sciences, Darmstadt, Germany
{dirk.burkhardt,kawa.nazemi}@h-da.de
[2] Riga Technical University, Kalku Str. 1, Riga 1048, Latvia
egils.ginters@rtu.lv

Abstract. In the domain of mobility and logistics, a variety of new technologies and business ideas are arising. Beside technologies that aim on ecologically and economic transportation, such as electric engines, there are also fundamental different approaches like central packaging stations or deliveries via drones. Yet, there is a growing need for analytical systems that enable identifying new technologies, innovations, business models etc. and give also the opportunity to rate those in perspective of business relevance. Commonly adaptive systems investigate only the users' behavior, while a process-related supports could assist to solve an analytical task more efficient and effective. In this article an approach that enables non-experts to perform visual trend analysis through an advanced process support based on process mining is described. This allow us to calculate a process model based on events, which is the baseline for process support feature calculation. These features and the process model enable to assist non-expert users in complex analytical tasks.

Keywords: Adaptive visualization · Transportation and logistics · Process mining

1 Introduction

The digitalization is a challenging task for almost every economic area. New upcoming technologies coupled with novel business ideas lead to innovative and revolutionizing business solutions. Many of them have the potential to change markets fundamentally and imperil primarily smaller players like small and medium sized enterprises (SME). The transportation and logistics domain is one of these fields, where a variety of new technologies and business ideas arise. However, in the current state it is rather difficult to estimate which of these innovations will be important and successful [10]. Beside technologies that aim on ecologic and economic transportation, such as alternative engines e.g. for parcel trucks, there are also fundamental different approaches like central packaging station or deliveries via drones. Yet there is a growing need for tools that enable identifying new technologies, ideas and furthermore, give also the opportunity to rate those in perspective of (future) relevance. The analytical challenges that

E. Ginters et al. (Eds.): ICTE ToL 2019, LNITI, pp. 319–327, 2020.
https://doi.org/10.1007/978-3-030-39688-6_40

have to be faced particularly are: (1) What technologies, business ideas, strategies etc. are coming up? (2) How have these upcoming technologies, business ideas, strategies etc. to be rated in perspective of future relevance? And (3), what of these upcoming technologies, business ideas, strategies etc. are relevant for the own market and more important the own business? These narrowed questions are substantial to answer as early as possible to be able to react on market changes.

Indeed, solutions for trend analysis, especially on patent data are still standard and often used. However, these tools use patent data only, which limits the gathered insights significantly on those that can be IPR protected. New business ideas or technological procedures cannot be registered and can therewith not identified with this data fundament. Even more, the patent registration procedures take approx. two years, which results in identifying trends that are at least two years old as well. Next to the data side, most of these tools are expert tools and require knowledge about the domain, technical understanding as well es economic experience to be able to perform useful analysis. Such kinds of experts are rare in SMEs and are therewith critical, so that most of these tools are not suitable for SMEs.

Therefore, in this chapter a different solution is proposed. First of all, a modular and scalable processing pipeline is defined that enables visual trend analysis on a variety of data fundaments for innovation management. Standard in that field is patent analysis, but due the long patent registration process and restriction on patents on technologies only, the focus will be on other data sources. Very important are research data from e.g. (open access) digital libraries and web data, such as news from enterprises or market magazines/blogs/portals. With this data and a highly interactive visual analytical solutions for trend analysis [3–5], technology and innovation foresight is possible to identify early market trends. However, to perform sufficient analysis professional knowledge about how to do an analysis would be still required. Therefore, in the chapter is described a novel graphical assistance feature that supports users via process mining [1, 2] in the analysis process. This allows even non-analysts or non-experts to observe the market for relevant innovations and rate them in regards of relevance. This is, as already mentioned, in particular required in the transportation domain, where a high number of small and medium-sized players are active and do not have the budget for intensive analytical investigations. A system that allows an assisted analysis enables these enterprises to be aware of market trends and support strategic decision making. The main contribution is a general model for supporting users based on the events by the mined process and the support features. This model is further implemented in a visual trend analytics system to assist users.

2 Design of a Process-Based User Supporting System

Common system that enable a process support consider a statically defined process where users are guided through the different process steps. A very basic implementation are installation wizards, which run during setup of e.g. windows programs [7]. The advantage of such process-driven programs is that the user just needs to follow each step so that there is only a low risk of failing it. In complex visual analytical system with high number of analysis features a static process model would not work, however, a process support would enable especially non-expert users in using those systems.

To enable process support, a dynamic approach is described that is based on process mining [1]. Van der Aalst originally introduced process mining to structure and optimize business processes, but the principle idea is modified to apply it on visual analytical system with the goal of providing adaptation functionality to assist the user.

2.1 General Process Support Methodology

The general process support methodology owns three major phases: (1) The training phase by analyzing data, (2) the process and support feature mining phase and (3) the supported data analysis phase. Thereby the first and third phase are identical and use overall the same system and data, but due to trained process model and support feature model the user could be supported by system meanwhile performing the data analysis. But the general analysis capabilities are still the same. The methodology overview is given in Fig. 1.

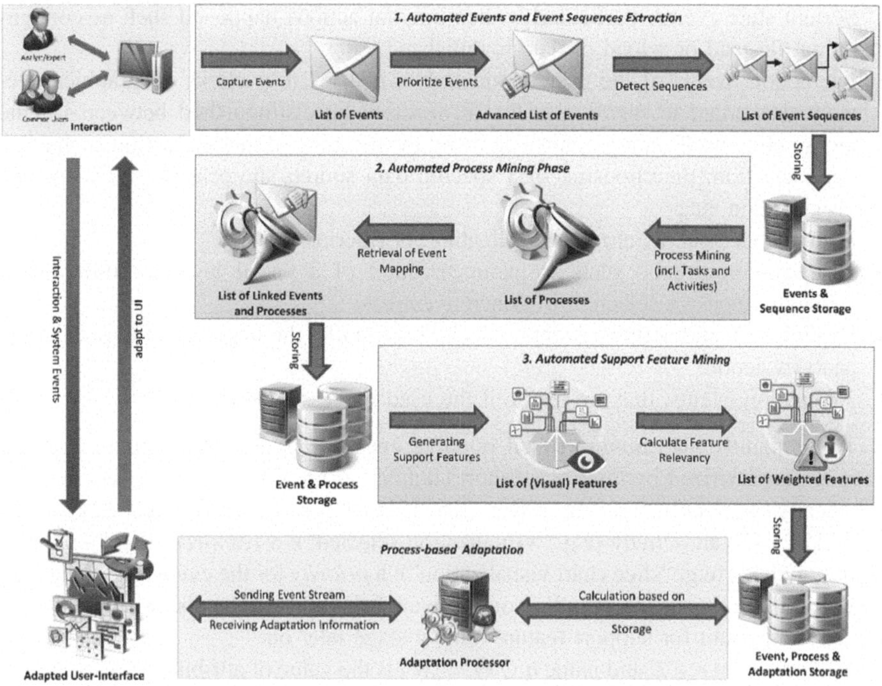

Fig. 1. Generalized interaction and activity processing model to support the user during his analysis tasks.

In the first phase, the training phase, the user already can analyze data with all analysis capabilities, but without any system support. In this phase, the potential users are restricted to analysts and experts who know how to use the system and how to perform a sufficient analysis. While the users are working with the system, the system logs any actions – the major focus lays on interaction events, such as mouse clicks and keyboard input, but also internal system states are logged.

In the second phase, the system performs the mining. These phase can be divided in three major steps: (1) Event extraction and event sequence determination, (2) the process mining to generate a process model and (3) the support feature mining where, relating to the generated process, concrete support features are determined and generated, such as recommendation and interface adaptations to certain databases, visualizations or filter criteria while performing a specific process task.

2.2 General Model Definition

The general Processing mining is majorly based on the definition of van der Aalst [1] with just some minor modification for different usage purpose. To perform the process mining, first an even log L is defined. The event log consists of a variety of structured events of the system, such as:

- System state events, particularly when internal actions happened such as configuration file read or reload of data is initialized;
- Interaction events of the user, particularly when the user clicks on graphical element. Important to mention is that it needs to be distinguished between i.e. the mouse click itself on a graphical spot and the meaning of the interaction such as an item selection, the choosing of a specific data source, the selection of a specific visualization etc.;
- Data events, such as changing, reloading, or filtering of data;
- Visualization events, such as parametrization of a visual layout, reordering of elements, focus or selection of concrete entities;
- Explicit task and activity events, which "bookmark" the beginning of specific real analysis action;
- Adaptation events, that are result of the used process support.

Consequently, the amount of all possible events $\mathcal{E}$ is defines, whereas an event $e \in \mathcal{E}$ is characterized by a set of attributes defined as Θ with a set of attribute names, specifically: an iteratively given *id* (e.g. "151451519"); a *timestamp* (e.g. "2019-09-24 08:32:12.250"); an *activity* (e.g. "visualization selected"); a *resource* (e.g. "icon slice chart"); a *target* (e.g. "slice chart visualization"); a *priority* for the event (e.g. "high"); a list of further *properties*, demanding on the event behavior such as document ID, which are only important for support feature mining stage later on.

For any event $e \in \mathcal{E}$ and name $n \in \Theta$, $\#_n(e)$ is the value of attribute n for event e. If event e does not have an attribute named n, then $\#_n(e) = \perp$ (null value). For convenience the following standard attributes are assumed: $\#_{id}(e)$ is the *id* associated to event e; $\#_{timestamp}(e)$ is the *timestamp* associated to event e; $\#_{activity}(e)$ is the activity associated to event e; $\#_{resource}(e)$ is the resource associated to event e; $\#_{target}(e)$ is the target associated to event e; $\#_{priority}(e)$ is the priority associated to event e; $\#_{properties}(e)$ is the *list of further properties* associated to event e.

To name the events for any event $e \in \mathcal{E}$ is defined, whereas $\underline{e}$ is the name of the event that $\underline{e} = \#_{activity}(e)$. Additionally, it is defined that $\mathcal{A}$ is the amount of all possible activities, so that counts $a \in \mathcal{A}$.

Van der Aalst [1] mentions that a process consists of cases and cases consist of events, which finally would mean that an event log consists of cases, and cases consist of event. But due the fact that annotations for cases are not exist, is defined that a process can consists of logical processes (for a better distinguishing they can be named as subprocesses) and those consist of events. A logical process helps only to structure/cluster a bigger process in smaller units, but basically this alignment is from the principle processing point of view not existing and not required. Finally, this results in the definition that a process consists of events only.

Therewith the event log can be defined with let $\mathcal{A}$ be a set of activity names. The trace σ is a sequence of activities, i.e. $\sigma \in \mathcal{A}^*$. The event log L is a multi-set of traces over $\mathcal{A}$, i.e. $L \in \mathbb{B}(\mathcal{A}^*)$ while $\sigma \in L$ implies $\sigma \neq \emptyset$. In regards of the process mining, the log-based ordering relations are defined first. Therefore, $a, b \in \mathcal{A}$ and has to count:

- $a >_L b$ if and only if there is a trace $\sigma = \langle t_1, t_2, t_3, \ldots, t_n \rangle$ and $i = \{1, \ldots, n-1\}$ such that $\sigma \in L$ and $t_i = a$ and $t_{i+1} = b$;
- $a \rightarrow_L b$ if and only if $a >_L b$ and $b \not>_L a$;
- $a \#_L b$ if and only if $a \not>_L b$ and $b \not>_L a$ and $a \parallel_L b$ if and only if $a >_L b$ and $b >_L a$.

For the process mining, the definition of the α-algorithm [1, 2] is considered and L is an event log over $T \subseteq \mathcal{A}$. Therefore $\alpha(L)$ is defined as follow:

$$
\begin{aligned}
&T_L = \{t \in T \mid \exists_{\sigma \in L}\, t \in \sigma\}, T_I = \{t \in T \mid \exists_{\sigma \in L}\, t = first(\sigma)\},\\
&T_O = \{t \in T \mid \exists_{\sigma \in L} t = \text{last}(\sigma)\},\\
&X_L = \{(A, B) | \mathrm{A} \subseteq T_L \wedge \mathrm{A} \neq \emptyset \wedge \mathrm{B} \subseteq T_L \wedge \mathrm{B} \neq \emptyset \wedge \forall_{a \in A} \forall_{b \in B} \mathrm{a} \rightarrow_L \mathrm{b} \wedge\\
&\forall_{a_1, a_2 \in A} a_1 \#_L a_2 \wedge \forall_{b_1, b_2 \in B} b_1 \#_L b_2\},\\
&Y_L = \left\{(A, B) \in X_L \mid \forall_{(A', B') \in X_L} A \subseteq A' \wedge B' \subseteq B' \Rightarrow (A, B) = (A', B')\right\} \quad (1)\\
&P_L = \{ \mid (\mathrm{A}, \mathrm{B}) \in Y_L\} \cup \{i_L, o_L\},\\
&F_L = \left\{\left(a, p_{(\mathrm{A},\mathrm{B})}\right) \mid (\mathrm{A}, \mathrm{B}) \in Y_L \wedge a \in A\right\} \cup \left\{\left(p_{(\mathrm{A},\mathrm{B})}, b\right) | (\mathrm{A}, \mathrm{B}) \in Y_L \wedge b \in B\right\} \cup\\
&\{(i_L, t) \mid t \in T_I\} \cup \{(t, o_L) \mid t \in T_O\}, \text{and}\\
&\alpha(L) = (P_L, T_L, F_L).
\end{aligned}
$$

2.3 Explicit Versus Implicit Task Selection

The idea is majorly based of a passive system that most of time only observes the user's interaction and system status and presents support wherever it may help the user. However, for major task, particularly at the very beginning of the analysis, it is almost impossible to detect the analysis purpose automatically or gather the information implicitly and therewith a guidance is almost impossible as well. To solve this beginning challenge, an explicit task selection as well is provided. Figure 2 illustrates the explicit task selection at the very beginning, but even during the analysis the system can request if different activities may be possible and cannot be identified via the implicit event observations.

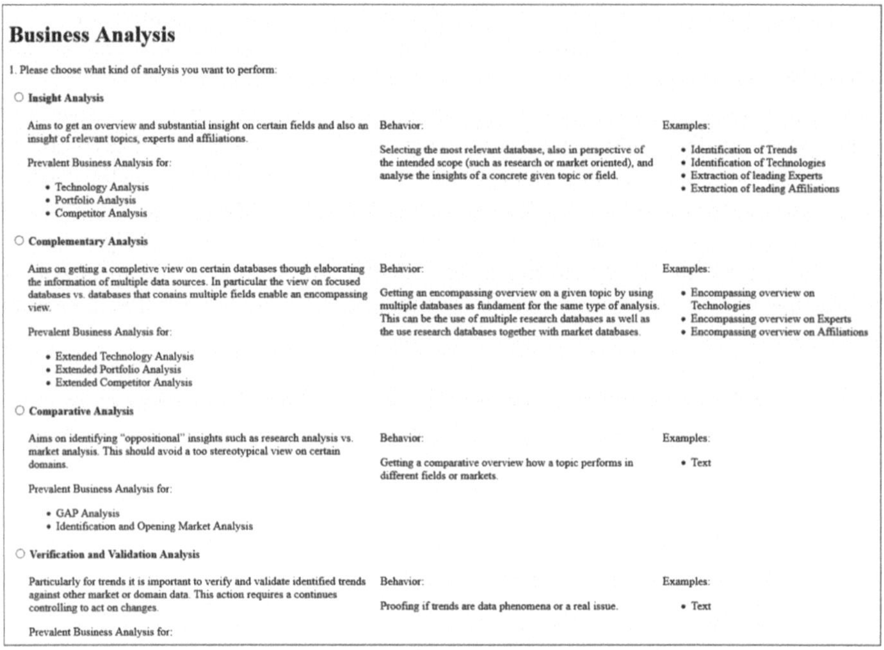

Fig. 2. Example screenshot of an explicit task selection.

A challenge is that the cases (see [1]) are ignored, which could bring a better structuring in the process, but yet it is aimed on an almost automatic approach where not stringent such tasks are explicitly named. It would be impossible to generate a solid process model when users are not giving regularly their intended analysis task or activity. In consequence, additional special events are defined, that are used as bookmarks for concrete beginnings of tasks and activities.

2.4 User Support Features

To enable an advanced analysis based on a process model, different support features are considered. So, the process mining is not the major result, it is more the fundament to use it as baseline to support the user in solving tasks, since the system "knows" the analysis procedures. The following listing is not complete, it should only outline the most important support features, which specifically make use from of the mined process model.

User Guidance. One of the major advancements of the system is to support users in their analysis work. Therefore, the mined process is used as baseline, to identify the current activity. Based on previous analysis activities, it could be furthermore extracted what aspects (i.e. data source, visualization, filter criteria etc.) are most often used and can now be suggested. If a concrete support feature could be identified toward the current user activity, a hint is shown to the user as suggestion – here the user can accept or decline it (see Fig. 3). If an aspect has a high probability, the system automatically performs the change – but, this function is very restrictively used, since it can confuse the user if e.g. visualization changes within a session without any notice.

The major goal of the user guidance features is to provide the user an invisible assistance through the analysis routine until the analysis goal is achieved. In previous works a workflow visualization based on BPMN [6, 7] was used to show the procedure as orientation, but first this requires extra space on the screen and second it was mostly hard understood due to the unknown notation and presentation. As future idea it will be tested to use abstracted process visualizations that just show e.g. the five most important analysis steps, which normally means that the mined process has to be simplified and abstracted manually.

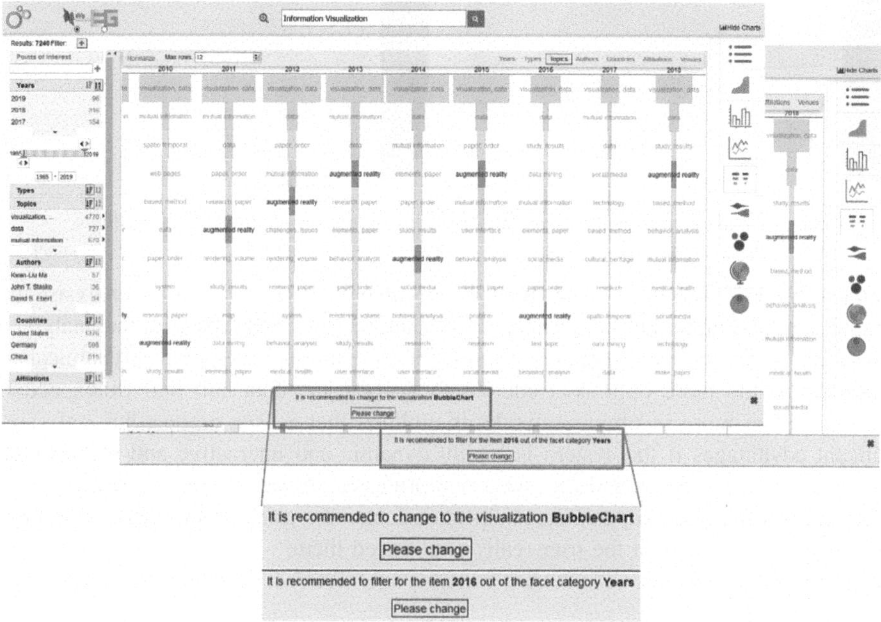

Fig. 3. Guiding the analyst through showing hints.

User-Interface Adaptation. The more the current analysis fits to an already learned analysis, the better the system can support the user via user-interface adaptation. Hereby the system notices the "intention" of the user and initiate changes on visual variables, visual layouts or the entire interface on behalf of the current performed activity. Thereby the functionalities are similar to those that are also currently used to guide the user (see section before), but the focus is more on changing visual aspects to gain a specific view on the data. Two examples are shown in Fig. 4, where the adaptation switches from a single view to a comparative view, because the system recognizes that the user wants to compare different data. And in Fig. 3, the system notices to change visualization layout or rather a specific filter, to show a specific data piece that most likely will be of interest for the user.

It is to mention that there are two different adaptation capabilities to consider: (1) a soft switch or rather a hint and (2) a direct adaptation. The soft switch is still too low to estimate that the change will almost face the user's intention. The second method is used, when the system could predict with a close to total probability what the user intents to. This last method is often used only at the beginning of an analysis. In future research, the individual user behavior should be given greater consideration by including user-based adaption capabilities (likewise described in [8, 9]).

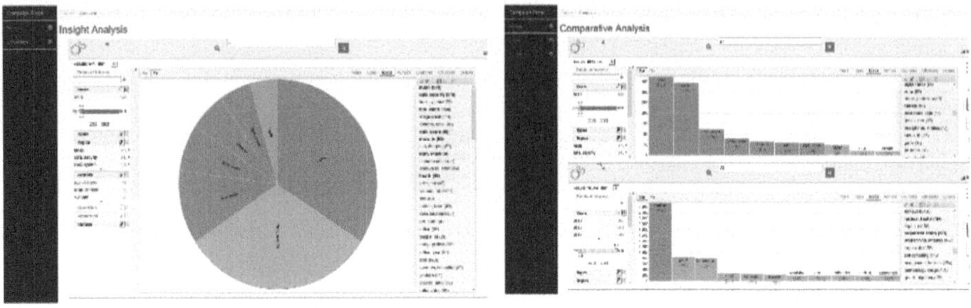

Fig. 4. Adaptation of the User-Interface demanding on required functionalities.

Interactive and Dynamic Interface. To be able to perform detailed analysis, the interface has an important role, since a solid analysis stays and falls with the analytical possibilities. So, the more interactive and from the analytical perspective dynamical an interface is, the more options it enables to analyze the data and find (unexpected) insights. Even more in previous works is outlined that a user support has only significant advantages if the system is highly dynamic and interactive and provides so many ways to e.g. analyze data, that a non-professional user is overstrained quickly. Here a process-driven support can help to structure the analysis process and recommends useful tools when the user really could need them.

3 Conclusion

The digitalization is a challenge in many domains where upcoming new technologies, business ideas, strategies etc. can revolutionize entire businesses. Particularly the transportation and logistics domain have huge variety of changes – to name just a few, there are significant optimizations of the transportation processes, new upcoming logistic ideas such as the delivery via drones or alternative delivery via packaging stations, but also the development of new transportation vehicles that more sustainable. To identify such trends, it is important to use analytical systems, that enable the identification of new trends, next to a rating in perspective of relevance and prediction of the impact on the own business. Even if those tools do exist, they are almost expert tools for professionals, whose are rare in SME. Therefore, in this chapter a new approach was described to assist and support particularly non-expert users in performing such (visual) trend analysis. As major fundament, process mining is used to generate common analysis processes and additionally extract supporting features that are commonly used by users. On behalf of these processing, a process support can be provided that guides and assist especially non-experts through the complex analysis tasks.

Acknowledgements. This work was partially funded by the Hessen State Ministry for Higher Education, Research and the Arts within the program "Forschung für die Praxis" and was conducted within the research group on Human-Computer Interaction and Visual Analytics (https://www.vis.h-da.de).

References

1. van der Aalst, W.: Process Mining: Data Science in Action. Springer, Heidelberg (2016)
2. Weijters, A., van der Aalst, W.: Rediscovering workflow models from event-based data using little thumb. Integr. Comput. Aided Eng. **10**(2), 163–190 (2003)
3. Nazemi, K., Burkhardt, D.: Visual analytics for analyzing technological trends from text. In: Proceedings of 23rd International Conference Information Visualisation (IV2019), pp. 191–200. IEEE (2019)
4. Nazemi, K., Retz, R., Burkhardt, D., Kuijper, A., Kohlhammer, J., Fellner, D.: Visual trend analysis with digital libraries. In: Proceedings of the 15th International Conference on Knowledge Technologies and Data-driven Business, pp. 14:1–14:8. ACM (2015)
5. Nazemi, K., Burkhardt, D.: A visual analytics approach for analyzing technological trends in technology and innovation management. In: Advances in Visual Computing, pp. 283–294. Springer, Cham (2019)
6. Burkhardt, D., Nazemi, K, Kohlhammer, J.: Visual process support to assist users in policy making. In: Handbook of Research on Advanced ICT Integration for Governance and Policy Modeling, pp. 149–162. IGI Global, Hershey (2014)
7. Burkhardt, D., Ruppert, T., Nazemi, K.: Towards process-oriented information visualization for supporting users. In: Proceedings of 15th International Conference on Interactive Collaborative Learning (ICL), pp. 1–8. IEEE (2012)
8. Burkhardt, D., Nazemi, K.: Dynamic process support based on users' behavior. In: 15th International Conference on Interactive Collaborative Learning (ICL), pp. 1–6. IEEE (2012)
9. Nazemi, K.: Conceptual model of adaptive semantics visualization. In: Adaptive Semantics Visualization. Studies in Computational Intelligence, vol. 646, pp. 193–297. Springer (2016)
10. Kayikci, Y.: Sustainability impact of digitization in logistics. In: Procedia Manufacturing, vol. 21, pp. 782–789. Elsevier (2018)

Logistics of People and Information in Employers-Based Virtual Test for University Students

Karlis Valtins(✉), Igors Tipans, and Natalja Muracova

International Cooperation and Foreign Students Department, Riga Technical University, Kronvalda Boulevard 1, 1010 Riga, Latvia
{karlis.valtins,natalja.muracova,igors.tipans}@rtu.lv

Abstract. Article discusses logistics around multifunctional and widely reproducible model for testing university student's practical knowledge and skills, matching them to the labor market needs, contributing to a university business cooperation, providing international outreach, increasing student's employability, providing quality assurance for curriculum content and more. Authors argue that all these things can be accomplished by a trivial tool – virtual test that is voluntary, cannot be retaken, is highly applicable to the engineering sciences and does not require capital investments. Article describes logistics of people and information within the proposed solution, including virtual and physical conditions, feedback loops, input/output sections and more. Article consists of literature review, expert discussions, model developed by the authors and analysis of previous initiatives in this area.

Keywords: Virtual testing · Data logistics · Systems modeling

1 Introduction

Perspective of logistics within this paper has a meaning of detailed managing and implementation of a complex operation – international employers test, which requires significant managerial efforts for people and information to go from one point to another. Worldwide university students are highly concerned about their future employment when they are approaching graduation from universities and authors of this paper argue that with a simple yet effective solution of testing environment that would be based on problem solving of real-life tasks rather than having theoretical questions could help students in a transition from university to labor market. Statistics show that 14.3% of young adults (aged 18–24) on average among OECD countries are neither employed nor in education and training [1], this is a significant number that indicates problems in the university to work transition. The overarching aim of proposed innovation would be to ensure joint efforts into building capacity for universities and industry representatives to upgrade academic and labor market performance, but foremost – to address student employment issues and check if the knowledge they have gained at the university is applicable to the real-life situations in the labor market. It is a normal practice for a university to engage into cooperation with companies

E. Ginters et al. (Eds.): ICTE ToL 2019, LNITI, pp. 328–336, 2020.
https://doi.org/10.1007/978-3-030-39688-6_41

representing labor market in terms of joint projects, guest lectures from industry experts, and participation in the designing of curriculum, commercialization of research and technology transfer processes. At the same time, majority of regional policy planners are still addressing the issue of matching the study process to the needs of labor market. In a recent report European Commission outlines, that majority of academics and businesses still does not engage into cooperation, because of the lack of cooperation mechanisms, and that creates a gap between academia and industry [2]. In addition, outside EU, this issue has been put into agenda and almost all countries are trying to become more embedded with the industry to promote entrepreneurship, start-up businesses, and technology transfer. But very few higher education institutions are assisting with a direct transition of students to the labor market, presumably because of the primary university mission (teaching/learning), efforts of keeping students in the higher education (at master or PhD levels) and lack of time and resources. Third mission of universities typically is addressing technology transfer and valorization of research, but literature shows that technology transfer processes are not fostering academic engagement [3]. Another important factor is student motivation for learning, it has been proven that most effective learning outcomes are related to the self-motivation of students and potential employment opportunity could provide a higher drive for students to be more engaged into the learning process, as they would see more merit for themselves [4]. It becomes evident that using existing forms of university business cooperation is not enough and additional cooperation mechanisms and social innovations are needed. Authors of this paper suggest, that with a trivial tool of a virtual online test, in which companies from different countries world-wide would help to prepare and give students questions/tasks, simulating real life situations, there could be a high possibility to address multiple issues at once – employment opportunities for students, motivation of students for learning, aligning study curriculum with the labor market, enhancing university teacher engagement and knowledge as well as utilizing existing or establishing new clusters for industry companies. At the same time, this tool would require well thought through logistics and coordination efforts coming from the university.

2 Current Landscape

It has become a regular practice for universities all around the globe to test their students at the end of each study subject and a relevant teacher usually makes these tests, this approach does not require advanced logistics and is common in the higher education field. Industry companies also often do testing in-house, establishing own criteria, priorities and tools to check knowledge of their employees. Only common ground in terms of testing is related to third parties that provide testing tools, software, certified tests. In online testing environment, like any other software, the attitudes of users are in line with their computer literacy [5, 6]. Here is a closer look to the most commonly used testing environments:

- University made subject based tests are relevant to the curriculum within the particular study subject, form of assessment varies from a presentation, project, written or online test, discussion, that helps university teacher to evaluate the knowledge of the student and often it results in a mark. Connectivity with industry in this category may happen if there is some advisory board that helps to check and amend curriculum in the whole study programme. All the data input and output elements in this category are closely related to the university teacher, his/her decisions and actions;

- Standardized tests with a certificate at the stage of completion such as Microsoft certification exams for specific software, language tests that approve certain level of language skills, i.e. TOEFL, IELTS. There tests are usually franchised and organized with the assistance of local partners globally. Logistics for these tests are complex and usually it takes a lot of coordination to prepare instructors, automate assessment etc.;
- Specific local tests made by companies prior to the employment of the personnel, usually 3rd party software is tailored to each individual institutional need;
- Third party made online tests for various knowledge assessment, plus software that can be used to create and deploy new tests using the same online software.

Advanced data gathering and cloud storage has become a standard within the virtual online environment and that has allowed universities to keep track about each student learning dynamics more comprehensively. It is also discussed in the literature that it is highly relevant to think about the employability aspects of students while they are studying [7]. It is often suggested that university teachers should come from industry in order to provide more in-depth view about the labor market needs, but that still have significant limitation because of being one person's opinion and information on the matter could be narrow or outdated. Online environment is becoming vaster and massive open online courses have become a niche in terms of mainstream as well as lifelong learning and specialized courses. Unfortunately, there is lack of evidence that these courses have close connectivity with the labor market and that they present real life questions/tasks. Still the concept of open access is emerging and provides high multiplication, transparency and accessibility.

3 Requirements Analysis

In order to support identification of the needs of three main target groups: university students, university teachers and companies, three international discussions were held in Riga, (Latvia), hosted by Riga Technical University during the period of 2015–2019. Experts from various fields representing all three target groups participated expressing their views about what is most important issues we need to address in the future of higher education, emphasizing university business cooperation, transition to labor market and online testing methods. First discussion cycle was held in 2015 with approximately 30 companies from Latvian Employers Confederation and Latvian Chamber of Commerce in the framework of VET2WORK project under the Lifelong Learning programme (project number: 550462-LLP-1-2013-1-LV-KA1-KA1ECETA). Second discussion was held during the international staff week of Riga Technical University in 2018 and approximately 50 representatives from various EU higher education institutions participated coming from university administration offices and faculties. Third discussion in 2019 included approximately 100 students from Japan, Georgia, Russia, Turkey, Finland and Latvia, participating in Riga Technical University organized summer schools. Discussions revealed the following needs for each target group:

University student's needs:

- To check if they are ready to enter labor market in terms of knowledge, skills and competences that they are gaining during their studies at the university;
- To test their knowledge in international context, not only matching for local, home country companies, but also international ones;
- Rise their possibility for employment after the graduation of the university, making the transition to labor market swifter and sounder.

University teacher's needs:

- To verify study course curriculum with external practitioners from industry, check if it is up to date, relevant and applicable in real life situations;
- To update curriculum based on the newest trends, equipment and methods used in the industry;
- Learn together with students about the specifics of industry, changes that happen over time to have most effective learning outcomes.

Company's needs:

- To recruit new employees, that have appropriate knowledge in the field company is operating, without requiring additional trainings;
- To have international cooperation collaborates, that could contribute to the opening of new markets, services, products or research findings;
- To be as efficient as possible with the time that is spent on things that are not directly connected with the business-related operations, including university business cooperation.

4 Proposed Model

Model established by authors of this paper comes from empirical knowledge obtained in the discussions, international projects and overall trend analysis for higher education. Every new IT system should be first conceptualized, and logistics of people and information should be analyzed, before the actual implementation (usual steps – concept, development, testing, and refinement) and this paper provides such a concept. TESTDRIVE solution takes approach of field or study programme based testing (for example: civil engineering, architecture, electrical engineering, computer sciences etc.), where both bachelor and master level students can take the test together during their final semester of studies. Test can be taken only once, by all study field students together, is voluntary and cannot be re-taken. There are four main stakeholders attributed to the IT system hosting the test, from which three correspond to the main beneficiary target groups – industry companies (local and foreign companies working in the particular study programme related industry, existing networks or clusters can be engaged as well as new ones can be created, group that will provide the tasks for the test), university students (enrolled in to the particular study programme, group that will take the test) and university teachers (all teachers of the particular study programme, group that will access the test after students have completed it in order to discuss together with students the questions and make adjustments to the course curriculum). In addition, another stakeholder is presented in the model (see Fig. 1) – TESTDRIVE instructor, person that should be appointed by the university implementing the proposed solution. Instructor should not be a teacher from the study programme and he/she should retain a neutral position in the testing process. Test in each study field should have different, adjusted questions/tasks every year, therefore making the testing environment, plus logistics dynamic and highly relevant. Significant factor for success is participation of international companies, it will allow students to test their knowledge in a global setting and teachers will be able to have broader outlook to the routine of industry companies. Sequence of the operations related to the TESDRIVE solution is a key element and both input/output operations are shown in the Fig. 1. Here is a detailed look at the logistics of people, resources and information:

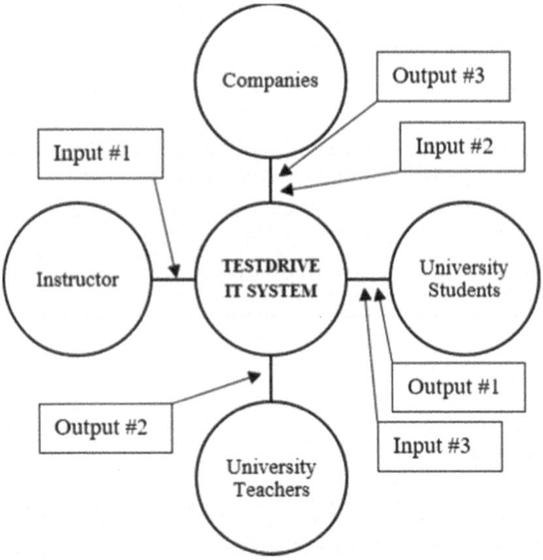

Fig. 1. TESTDRIVE infological model.

Input #1, instructor to the TESTDRIVE system: instructor appointed to administer the test created a table framework for the test; typically, it refers to approximately 20-30 questions that industry companies will fill in collectively. Instructor then assigning access permissions to the form enters the data about company (logo, e-mail addresses, name, location, field of operation). Framework consists of three parts – question/task, variables used and correct answer/solution. Upon completion, instructor sends the link to the framework to all companies in the cluster. This step happens typically one month before the scheduled test.

Input #2, companies to the TESTDRIVE system: industry companies receive the link to participate in the development of the test, employees are nominated that will enter the questions/tasks; these employees should correspond to the practitioners of the study field. These employees fill in online three sections that correspond to the real-life situation simulation, encompasses question/task, variables used in the task, and the correct answer/solution to the problem. They can enter text, numerical data and pictures, if needed. Each company is invited to submit 1–3 questions, depending on their availability and willingness to participate. After successful completion of the question, data is saved and uploaded to the TESTDRIVE system and instructor at the university can check and verify the task. If possible, instructor should instruct companies prior to the input #2 about the item sensitivity and cognitive activation of students, because it could lead to higher learning outcomes [8].

Input #3, university students to the TESTDRIVE system: students studying in particular study field are invited to a computer class, where instructor gives them a briefing about the test. Students have one hour to complete the test and system automatically shuffles the variables of the tasks so that questions that appear in the system for each student are different.

Output #1, TESTDRIVE system to the university students: Upon completion of the test, students immediately see the final score and right/wrong answers. Score is calculated by correct answers and time used for completion of the test.

Output #2, TESTDRIVE system to the university teachers: After all students have competed the test, instructor administers deployment of the test questions to the university teachers of the study programme. Teacher can then in the upcoming lectures review and discuss the questions together with the students.

Output #3, TESTDRIVE system to the companies: After all students have completed the test, instructor administers deployment of the test results to all companies that participated in the development of the test. Top ten highest ranked students and their contact information is sent to all companies and then companies if they want can contact these students with an internship or job offer.

Simulation of real-life situations could lead to the next version of the TESTDRIVE that could already include emulation problems industry companies are facing, as there is a high potential that companies also could find some relevant knowledge and artifacts from student given answers in the test. This future model would require even more significant logistical issues, including physical hardware and equipment used during the test. New approaches to the problems could be discovered using simulation. [9].

5 Data Flow

This section provides description about the data flow within the proposed model. As the data can be both – structured (numerical values, closed questions) and unstructured (open text, photos), it becomes more challenging to automate tasks of communication between the stakeholders, assessment part of giving test results immediately after the end of the test. Visual pattern recognition software is recommended to be used together with the TESTDEIVE IT system to advance the unstructured data pool; this would advance the data logistics significantly. For convenience it is recommended to use cloud-based storage, amount of data exchanged between the stakeholders is highly dependent on the study field, nature of the question/tasks given by the companies, file format used in the communication as well as amount of data inserted by the students. Three main sections of the data flow correspond to the cloud-based storage, user interface and system communication (see Fig. 2).

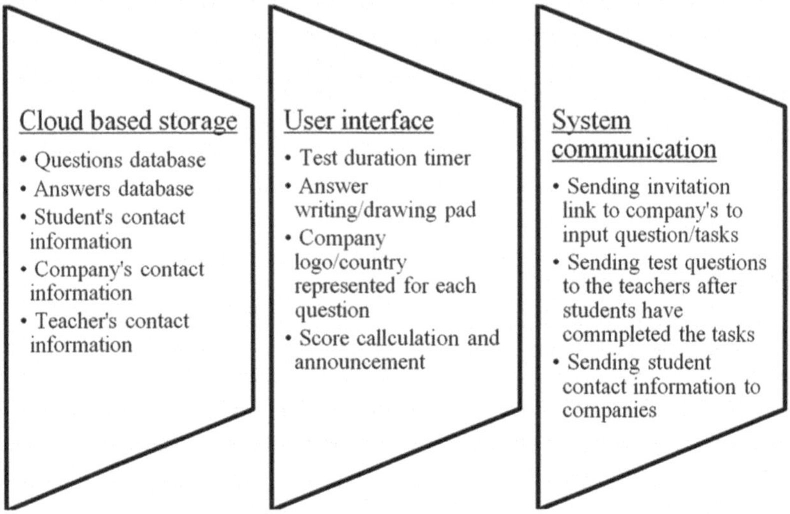

Fig. 2. Data flow structure in TESTDRIVE model.

It becomes apparent that developed testing software solution should address these three modules for successful implementation of the TESTDRIVE solution. In addition, several modules can be considered depending on the scope and budget available – interface languages, statistics and analytics of the test results, rating system for companies and teacher within the subject.

6 Conclusion

By conceptualizing TESTDRIVE IT system, authors have acknowledged that there are many added value elements or horizontal issues that proposed solution is addressing, besides the obvious benefit of benchmarking students' knowledge against the labor market. It could provide broader educational experience that could also contribute to development of human resources and entrepreneurial skills [10]. It can be forecasted that proposed solution will also tackle the following issues:

- Rising level of student motivation for learning and making preparation period for examination more effective and fruitful. Literature suggests that even if students are engaged in active learning and are enthusiastic, it does not automatically mean that they will learn [11], therefore external factors provide different kind of motivation;
- Introduction of the academic staff to the latest trends happening in the industry, which can lead to the upgrade of teacher's skills and knowledge. If teachers are also involved in the research process and are working on advancement of some technologies, this could be also a platform for discussion between industry and technologists [12];
- Adding international perspective to the study process and examination of students by inviting foreign companies to participate in the development of the test questions/tasks;
- Prevention of the brain drain by directly engaging students in to the local labor market. Although only top ten best performing students will be revealed to the industry companies, additional dialogue can be maintained with local and not international companies;
- Reducing university student dropouts and lowering unemployment rates;
- Making certain careers (for example engineering) more appealing to potential students by demonstrating the interaction with industry.

It can be argued that proposed solution will check if student skills match labor market needs and with the help of TESTDIVE IT system, it will be possible to verify student readiness to transit from university to the labor market. Activation of existing or establishment of new joint clusters for employers is a key element for this solution to work; therefore, institution implementing the solution should have a vast external outreach. Significant efforts also should be directed to the revision of curriculum and implementation of necessary amendment, based on the information that comes from the industry. This would provide a high-level quality assurance for any study programme. Finally, it is possible to say that proposed model would strengthen university business cooperation in particular regions, by introducing new mechanism for cooperation. This model can be applicable to any study direction and the intellectual material would be the questions that companies create, not the system itself.

Future research directions could be related, but not limited to: assessment of adaption of existing software to the TESTDRIVE model, elaborating on emulation-based tasks (with external hardware elements), comparison of logistics between proposed model and advances testing systems (for example Microsoft), financial feasibility analysis for the development of IT system. It is also suggest looking at the cooperation mechanisms for universities to join efforts establishing joint test by merging their existing industry networks and clusters.

References

1. OECD, Education at a Glance 2019: OECD indicators, p. 54. OECD Publishing, Paris (2019)
2. European Commission: The state of university-business cooperation in Europe final report, ISBN 978-92-79-80971-2, p. 5 (2018)
3. Perkmann, M., Tartari, V., McKelvey, M., Autio, E., Broströmc, A., D'Este, P., Fini, R., Geuna, A., Grimaldi, R., Hughes, A., Krabel, S., Kitson, M., Llerena, P., Lissoni, F., Salter, A., Sobrero, M.: Academic engagement and commercialisation: a review of the literature on university-industry relations. Elsevier Res. Policy **42**(2), 432–433 (2013)
4. Wang, C.K.J., Liu, W.C., Kee, Y.H., Chian, L.H.: Competence, autonomy, and relatedness in the classroom: understanding students' motivational processes using the self-determination theory. Heliyon **5**(7), 2 (2019)
5. Parhizgar, S.: Testing and technology: past, present and future. Theor. Pract. Lang. Stud. **2** (1), 174 (2012). ISSN 1799-2591
6. Backes, B., Cowan, J.: Is the pen mightier than the keyboard? The effect of online testing on measured student achievement. Elsevier Econ. Educ. Rev. **68**, 91 (2019)
7. Lane, J.E.: Building a Smarter University: Big Data, Innovation, and Analytics, p. 278. State University of New York Press, Albany (2014)
8. Naumannad, A., Rieserb, S., Musowcd, S., Hochwebercd, J., Hartig, J.: Sensitivity of test items to teaching quality. Elsevier Learn. Instr., p. 42 (2019)
9. Buckova, M., Skokana, R., Fuskoa, M., Hodon, R.: Designing of logistics systems with using of computer simulation and emulation. Transp. Res. Procedia **40**, 978–985 (2019)
10. Ferreira, J.J.M., Raposo, M., Rutten, R., Varga, A.: Cooperation, Clusters, and Knowledge Transfer: Universities and Firms Towards Regional Competitiveness, p. 249. Springer, Heidelberg (2013)
11. Nemova, D., Bagautdinov, R., Mushinskiy, A.: University-business cooperation based on virtual research information service. Procedia Eng. **117**, 1116 (2015)
12. Barkley, E.F.: Student Engagement Techniques: A Handbook for College Faculty, p. 6. Wiley, Hoboken (2009)

Improving the Energy Efficiency of Railway Electrical Traction by Creating Smart Grid System

Oleksandr Plakhtii(✉), Karina Trubchaninova, Vladyslav Panchenko, and Olga Anannieva

Ukrainian State University of Railway Transport, Kharkiv 61001, Ukraine
a.plakhtiy1989@gmail.com,
karyna.trubchaninova@kart.edu.ua,
vlad_panchenko@ukr.net, romashka13052015@gmail.com

Abstract. The article presents technical solutions to improve the energy efficiency in the system of traction rail power supply. Railway transport is a powerful consumer of electricity. A rather high part of the cost of railway transportation is determined by the cost of electricity consumed. In connection with the constant increase in the cost of electricity, an urgent task is to increase the energy-efficiency of electric transport systems, which will ensure a reduction in the cost of rail transportation. To improve the energy efficiency of electric traction, the concept of an intelligent traction power supply system is proposed, which provides monitoring and control of the transmission of electricity to the traction network. This system ensures the rational use of the recovery mode at the traction substations and in the electric rolling stock, work with powerful electrical energy storage devices, as well as connecting alternative energy sources to the traction power supply network. The Matlab/Simulink is used for developing a simulation model of the proposed intelligent traction power supply system. The research audience is engineers and scientists who work in the field of energy and railway power supply systems.

Keywords: Energy-efficiency · Smart grid · Alternative energy

1 Introduction

In recent years, there has been a tendency in the general industrial energy sector to switch from classical to intelligent electric networks. This is due to the need to save energy resources, namely the increase in the need to account for electricity, the need to compensate for reactive power and higher harmonics with active power filters and an increase in the amount of electricity that is generated from alternative power sources. Smart power networks use information and communication networks and technologies to gather information and control power generation and energy consumption, which automatically improve efficiency, reliability, economic benefits, and sustainability of power generation and distribution [1–7].

Intelligent electrical networks (smart Grid) is positioned as a priority in the development of energy in the near future in the EU, USA, China and other countries of

E. Ginters et al. (Eds.): ICTE ToL 2019, LNITI, pp. 337–343, 2020.
https://doi.org/10.1007/978-3-030-39688-6_42

the world. In the first place for investments in Smart Grid for the next 5 years - China ($7 billion), the second - the United States ($7 billion), the third - Japan ($900 million) [8].

The use of the Smart Grid concept in railway traction networks is promising. Railway traction substations are quite powerful consumers of electricity and the need to improve the energy efficiency of these systems is a priority. This can be achieved by the combined use of alternative power sources, high-power energy storage and semiconductor converters: active power filters, active rectifiers, regulators, charging drives, etc. [9–11]. Let's consider examples of using alternative power sources on the railways of the world:

- Netherlands. The Netherlands Railways in cooperation with the major railway company NS has entered into a contract with Eneco company, according to which traction power supply is gradually switching to renewable power sources. So back in 2016, 75% of NS traction costs were consumed from Eneco's wind power plants, which is 1.4 billion kWh per year. In addition to significant energy savings due to the use of alternative power sources, the influence of the environmental factor is quite important. Based on 2014 data, prior to the partnership with Eneco, NS company used 1.2 TWh of electricity and emitted 550 kt of CO2, while the full VIVENS contract emitted 640 kt (kilotons) of CO2 [12];
- Great Britain. On UK Railways, the use of alternative sources of electricity began in 2011 with the use of solar panels with a total capacity of 30 kW to power the railway signalling and lighting systems at Wessex station. In the course of further development of the project "Aldershot" is expected to increase the power of solar panels to power traction consumers. For example, the Riding Sunbeams project research team believes solar power could provide 20% of Merseyrail's traction network in Liverpool, as well as 15% of commuter routes in Kent, Sussex and Wessex. By the end of 2020, Riding Sunbeams hopes to build and connect the world's first full-scale community-owned and suburban solar farm with British Railways. [13];
- India. At the moment, there are about 250 electric trains in India, additionally powered by solar panels attached directly to the roof of the train. The concept of creating solar farms along the railway tracks is in development. Indian Railways, the country's largest energy consumer, has announced its intention to install solar farms with a total capacity of 30 GW in 10 years. This plan is part of India's drive to replace polluting coal-fired energy with cleaner energy sources, including up to 100 gw of solar power. By 2025, Indian Railways expects to meet a third of its electricity demand from renewable energy sources by building 1.1 GW of rooftop projects and large-scale track solar farms total 3.9 GW. Thus it is possible to draw a conclusion about wide development of use of alternative sources of the electric power in systems of traction power supply.

The proposed scheme that allows the use of electricity from alternative energy sources in both the general industrial and traction network (see Fig. 1). At the same time, the power active filter ensures high quality of electricity transmitted from alternative sources, as well as compensation of reactive power and higher harmonics of the railway traction substation.

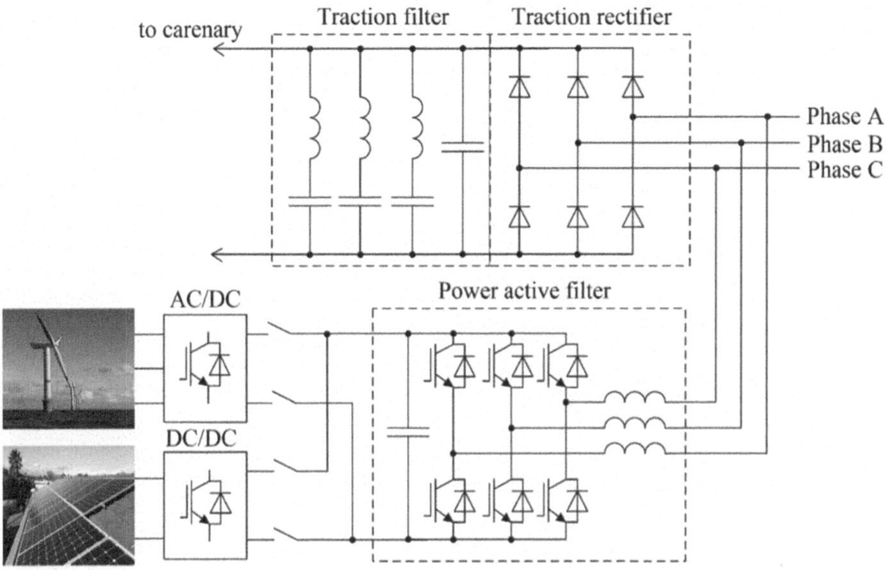

Fig. 1. Proposed block diagram of connection of alternative power sources to traction substation.

The aim of research is to study the energy processes in the proposed Smart-Grid system for railway application, namely, to evaluate the possibility of correcting the power factor in traction DC substation with power active filter and wind generator.

2 Matlab Simulation of the Proposed Smart Grid Scheme

To confirm the energy-saving effect of the use of a wind generator and a power active filter, as well as the analysis of the quality of electricity consumed by a DC traction substation in the program Matlab/Simulink simulation was carried out. The model consists of a six-pulse diode rectifier, a power active filter, a wind generator with its rectifier (see Fig. 2). During the simulation, the waveforms of the input current of the three-phase diode rectifier, the current generated by the power active filter and the resulting current in the general industrial network were obtained (see Fig. 3).

For numerical estimation of parameters of quality of input current of traction substation at application of the power active filter the Fourier analysis of corresponding signals is carried out (see Fig. 4).

As seen in Fig. 4, the use of the power active filter in the traction substation can reduce the harmonic distortion of the input current from 37.05% to 2.15%. At the same time, by reducing the content of the higher harmonics of the input current, the mean square value of the current is reduced from 2138 A to 2019 A, which accordingly leads to a decrease in the total power consumption.

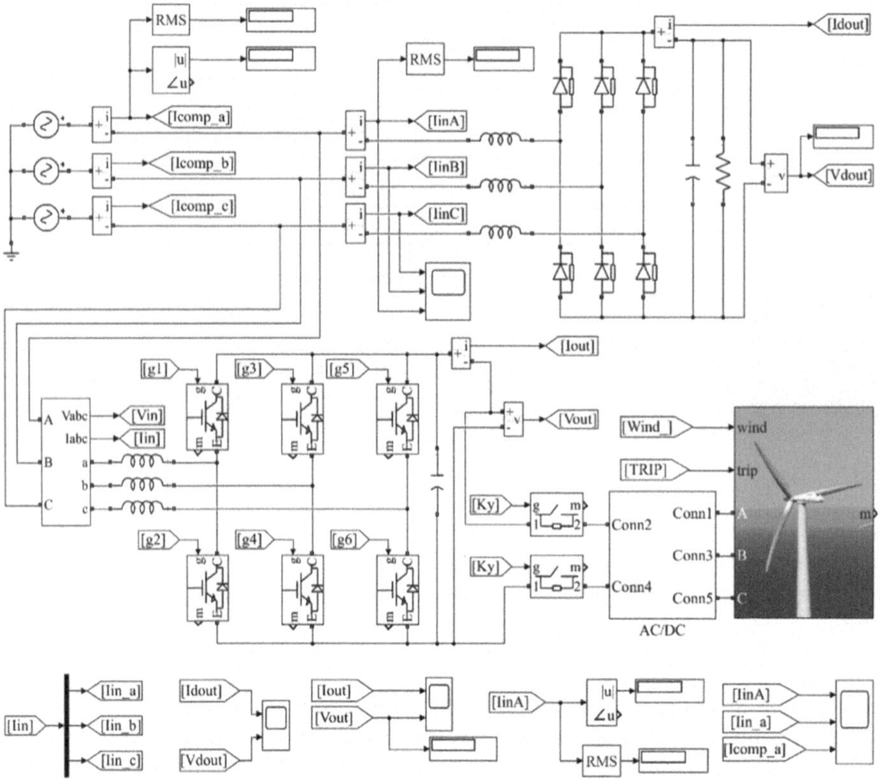

Fig. 2. Model of six-pulse diode rectifier with power active filter with wind generator.

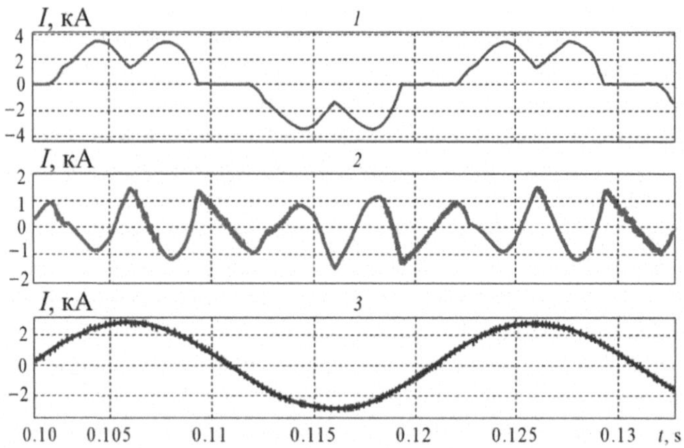

Fig. 3. Simulation results: 1 - input current of the three-phase diode rectifier; 2 - current generated by the power active filter; 3 - the resulting current in the general industrial network.

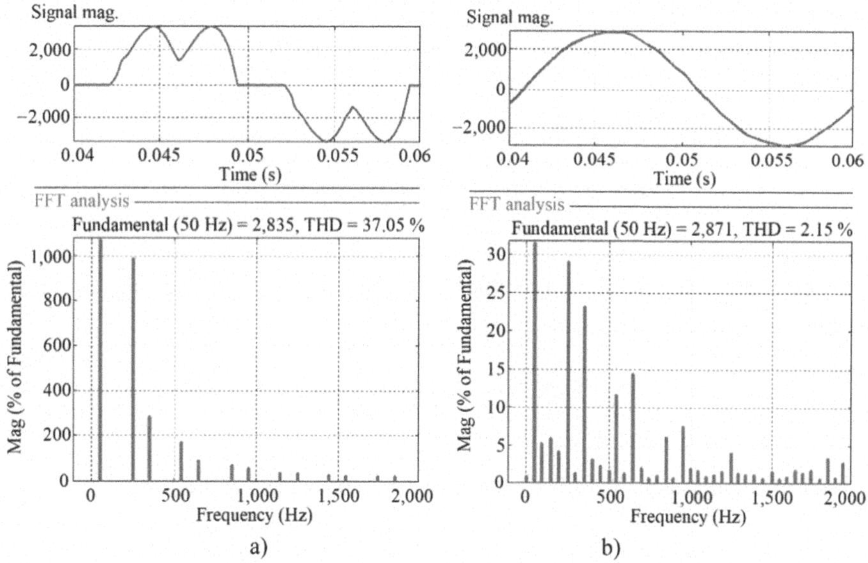

Fig. 4. Fourier-analysis results: (a) traction substation current (before compensation); (b) grid current after compensation.

3 Analysis of the Energy Saving Effect from Compensation of Higher Current Harmonics

The use of Smart Grid system, alternative sources of electricity in combination with power active filters will almost completely compensate. A method for determining the energy-saving effect from higher harmonics compensation is proposed. Additional power losses in electrical networks from higher harmonic currents, which are uniquely determined based on the resulting value of the total harmonic distortion (THD) of the network current [14].

This method is valid in the case when the influence of the skin effect on the resistance of the network with a limited spectrum of higher harmonics is insignificant. In this case, additional losses in the electrical network from higher harmonics can be calculated based on an increase in the root mean square value (RMS) of the current, and, consequently, an increase in losses in the quadratic dependence on the RMS value of the current. The harmonic distortion coefficients of alternating current THD_I are defined as:

$$THD_I = \frac{\sqrt{\sum_{m=2}^{m=\infty} I_m^2}}{I_1} \quad (1)$$

where I_m – is the RMS value of the m-th harmonic. Note: THD_I in formulas are given in relative values [0 to 1]. The root mean square value (RMS) of the current can also be expressed:

$$I_{RMS} = \sqrt{\frac{1}{T}\int_{0}^{T} i^2(t) \cdot dt} = \sqrt{I_1^2 + \sum_{m=2}^{m=\infty} I_m^2} \tag{2}$$

From expression (1) it is possible to express the sum of squares of higher harmonics:

$$\sum_{m=2}^{m=\infty} I_m^2 = (THD_I \cdot I_1)^2 \tag{3}$$

From the expressions (1) till (3) I_{RMS} can be represented in the form:

$$I_{RMS} = \sqrt{I_1^2 + (THD_I \cdot I_1)^2} = \sqrt{I_1^2 \cdot (1 + THD_I^2)} \tag{4}$$

The relative increase in power loss caused by higher harmonics can be expressed:

$$\Delta P = \frac{I_{RMS}^2 \cdot R}{I_1^2 \cdot R} = \frac{I_1^2 \cdot R \cdot (1 + THD_I^2)}{I_1^2 \cdot R} = 1 + THD_I^2 \tag{5}$$

Thus, an unambiguous dependence between the harmonic distortion coefficient of the consumed current and the percentage of additional power losses is established.

4 Conclusion

The Smart-grid concept of increasing the energy-efficiency of the railway power supply system through the use of a power active filter and alternative power sources is proposed. The Matlab-modeling of the proposed concept was conducted, which confirmed the possibility of generating high-quality electricity. The energy-saving effect from compensation of higher harmonics is determined. On the basis of the given expressions it is possible to define power saving effect in a network from application of the power active filter. The typical THD value of the input current of a three-phase rectifier, depending on the output filter capacitance and input inductance, lies in the range of 30% to 90%. Thus, the power active filter allows to reduce the value of losses in the network caused by higher harmonics from 10% to 48%.

The advantages of the proposed approach of the smart grid system include the possibility of simultaneous transmission of electricity from alternative energy sources and the ability to compensate for higher harmonics of currents and reactive power. The disadvantage of this study is the lack of analysis of the efficiency of electric power transmission from a wind generator to the network, which is planned to be investigated in the future. The benefit of the research is the possibility of increasing the energy efficiency of power supply system's due to the energy of alternative sources and reducing additional power losses from higher harmonics and reactive power.

Acknowledgements. For support in carrying out scientific researches the gratitude to the management of the Ukrainian state railway transport is expressed.

References

1. Larik, R.M., et al.: Smart grid technologies in power systems. J. Appl. Sci. Eng. Technol. **11**(6), 633–638 (2015)
2. Zhemerov, G., Tugay, D.: Energy saving in smart grid systems with trunk mode of customer connection. In: IEEE International Conference on Intelligent Energy and Power Systems (2014)
3. Plakhtii, O., et al.: Analyses of energy efficiency of interleaving in active voltage-source rectifier. In: IEEE International Conference on Intelligent Energy and Power Systems (2018)
4. Chakrabortty, A., Bose, A.: Smart grid simulations and control. In: Power Electronics in Renewable Energy Systems and Smart Grid, pp. 585–624 (2019)
5. Fomin, O.: Improvement of upper bundling of side wall of gondola cars of 12-9745 model. J. Metall. Min. Ind. **1**, 45–48 (2015)
6. Hossain, M.R., et al.: Smart grid. In: Green Energy and Technology, pp. 23–44 (2013)
7. Tugay, D., et al.: Energy efficiency of microgrid implementation with solar photovoltaic power plants. In: IEEE International Conference on Intelligent Energy and Power Systems (2018)
8. https://ses.jrc.ec.europa.eu/sites/ses/files/documents/smart_grid_projects_in_europe.pdf. Accessed 21 Oct 2019
9. Cui, S., Yu, Q., Gu, G., Gang, Q.: Research on the architecture of electric power information communication network for smart grid. In: IEEE Conference on Energy Internet and Energy System Integration (2017)
10. Plakhtii, O., Nerubatskyi, V., Sushko, D., et al.: Improving energy characteristics of AC electric rolling stock by using the three-level active four-quadrant rectifiers. East.-Eur. J. Enterp. Technol. **4**(8), 6–14 (2019)
11. Plakhtii, O.A., Nerubatskyi, V.P., et al.: Analysis of the energy efficiency of a two-level voltage source inverter in the overmodulation mode. Sci. Bull. National Min. Univ. **4**(172), 68–72 (2019)
12. https://www.railway-technology.com/features/featuremichel-kerkhof-of-eneco-discusses-the-dutch-rail-networks-renewable-revolution-4647194. Accessed 21 Oct 2019
13. https://www.theguardian.com/business/2019/aug/22/rail-line-in-hampshire-is-worlds-first-to-be-powered-by-solar-farm. Accessed 21 Oct 2019
14. Plakhtii, O., Nerubatskyi, V., Ryshchenko, I., et al.: Determining additional power losses in the electricity supply systems due to current's higher harmonics. East.-Eur. J. Enterp. Technol. **1**(8), 6–13 (2019)

Bicycle Path Network Designing and Exploitation Simulation as a Microservice Architecture

Artis Aizstrauts[1], Egils Ginters[1,2(✉)], Dirk Burkhardt[3], and Kawa Nazemi[3]

[1] Sociotechnical Systems OU, Sakala Str. 7-2, 10141 Tallinn, Estonia
artis.aizstrauts@gmail.com, soctech@soctech.net
[2] Riga Technical University, Kalku Str. 1, Riga 1658, Latvia
egils.ginters@rtu.lv
[3] Darmstadt University of Applied Sciences, Haardtring 100, 64295 Darmstadt, Germany
{dirk.burkhardt,kawa.nazemi}@h-da.de

Abstract. Simulation is recognized as a suitable tool for sociotechnical systems research. But the variety and complexity of sociotechnical systems often leads to the need for distributed simulation solutions to understand them. Models that are built for infrastructure planning are typical examples. They combine different domains and involve variety of simulation approaches. This article proposes an easy management environment that is used for VeloRouter software – a multi agent-based bicycle path network and exploitation simulator that is built as a microservice architecture where each domain simulation is executed as a different microservice.

Keywords: Sociotechnical systems simulation · Bicycle path network planning · Easy Communication Environment

1 Introduction

Within FP7 FLAG-ERA FuturICT 2.0 project "Large scale experiments and simulations for the second generation of FuturICT" (2017–2020) the authors of this article worked on a sub-topic "Intermodal bicycles network as efficient kind of urban green transport (VeloRouter 2.0)". That included development of prototype of intermodal bicycle network modeling tool, called VeloRouter, to be used both by policy planners and decision makers, as well as cyclists for bicycle infrastructure improvement and everyday route planning. Article presents its solution as a microservice architecture.

Rapid technological development has brought changes to many realms of life. With that, also the notion of sociotechnical systems has become widely accepted in many domains as sociotechnical factors are at the foundation of nearly all forms of contemporary human activity. Education, healthcare, entertainment and social media are among the many current, and frequently intersecting, examples of complex sociotechnical systems [1]. Transportation system is no exception. In fact, Schmitt argues that cities are the largest, most complex and most dynamic man-made systems [2].

E. Ginters et al. (Eds.): ICTE ToL 2019, LNITI, pp. 344–351, 2020.
https://doi.org/10.1007/978-3-030-39688-6_43

Socio-technical approach enables us to systematically identify, describe, and study the more obscure, non-linear effects of multiple, dynamically shifting interactions among large collections of socio-technical system components [1]. Hettinger et al. conclude that sociotechnical approach can drive improvements in system performance and safety [1] therefore this approach fits the needs of a research focusing on transportation and planning issues in cities.

The next section of this article introduces the underlying assumptions and understanding of VeloRouter. The authors argue how introduction of sociotechnical approach, participatory planning methods and simulation modeling can lead to better results. The third section describes the main functionality of VeloRouter. And finally, the fourth section of this article outlines the building blocks and communication mechanisms of VeloRouter as a microservice architecture.

2 Sociotechnical Approach to Urban Planning

According to the World Bank data urban population in the world exceeds 50% since 2007 [3] and for more than ten years more than half of the world's population live in cities. With growing cities and increasing environmental and climate challenges urban planning is more important than ever before. For policy planners, decision makers and engaged citizens technology development has enabled large scale data collecting, excessive computation abilities and also direct citizen participation. Rapidly spreading and evolving technologies led to development of fields like urban informatics and participatory planning etc.

At the same time research points out that traditional methods of planning and managing large cities have reached their limits and need a radical re-thinking. On the computational side, this necessitates the integration of new methods and instruments. On the planning and design side, this requires an involvement of stakeholders and decision makers much earlier than usually done in the past [2].

Architect and urban design consultant Gehl [4] has defined four key objectives in urban planning – lively cities, safety, sustainability, and health. All four objectives can be strengthened by increasing the concern for pedestrians, cyclists, and city life in general:

- Lively cities are promoted when more and more people are invited to walk, bike and stay in city space;
- Safe cities are promoted when more people move about and stay in city space;
- Sustainable cities are promoted if a large part of the transport system can take place as what Gehl calls "green mobility", that is travel by foot, bike or public transport;
- Healthy city is strengthened dramatically if walking or biking can be a natural part of the pattern of daily activities.

In other words, there is a need for diversified infrastructure, that is aimed at diverse needs within the society and leads to vivid, secure, sustainable cities and healthy societies. In particular, cycling should be viewed among the pivotal nodes of diversified transportation infrastructure. Cycling promotes healthy lifestyle, reduces CO2 emissions, improves urban public space and cyclers infrastructure can also be used at least partly by skaters, skateboarders, scooter drivers, etc. Cycling should be encouraged by municipalities – by building infrastructure that corresponds to the needs of society, by enabling cyclists to participate in planning process, etc.

For cycling to become a fully integrated element within a multimodal transport system and not just a type of tourism, an appropriate infrastructure is necessary – cycling routes, lighting, bike rental, parking, repair services, as well as alignment with other types of transport. Cyclists need route planning tools, information about road surface quality, relief, and usage patterns during different weather conditions and over different days of the week [5].

But cities lack instruments to gather and analyze rapidly changing trends and to plan and build the necessary infrastructure in an effective manner. Decision makers and policy planners have to respond to the actual needs, make data driven projections, adjust the society's habits according to strategic vision of the city. This aim can not be achieved without appropriate tools. It is necessary to combine modern instruments and methods or to invent new ones to allow for the simulation of successful scenarios for sustainable future cities [6]. Special care must be taken to incorporate the local stakeholders, the legislative authorities, politics and economy, as well as the geographic, social and environmental context in the process [2].

At the same time the citizens lack instruments to interact with the decision makers and express their needs in an easy, inclusive way (without "red tape" burden), to give feedback about existing infrastructure, to submit their own proposals and to plan their routes based on distance, terrain, weather conditions, etc. Available tools – like petitions.net for participation and bikemap.net or mapmyride.com for bicycle route planning – focus on these things separately and usually on a different scale. Petitioning tools are meant for any type of policy proposals and need additional filtering, but route planners usually calculate the shortest distance without taking into account other factors and these. No existing tools incorporate both – route planning and functionality necessary for municipalities [15].

The authors of this article view VeloRouter as a viable sophisticated alternative for municipal decision making, besides it provides higher level of citizen involvement in local infrastructure development. End-user involvement in early planning phases gives much higher acceptance of end result and can be an important instrument for democratic governance.

3 VeloRouter – Bicycle Path Network Design and Exploitation

VeloRouter is a web-based tool for participatory urban planning that integrates live-data based services and simulation modeling and allows stakeholder interaction. Previous research concluded that existing products aimed at cycling routes design mainly offer the capabilities of planning and publishing the routes, but do not provide functionality necessary for municipalities to build justified bicycle path network [5]. Three main groups of stakeholders or target audiences for VeloRouter are: decision makers, infrastructure developers and cyclists. Their main benefits are as follows:

- Decision makers (municipalities) can determine the most popular routes, receive feedback and communicate with infrastructure users, analyze different possible bicycle route infrastructure projects and compare them, if needed also against a fixed budget;
- Infrastructure developers maintain the bicycle infrastructures digital twin in VeloRouter in order to update the information about the actual state of the infrastructure, keep track of real-life situation and plan maintenance works;
- Cyclists can plan their routes, find out the potential occupancy of it and participate in infrastructure development by submitting their own projects and giving feedback for other submitted projects and plan their trips.

There are four main services that are integrated in VeloRouter and provide its wide functionality: weather service, bicycle usage service, traffic service and infrastructure service. Weather service provides forecast for each given date and time. This information is used for simulation and user bicycle path navigation. Bicycle usage service calculates how many cyclists will be riding on each given time. This information is used for simulation and user bicycle path navigation. Traffic service provides prediction on traffic load on the streets, this information is critical for low level bicycle infrastructure parts of the route. Infrastructure service has all the information about the actual state of each edge of the route. This information is used for simulation and user bicycle path navigation.

Simulation modeling is widely accepted as a method for sociotechnical systems research [1]. In addition, Lawson et al. point out that computer-based simulation modeling is an effective tool for helping to address the challenges associated with managing visitor use in wilderness settings [7]. Their study shows, how simulation modeling can be used as a tool for understanding existing visitor use patterns within a specific area, estimating the impact of increasing visitor use levels on management objectives, and evaluating the effects of alternative policy decisions.

VeloRouter is a multi agent-based bicycle path network and exploitation simulator. It is, in fact, a multi-model simulation system that contains several sub-models. Each sub-model is built with domain specific technologies and simulation approach. For example – multi agent-based bicycle path network and exploitation simulator. It is designed in the Repast Symphony environment and it uses OpenStreetMap spatial data. Multi-model simulation systems are built as multi service applications. VeloRouter technological infrastructure enables simulation services to communicate among each other and also provides information services – allowing different services to co-exist in one communication environment.

4 Communication Architecture

As mentioned in previous section, VeloRouter consists of several services that comprise microservice architecture. Microservice architecture is an approach to developing a single application as a suite of small services, each running in its own process and communicating with lightweight mechanisms, i.e. AMPQ. These services are built around business capabilities and independently deployed and developed. As for the key benefits of such architecture scalability (relatively easy to duplicate services to gain more computing power), availability/resiliency (if one service breaks down, general application can still provide functionality) and independency (each service can be built with the tools and technologies that suits the best for each case, regardless of technologies used for other services) can be mentioned [8].

Communication among such independent deployable components thus becomes crucial. Authors have developed Easy Communication Environment (ECE) for VeloRouter sub-model communication [9]. ECE architecture can be divided into four levels of abstraction (see Fig. 1).

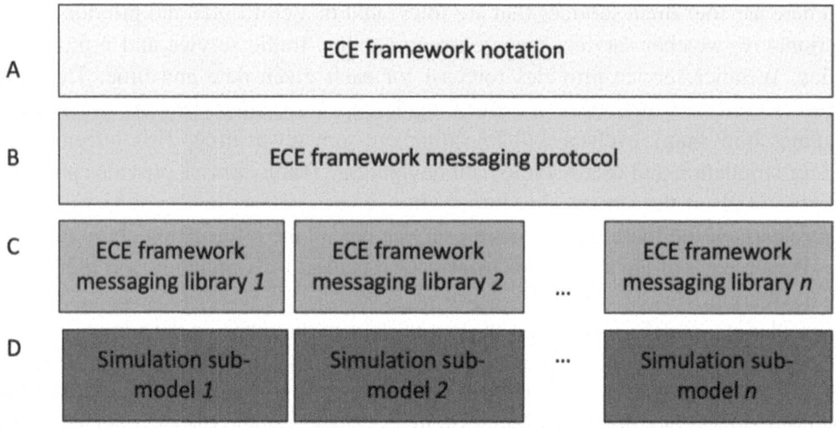

Fig. 1. Easy Communication Environment architecture abstraction layers.

D – Simulation Sub-model. Sub-model is any program or algorithm that consumes or produces any data. Distributed simulation approach highly depends on intercommunication among sub-modules [9]. In such approach sub-models are distributed by various reasons, i.e. by domain specific, to manage load balance, to parallelize simulation process, etc. In any case communication is crucial to maintain distributed simulation and achieve main simulation purpose.

In VeloRouter use-case, sub-module communication is used in many cases, for example, between bicycle usage simulation service and weather forecast service, between route infrastructure service and bicycle usage simulation service. Each of the service may publish several information endpoints for other sub-modules to use. For example, route service is responsible for the route calculations and may have such endpoints as – *closes_path*, *best_path*, *less_occupied_path*, etc. It is up to simulation designer to decide what information should be public and used by other simulation sub-models. ECE by itself does not put any limitations on it.

B – ECE Framework Messaging Protocol. Messaging protocol consists of two functional parts – message transportation and message carrier format.

For message transportation ECE uses AMQP (Advanced Message Queuing Protocol) protocol. It is an open standard application layer protocol for message-oriented middleware. The defining features of AMQP are message orientation, queuing, routing (including point-to-point and publish-and-subscribe) and reliability [10]. Reliability is one of the core features of AMQP, and it offers two preliminary levels of Quality of Service (QoS) for delivery of messages – At most once and At least once [11].

At most once service level guarantees a best-effort delivery. There is no guarantee of delivery. The recipient does not acknowledge receipt of the message and the message is not stored and re-transmitted by the sender. QoS level 0 is often called "fire and forget" and provides the same guarantee as the underlying TCP protocol [12].

At least once service level guarantees that a message is delivered at least one time to the receiver. The sender stores the message until it gets a PUBACK packet from the receiver that acknowledges receipt of the message. It is possible for a message to be sent or delivered multiple times [12].

And for the message carrier, ECE uses JSON (JavaScript Object Notation) data format. JSON is an open-standard file format that uses human-readable text to transmit data objects consisting of attribute–value pairs and array data types (or any other serializable value). It is a very common data format, with a diverse applicability [13].

All to gather, ECE combines both of these technologies to deliver sophisticated communication architecture that is highly reliable and easy maintainable [14]. AMQP ensures that message is delivered to the right client/subscriber (and resilience) and JSON allows to transmit complicated data that can be parsed and understood by most of the programming languages [13].

A – ECE Framework Notation Level. Notation level is the highest level in ECE communication architecture. This level ensures that each node (sub-model) understands each other and speaks the same language. ECE notation is based on JSON data format, which is used to format the message, the notation defines mandatory fields in each communication part and communication behavior. ECE communications consist of three communications events – request, response, acknowledgment (see Fig. 2, see Fig. 3).

Communication pattern is described in Fig. 2, the first action is a request that is made by Sub-module A with message that is formed according to ECE communication notation (see Fig. 3 as a request example).

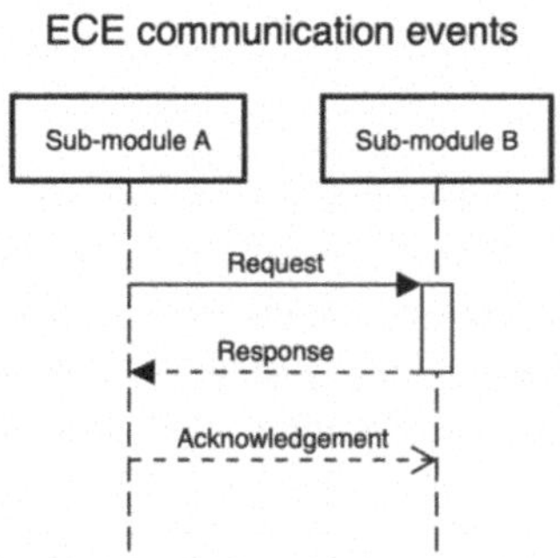

Fig. 2. ECE communication events.

This process is asynchronous, when the Sub-module B is ready to respond, it sends back the response message. Afterwards Sub-module A sends back the acknowledgment message to notify Sub-module B, that the response is received and understood.

```
{
    "type": "request",
    "request_guid": "2fbec59a-7686-4f96-9686-88aaff7ef3a5",
    "service_guid": "9b0ae71b-aeb6-42db-bbcf-2875cafb0a39",
    "service_endpoint": "closes_path",
    "payload": {
        "location_node_start": 847584,
        "location_node_end": 234903
    }
}
```

Fig. 3. ECE request example.

Figure 4 gives example on how ECE architecture abstraction layers are represented in communication between two sub-models. In this figure one can see all the main parts that are required to deliver a message from one entity to another. In such case message is constructed affirmative with ECE notations and encoded in JSON format. Sub-module has a library for its specific programming language that provides AMQP communication. AMQP requires message broker for communication consistency it ensures that messages are delivered at the right recipient in the right order.

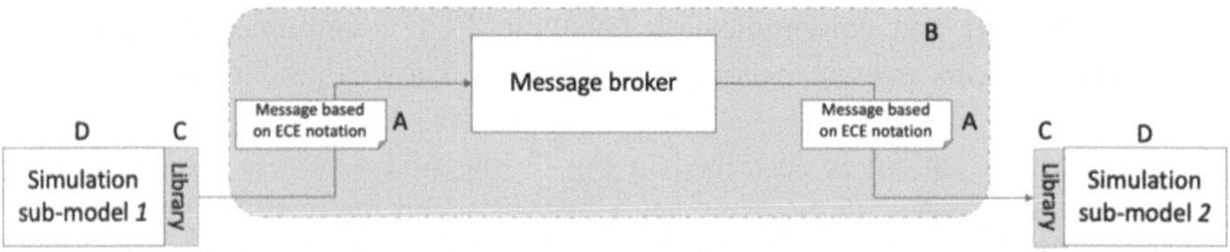

Fig. 4. ECE communication example between two sub-modules.

5 Conclusion

Development of a diversified and inclusive transport scheme that corresponds to the needs of all stakeholders is a challenge in many modern cities. Socio technical approach used in VeloRouter addresses many of those challenges.

VeloRouter combines functionality meant for decision makers, infrastructure developers and cyclists, that ensures up-to date and comprehensive information exchange among stakeholders. For such operability VeloRouter consists of several services or microservices. Microservice approach is beneficial for this kind of system as it enables to separate different functions, it is easier to maintain and they are reusable. The communication among the microservices has to be defined to be simple and universal at the same time. The authors propose Easy Communication Environment based on JSON and AMQP to ensure simplicity and resilience. Further development of ECE will focus on time advancement in such simulation environment.

Acknowledgements. The article publication is initiated by FP7 FLAG-ERA FuturICT 2.0 (2017–2020) "Large scale experiments and simulations for the second generation of FuturICT" STIMUL project "Intermodal bicycles network as efficient kind of urban green transport (VeloRouter 2.0)".

References

1. Hettinger, L.J., Kirlik, A., Goh, Y.M., Buckle, P.: Modelling and simulation of complex sociotechnical systems: envisioning and analysing work environments. Ergonomics **58**(4), 600–614 (2015)
2. Schmitt, G.A.: Planning environment for the design of future cities. In: Muller, A.S., Aschwanden, S., Halatsch, J., Wonka, P. (eds.) Digital Urban Modeling and Simulation. Springer (2012)
3. United Nations Population Division, World Urbanization Prospects: 2018 Revision. https://data.worldbank.org/indicator/SP.URB.TOTL.in.zs. Accessed 27 Oct 2019
4. Gehl, I.: Cities or People. Island Press, Washington, D.C. (2010)
5. Ginters, E., Mezitis, M., Aizstrauta, D.: Sustainability simulation and assessment of bicycle network design and maintenance environment. In: Proceedings of the IEEE 2018 International Conference on Intelligent and Innovative Computing Applications (ICONIC), Mauritius, pp. 1–7 (2018)
6. Bettencourt, L., West, G.: A unified theory of urban living. Nature **467**(7318), 912–913 (2010)
7. Lawson, S.R., Itami, R.M., Randy, G.H., Manning, R.E.: Benefits and challenges of computer simulation modeling of backcountry recreation use in the desolation lake area of the John Muir wilderness. J. Leisure Res. **38**(2), 187–207 (2006)
8. Amundsen, M., McLarty, M., Mitra, R., Nadareishvili, I.: Microservice Architecture: Aligning Principles, Practices, and Culture. O'Reilly Media, Sebastopol (2016)
9. Aizstrauts, A., Ginters, E., Baltruks, M., Gusev, M.: Architecture for distributed simulation environment. Proc. Comput. Sci. **43**, 18–26 (2014)
10. OASIS Advanced Message Queuing Protocol (AMQP) Version 1.0. http://docs.oasis-open.org/amqp/core/v1.0/. Accessed 25 Oct 2019
11. Naik, N.: Choice of effective messaging protocols for IoT systems: MQTT CoAP AMQP and HTTP. In: Proceedings of IEEE International Systems Engineering Symposium (ISSE), pp. 1–7 (2017)
12. Message quality of service (QOS). https://www.ibm.com/support/knowledgecenter/en/SSFKSJ_9.1.0/com.ibm.mq.dev.doc/q123795_.htm#q123795___qos. Accessed 25 Oct 2019
13. JavaScript Object Notation. https://www.json.org. Accessed 25 Oct 2019
14. Vinoski, S.: Advanced message queuing protocol. IEEE Internet Comput. **6**, 87–89 (2006)
15. Ginters, E., Baltruks, M., Sakne, I., Merkurjevs, J.: Dual use bicycle path network designing and exploitation environment – VeloRouter. In: Proceedings of European Modeling and Simulation Symposium (EMSS 2016), Cyprus, pp. 10–14 (2016)

The Information and Analytical System for Monitoring Roadside Pollution by Traffic Flows

Vasyl Mateichyk[1(✉)], Mykola Tsuman[1], Ganna Weigang[2], Dmitry Kozodoy[3], Zura Sansyzbajeva[4], and Yuriy Grytsuk[5]

[1] National Transport University, Kiev 01010, Ukraine
{wmate, tsuman}@ukr.net

[2] Banking University, Kiev 04070, Ukraine
weigang.ganna@gmail.com

[3] Ukrainian State University of Railway Transport, Kharkiv 61001, Ukraine
dmitry_1980@ukr.net

[4] L.N. Gumilyov Eurasian National University, Munaitpasov Street 5, Nur-Sultan, Kazakhstan
zura_astana@mail.ru

[5] Donbas National Academy of Civil Engineering and Architecture, Kramatorsk 84333, Ukraine
yuri.gritsuk@gmail.com

Abstract. The intensive growth of the territory of modern cities is closely connected with the development of transport infrastructure. Digitalization of traffic management processes is the prerequisite for efficient organization of urban transport to ensure sustainable development of cities. Using artificial intelligence, Big Data and predictive analytics, new telecommunication technologies and the Internet of Things, computer modeling, blockchains, new "human-machine" interaction interfaces, and new technologies of security, including cybersecurity contribute to the expansion of modern transport infrastructure. These technologies were used for developing the information and analytical system to monitor the roadside pollution by traffic flows. The software designed enables to study the influence of the parameters of individual vehicles in a traffic flow, operational and climatic conditions, as well as the characteristics of transport infrastructure on roadside pollution. Information and analytical system involves the use of vehicle-to-infrastructure technologies for assessing and forecasting roadside pollution for possible options of transport infrastructure.

Keywords: Road transport · Traffic flow · Pollution monitoring

1 Introduction

Road vehicles are currently the most efficient means of passenger and freight transportation, individual trips, etc. At the same time, road vehicles are the largest consumers of fuel and other resources. They are the sources of noise, vibrations,

E. Ginters et al. (Eds.): ICTE ToL 2019, LNITI, pp. 352–359, 2020.
https://doi.org/10.1007/978-3-030-39688-6_44

electromagnetic radiation and environmental pollution by different solid and gaseous substances. The issues of road vehicle performance have been important throughout the history of automobiles. With increasing number of automobiles, the issues of environmental impact of road transport have become more urgent.

Many scientists have devoted their research to environmental impact of road network objects – separate vehicles and traffic flows – and the ways to improve their environmental safety [1–10]. The investigations also focused on the development of methods to study and predict environmental safety of separate vehicles and roadside pollution by traffic flows. The development of transport environmental monitoring systems was of great scientific interest as well. The paper [1] is devoted to the study of environmental pollution by road transport and methods for preventing such pollution. It results in the development of a "driver-vehicle-road-environment" system which enables to optimize the operation factors for reducing environmental pollution by road transport. The research [2] deserves particular attention. The results of their research include the systematized quantitative estimation of harmful emissions by different categories of vehicles, the estimation of urban environmental pollution by road transport. They also analyzed measures aimed at reducing air pollution by vehicles. The issues of improving vehicle environmental safety are described in [7]. The research includes systems analysis methods for developing a "road vehicle" system mathematical model. It enables to evaluate vehicle environmental safety and to find ways to improve it. The peculiarities of developing a "traffic flow-road" system mathematical model and creating specific elements for intelligent monitoring system of vehicle environmental safety in operation are described in [8, 9]. The paper [10] describes the peculiarities of vehicle-to-infrastructure (V2I) information system and the process of monitoring and evaluating the vehicle technical condition in operation with the possibility of forecasting. The system is based on the general approach to investigate the "Vehicle - Driver - Operating Conditions - Vehicle Operation Infrastructure" system.

The composition and level of hardware and software for intelligent monitoring systems significantly depend on monitoring objects and monitoring tasks. Vehicle propulsion systems, separate vehicles and traffic flows are considered to be monitoring objects in road transport. Depended on monitoring tasks intelligent monitoring systems can include monitoring systems of monitoring objects technical performance (such as the speed of motion, the mode of motion, operating process parameters, technical condition, traffic flow composition, etc.) [10]. On the other hand, the monitoring systems of environmental performance are more science intensive because it significantly depends on both operational technical performance and design parameters of monitoring objects [11]. Development of the information and analytical system (IAS) can solve these tasks of monitoring roadside pollution by traffic flows.

The aim of the work is to develop of the IAS for monitoring roadside environment pollution by traffic flows including possibility of its forecasting. It must consider the technical condition parameters and motion modes of separate vehicles in the traffic flow, road infrastructure parameters and weather conditions. This system can be used as an element of the intelligent transport system in the research and optimization of road traffic control processes and road infrastructure parameters.

2 The "Traffic Flow – Road – Roadside Environment" System

The theoretical basis of IAS functioning is the system "traffic flow - road - roadside environment". The structure of this system is presented in Fig. 1. The input of the system at the *A* level is the parameters of the traffic flow. This is the structure of the traffic flow by vehicle categories *k* (M1, M2, M3, N1, N2, N3), environmental classes *j* (Euro-0… Euro-6) and fuel types *p* (gasoline, gasoline and ethanol blends E5–E85, diesel fuel, compressed natural gas, liquefied petroleum gas); traffic flow intensity and vehicle motion modes. The vehicle motion modes are characterized by speed, acceleration, and performance of the vehicle propulsion system. Road and roadside environment (*Env*1… *Env*5) are characterized by the type and condition of the road surface, weather conditions. In the process at the *A* level, when the vehicles are moving in the traffic flow, depending on the parameters of the road and the roadside environment, the corresponding forces of air resistance P_W and road resistance P_ψ arise. The output of the system at the *A* level is the specific emissions m_{ikpj} and the noise levels L_k of separate vehicles polluting the roadside.

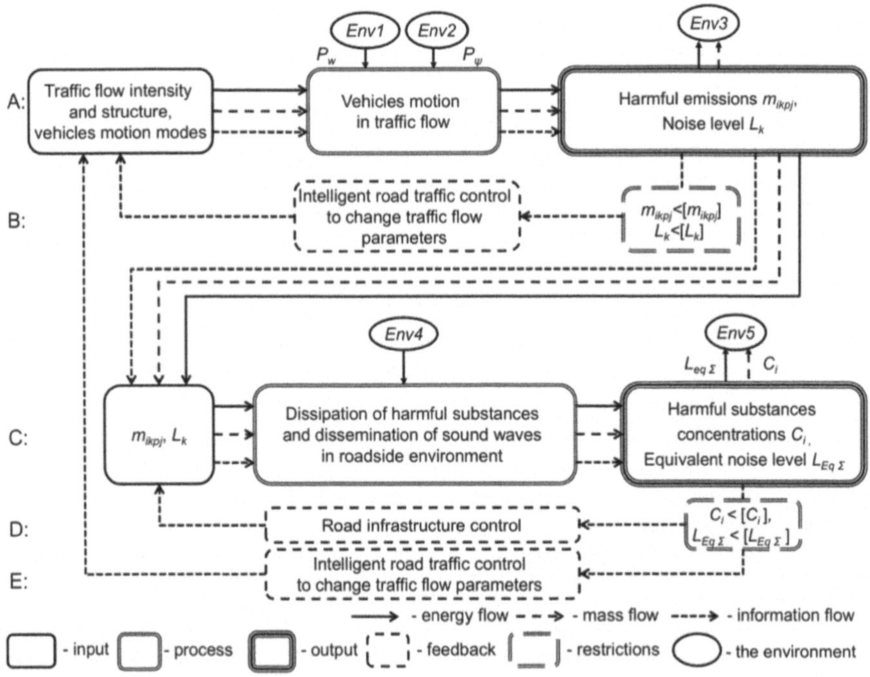

Fig. 1. The structure of the "Traffic Flow – Road – Roadside Environment" system.

The specific emissions m_{ikpj} of carbon monoxide *CO*, hydrocarbons C_mH_n, nitrogen oxides NO_x, particulate matters *PM* and specific fuel consumption are determined by simulation results depending on the current conditions, parameters and driving modes of the separate vehicles in flow. For this purpose, mathematical models of operating processes of vehicular engines are used, which are described in detail in [12–14]. Specific fuel consumption is used to determine also sulfur dioxides SO_2 and carbon dioxides CO_2 emissions.

Restrictions of the system at the *B* level are the specific emission limits of the pollutants $m_{ikpj} < [m_{ikpj}]$ and the noise levels $L_k < [L_k]$ of separate vehicles. Feedback at the *B* level enables to control the *A* level output to provide these restrictions. The control occurs by changing the intensity and structure of the traffic flow, vehicles motion modes in the flow. This can be done through intelligent road traffic control.

In the process of the *C* level, the dissipation of harmful substances and the dissemination of sound waves from separate vehicles in the roadside environment, depending on the road parameters and current weather conditions. As a result, at the output of the *C* level, the concentrations of harmful substances C_i in the air of the roadside and the equivalent noise level of the traffic flow $L_{eq\ \Sigma}$ are obtained.

Restrictions of the system at the *D* level are the maximum permissible concentrations of pollutants $C_i < [C_i]$ in the air and the maximum permissible noise level $L_{eq\ \Sigma} < [L_{eq\ \Sigma}]$ of the traffic flow. Feedbacks at the *D* and *E* levels enable to control the *C* level output to provide these restrictions. At the *D* level the road infrastructure is controlled to change the parameters of the dissemination of harmful substances and noise in the roadside environment. At the E level, intelligent road traffic control is used to change the traffic flow parameters.

3 Characteristics of the Information and Analytical System

The structure of developed IAS is shown in Fig. 2. Traffic flow parameters can be monitored and controlled according to the information from vehicle parameters database. The traffic flow structure is created according to the vehicle categories, fuel types, vehicle environmental classes and traffic flow intensity using the current On-Board Diagnostics (OBD) and Global Positioning System (GPS) data from separate vehicles. Route parameters and weather conditions are determined by the monitoring system according to the vehicle current location data. The simulation of vehicle environmental safety in the traffic flow and the evaluation of roadside pollution by the traffic flow are based on the determined traffic flow parameters and current OBD and GPS data from separate vehicles. The use of geographical information systems (GIS) enables to visualize and analyze roadside pollution by traffic flows. Thus, the concept of connected vehicles in modern transport infrastructure (known as Vehicle-to-infrastructure) is implemented.

Evaluation of the separate vehicle environmental performance in motion on the real routes is possible using OBD data and vehicle location current parameters based on GPS data. Figure 3 shows the enlarged functional scheme of vehicle performance monitoring under operating conditions.

The data set of real characteristics of the vehicle route is determined according to the vehicle current location data. This is the basis for determining real resistance forces of the vehicle motion on the route. Using the mathematical model of the "road vehicle"

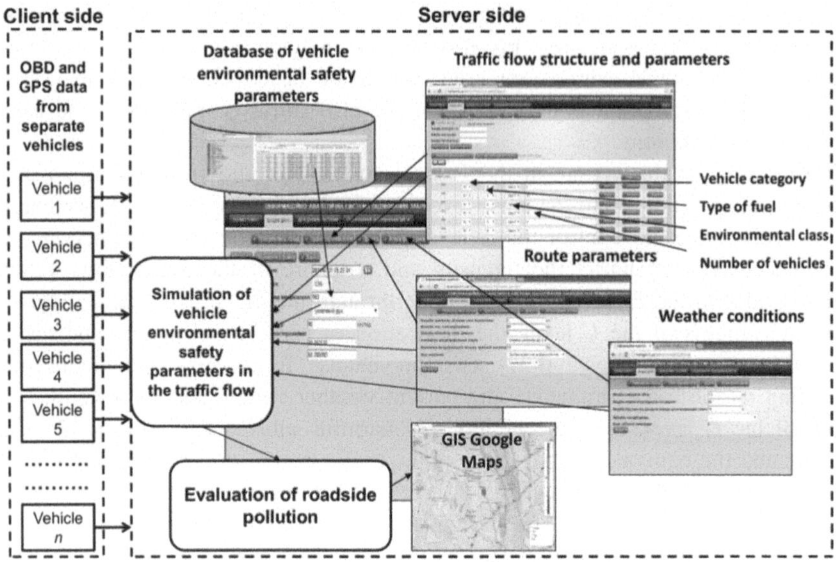

Fig. 2. The structure of the information and analytical system.

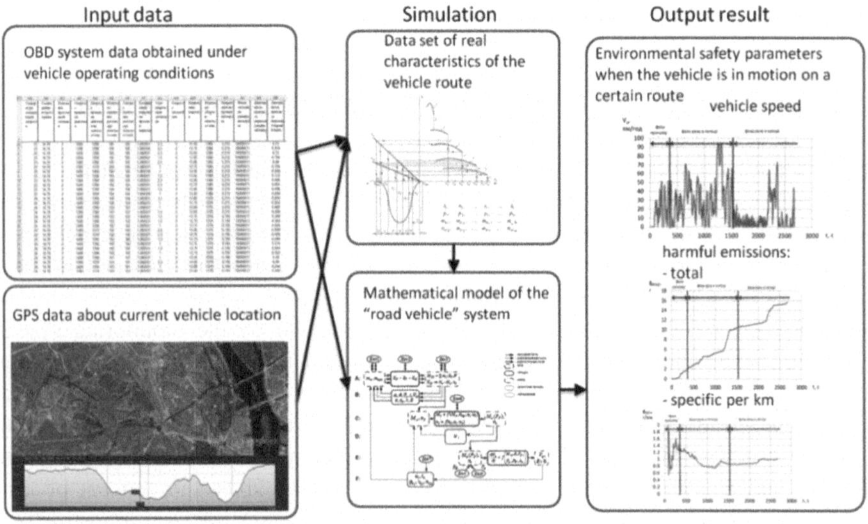

Fig. 3. Algorithm of data obtaining and harmful substances emissions modeling from separate vehicles.

system [7], the environmental safety is evaluated when the vehicle is in motion on a certain route. The obtained data of vehicle motion parameters and environmental safety parameters are transferred to remote databases for further analysis using data transmission systems.

4 The Monitoring Results

Figure 4 shows the examples of current parameters of the vehicle with a gasoline engine motion in the city traffic flow and a separate vehicle environmental safety.

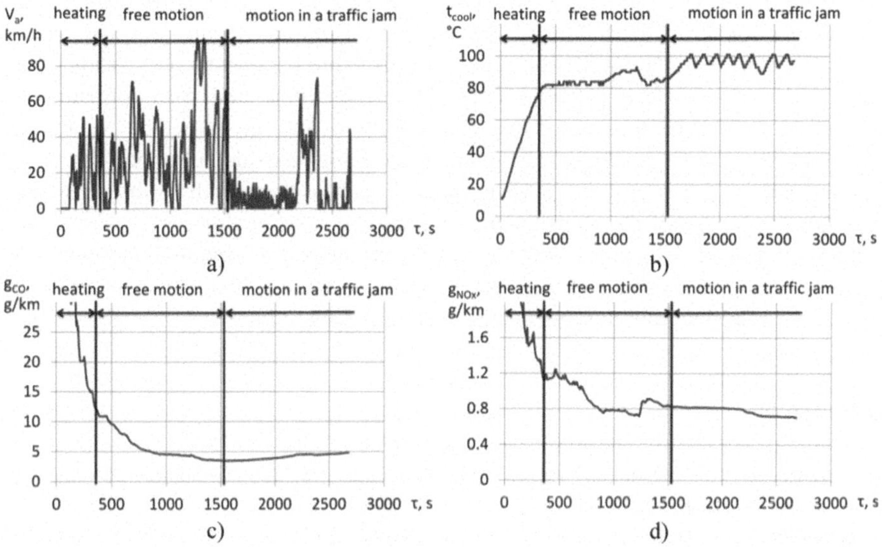

Fig. 4. Motion parameters and environmental safety performance of separate vehicle in the traffic flow: (a) speed; (b) coolant temperature; (c) CO emission; (d) NO_x emission.

The analysis of the obtained vehicle motion parameters (Fig. 4) enables to determine three typical motion phases, namely: heating, free motion and motion in traffic jam. The heating phase is characterized by low speed of motion V_a, increasing the coolant temperature t_{cool} to operating value, the increased values of specific harmful emissions, in particular, carbon monoxide emission g_{CO} and nitrogen oxides emission g_{NOx}.

The phase of free motion is characterized by increasing the speed mode (see Fig. 4a) and normal value of the coolant temperature (see Fig. 4b). This contributes to the decreased values of specific harmful emission (Figs. 4c and d).

The temperature mode of the cooling system in the phase of motion in a traffic jam is characterized by maximum values of the coolant temperature and periodic turning on the radiator cooling fan (see Fig. 4b). The values of specific harmful emission increase in this phase (see Figs. 4c and d).

Based on the processing of OBD and GPS data from separate vehicles, the IAS provides real-time simulation of harmful substance concentrations in the air of the roadside environment. It also can be visualized and analyzed depending on the geographic coordinates using a Google Maps application.

5 Conclusion

The developed information and analytical system enables to monitor and optimize the parameters of the "Traffic Flow – Road – Roadside Environment" system according to the level of the roadside pollution. The level of the roadside pollution is estimated by the concentrations of harmful substances in the air of the roadside environment and the equivalent noise level depending on the technical condition parameters and motion modes of vehicles in the traffic flow, road infrastructure parameters and weather conditions. This system also provides control of traffic flow and road infrastructure parameters to ensure minimum roadside pollution and involves the vehicle-to-infrastructure technologies for assessing and forecasting roadside pollution. The use of geoinformation system ensures to visualize and analyze the harmful substances concentrations in the air of the roadside environment.

Further research will focus to the use of the developed information and analytical system to study the influence of different algorithms of traffic flow management on the level of the roadside pollution in specific sections of real routes.

References

1. Gutarevych, Yu.F.: Reducing Vehicle Emissions under Operating Conditions: Monograph. Vyshcha shkola, Kyiv (1991)
2. Lukanin, V.N., Trofimenko, Yu.V.: Environmental burden decrease in motor transport operation. The outcomes of science and technology. Ser. Automobile and city transport. VINITI, Moscow (1996)
3. Zvonov, V.A., Kozlov, A.V., Kutenev, V.F.: Environmental Safety of a Car in the Complete Life Cycle. NAMI, Moscow (2001)
4. Kanilo, P.M.: Motor Transport: Fuel and Environmental Problems and Prospects: Monograph. KhNAHU, Kharkiv (2013)
5. Eastwood, P.: Particulate Emissions from Vehicles. Wiley, Chichester (2008)
6. Faiz, A., Weaver, C.S., Walsh, M.P.: Air Pollution from Motor Vehicles: Standards and Technologies for Controlling Emissions. The World Bank, Washington, D.C. (1996)
7. Mateichyk, V.P.: Evaluation Methods and Methods for Improving Environmental Safety of Road Vehicles: Monograph. NTU, Kyiv (2006)
8. Mateichyk, V.P., Tsiuman, M.P., Waigang, H.O.: Simulation of the "traffic flow – road" system. Inter-Univ. Collected Articles Sci. Notes **46**, 371–381 (2014)
9. Mateichyk, V.P., Tsiuman, M.P.: Creating the structure of intelligent monitoring system for environmental safety indicators of vehicles at various stages of life cycle. Herald Nat. Transp. Univ. **32**, 193–200 (2015)
10. Gritsuk, I.V., Volkov, V., Mateichyk, V., Grytsuk, Yu., et al.: Information model of V2I system of the vehicle technical condition remote monitoring and control under operating conditions. SAE Technical Paper 2018-01-0024 (2018)
11. Mateichyk, V.P., Waigang, H.O., Smieszek, M.: Information monitoring system of roadside pollution by traffic flows. Automobile Electron. Mod. Technol. **4**, 74–77 (2013)
12. Gritsuk, I., Volkov, V., Mateichyk, V., Gutarevych, Yu., Tsiuman, M., Goridko, N.: The evaluation of vehicle fuel consumption and harmful emission using the heating system in a driving cycle. SAE Int. J. Fuels Lubricants **10**(1), 236–248 (2017)

13. Gritsuk, I.V., Mateichyk, V., Tsiuman, M., Gutarevych, Yu., et al.: Reducing harmful emissions of the vehicular engine by rapid after-start heating of the catalytic converter using thermal accumulator. SAE Technical Paper 2018-01-0784 (2018)
14. Kuric, I., Mateichyk, V., Smieszek, M., Tsiuman, M., Goridko, N., Gritsuk, I.: The peculiarities of monitoring road vehicle performance and environmental impact. In: Stančeková, D., Vaško, M., Rudawska, A., Čuboňová, N., Sapietová, A., Mrázik, J., Szabelski, J., Tlach, V. (eds.) Innovative Technologies in Engineering Production ITEP 2018. MATEC Web of Conferences, vol. 244, p. 03003. EDP Sciences (2018)

Features of Environmental Diagnostics of Heat Motors and Boiler Plants by Information Methods

Andrey Polivyanchuk[1(✉)], Maksym Ahieiev[2], Artur Kagramanian[3], Andrejs Baranovskis[4], and Oleksandr Samarin[2]

[1] O. M. Beketov National University of Urban Economy in Kharkiv, Kharkiv 61001, Ukraine
ap3@ukr.net

[2] Kherson State Maritime Academy, Kherson 73009, Ukraine
maxageev73-73@ukr.net, samarinl62@gmail.com

[3] Ukranian State University of Railway Transport, Feuerbach Square 7, Kharkiv, Ukraine
kartal2@ukr.net

[4] Riga Technical University, Azenes Street 12, Riga 1048, Latvia
dr.baranovskii@gmail.com

Abstract. Environmental friendliness is one of the most important indicators of the quality of modern heat engines and boiler plants, due to the significant negative impact of the chemical and physical nature of these objects on the environment. When creating modern measurement information systems to determine the environmental performance of boiler plants and heat engines, the following goals are set: ensuring a wide scope of use and a long life of equipment, increasing its compactness and mobility, ensuring the high accuracy and cost-effectiveness of using these systems. Therefore, the task of creating a universal system for the ecological diagnosis of heat engines and boiler rooms, which take measures to improve the efficiency of measurements, is appropriate. The purpose of the work was to create and experimentally work out an automated, non-versatile system for ecologically diagnosing heat engines and boiler installations of various purposes according to the indicators characterizing the material and physical pollution of the environment. The technical solution of realization of system of ecological monitoring and diagnostics of heat engines and boiler rooms by information methods is offered. The structure, composition and functional purpose of the basic modules of this system are shown: instrumental, test-demonstration and laboratory.

Keywords: Automotive transport · Diagnostic system · Pollution monitoring

1 Introduction

Today, environmental friendliness is becoming one of the most significant indicators of the quality of modern heat motors (HM), mobile ones and boiler plants (BP) used in the energy sector, industry, communal and other areas. This is due to the significant

E. Ginters et al. (Eds.): ICTE ToL 2019, LNITI, pp. 360–367, 2020.
https://doi.org/10.1007/978-3-030-39688-6_45

negative impact of HM and BP on the environment: the emission of greenhouse gases and pollutants (pollutants) into the atmosphere, acoustic pollution of residential territories, etc.

When conducting environmental diagnostics, HMs and BPs face the problem of the need to increase universality and accuracy of measuring equipment, which is due to the phased improvement of environmental regulatory documents, the expansion of the scope of environmental standards, and a decrease in the levels of monitored indicators. In these conditions, the creation of effective systems for the environmental diagnosis of HE and BI, corresponding to current and perspective regulatory requirements, with a wide application area is an actual direction for research [1–3]. Wherein, this system should have such properties as: universality: the ability to diagnose HM and BP; high accuracy of measurements at the minimum cost of equipment; compactness, mobility and convenience in operation; ability to improvement.

The purpose of this work was to create a system for the environmental diagnosis of HM and BP (called further as a diagnostic system) using information methods to increase the effectiveness of monitoring indicators of chemical and physical pollution of the environment. For this, the following tasks were solved: (1) the creation of a system diagnostic in accordance with the requirements of regulatory documents - UNECE Regulation R-49, R-83, R-96, ISO 8178, etc.; (2) the realization of information methods increase efficiency environmental performance monitoring HM; (3) experimental studies of the diagnostic system and information methods on full-scale objects.

2 Description of the Diagnostic System

Based on the world and their own experience in the creation and operation of equipment for the environmental monitoring of energy installations reflected in works [4–8], by the authors developed a universal, multi-functional diagnostic system, allowing to determine the following environmental indicators: concentration of pollutants in the exhaust gases of HM, flue gases BP and atmospheric air; mass, specific and average operational emissions of pollutants; noise levels, thermal pollution, vibration; surface concentrations of pollutants in atmospheric air, etc.

The diagnostic system consists of 3 functional modules (see Fig. 1).

The instrumental module is designed for execution directly monitor the technical and environmental indicators of the studied objects using: - a multi-channel portable gas analyzer OKSI-5 M, a compact PM emission control system - MKT-2 [9], a sound level meter-registrar DT-6652, a thermal imager Testo 871, a vibration meter VM6360 and etc.

Testing and demonstration module is intended for testing measuring of appliances and equipment, simulation modeling of selection and analysis gas samples, demonstrating the principles of operation, technical capabilities and operating rules for environmental diagnostics of HM and BP in the preparation of qualified specialists.

The laboratory module is intended for carrying out the qualitative and quantitative analysis of gas samples taken during the study of sources of environmental pollution using photometric, gravimetric, ionometric and other measurement methods.

Fig. 1. Appliances and equipment of diagnostic systems.

The instrumental module is designed for execution directly monitor the technical and environmental indicators of the studied objects using: - a multi-channel portable gas analyzer OKSI-5 M, a compact PM emission control system - MKT-2 [9], a sound level meter-registrar DT-6652, a thermal imager Testo 871 and ect.

Testing and demonstration module is intended for testing measuring of appliances and equipment, simulation modeling of selection and analysis gas samples, demonstrating the principles of operation, technical capabilities and operating rules for environmental diagnostics of HM and BP in the preparation of qualified specialists.

The laboratory module is intended for carrying out the qualitative and quantitative analysis of gas samples taken during the study of sources of environmental pollution using photometric, gravimetric, ionometric and other measurement methods.

3 Information Methods to Improve the Efficiency of Control

In the diagnostic system are implemented information methods for monitoring and improving the accuracy of measuring the average operational emissions of gaseous pollutants and PM - indicators GAS and PT, measured in accordance with the requirements of the UNECE Regulation R-49, R-96 and standard ISO 8178.

3.1 Method for Assessing the Accuracy of Measurements of GAS and PT

This method allows you to determine the resulting measurement error of the indicated of these values – δGAS, δPT as the error of the result of indirect measurements [10]:

$$\delta y = \frac{1}{y} \cdot \left(\sum_{j=1}^{m} \left(\frac{\partial y}{\partial x_j} \cdot \partial x_j \cdot x_j \right) \right)^{0.5} \tag{1}$$

where: y - the value determined by calculation from the known dependence $y = (x_1, x_2, \ldots, x_m)$; $x_1, x_2, \ldots, x_m$ - values measured directly during the testing quantity m; δx_i - known error measurement of quantities of x_i.

As a result of applying this formula to the dependencies for determining the values of GAS and PT established by the ISO 8178 standard, the expressions:

$$\delta GAS = \sqrt{\delta M_{gasi}^2 \cdot \sum_{i=1}^{n} (WF_i \cdot k_{Mgas})^2 + \delta P_i^2 \cdot \sum_{i=1}^{n} (WF_i \cdot k_{P_i})^2} \tag{2}$$

$$\delta PT = \sqrt{\delta m_j^2 + \delta m_{sam}^2 + \delta G_{edf}^2 + \delta P_i^2 \cdot \sum_{i=1}^{n} (WF_i \cdot k_{P_i})^2} \tag{3}$$

where: δM_{gasi}, δP_{ei}, δm_f, δm_{sam}, δG_{edf} – errors of direct measurements of quantities: mass emission of gaseous pollutants, engine power, mass of PM collected on the filter, sample weight and equivalent mass flow in the sampling system, respectively; n – number of test cycle modes; WF_i – weight coefficient of the i-th test mode; K_{Mgasi}, K_{Pei} – coefficients taking into account the influence of M_{gas} and P_e values on the accuracy of measurements of GAS and PT, determined experimentally.

The dependencies (2) and (3) can be used in the analysis the influence of errors of the measuring equipment for the accuracy of determining the GAS and PT values.

3.2 Method for Improving the Accuracy of Measuring the Indicator PT

This method is implemented in the MKT-2 microtunnel and makes it possible to increase the accuracy of measurements of the indicator PT due to accounting the methodological error of measurement of a given quantity - δPT_t, which arises due to influence of the sample temperature in front of the sampling filter - t_f on the measured PM mass emission - m_{pm} [4, 6] (see Fig. 2). The temperature t_f during PT control depends on the temperature of the air entering the tunnel – t_{dil} and the maximum sample temperature in the tunnel - $t_{f(max)}$, which, in accordance with the requirements of standard ISO 8178, can change in the ranges: t_{dil} = 20–30 °C and $t_{f(max)}$ = 42–52 °C.

Based on the analysis of the results of experimental studies of error δPT_t, given in the works [4–6, 11–13], the authors proposed a dependence for its determination [14]:

$$\delta PT_t = \sum_{i=1}^{n} k_{modei} \cdot \cdot K_{Mpmi} \cdot (t_{fi} - t_{f0i}) \tag{4}$$

where: K_{modei} – a coefficient that takes into account the influence of the engine operating mode on the quantity δPT_t, determined by the method described in work [9]; K_{Mpmi} – coefficient equal to the ratio of the mass emission of PM in the i-th mode to the average per cycle of PM emission; t_{fi} – the actual value of the sample temperature in the i-th mode; t_{f0i} – sample temperature at the i-th mode, taken as the base temperature and corresponding to the values t_{dil} = 20 °C and $t_{f(max)}$ = 52 °C.

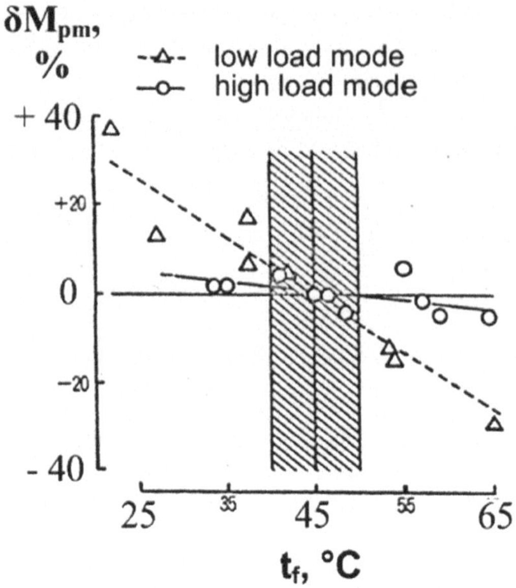

Fig. 2. Experimental data on the effect of t_f per measured mass emission PM [4].

4 The Analysis of the Diagnostic System Testing

The diagnostic system was tested on full-scale objects: transport diesels: automotive – 4CHN12/14, tractor - D65M, diesel locomotive - diesel train engine DEL-01 (see Fig. 3) and boiler units: gas - DKVR-20/13 and AOGV-100E and solid fuel - KCHM-2 M-4. As a result of tests of these engines in cycles established by the UNECE Regulation R49, R-96 and standard ISO 8178, the coefficients K_P, K_{Mgas} and K_{Mpm} used to assess the accuracy of measurements of GAS and PT (see Table 1) the resulting errors are investigated measurements of these quantities and the range of variation of the methodological error δPT_t was determined (see Fig. 4).

Analysis of the results shows the following.

The resulting measurement errors of the GAS and PT provided by the system diagnostic are: $\delta GAS = \pm$ 2.8%, $\delta PT = \pm$ 4.3%; at the same time, the greatest influence on the accuracy of measurements of the GAS has the error in measuring engine power - δP_{ei}, on the measurement accuracy of indicator PT - the measurement error of PM mass is δm_f; increase in the accuracy of measurements of P_{ei} and m_f 2 times values allows us to reduce the errors of δGAS and δPM by 17% and 21%, respectively.

The range of variation of the methodical error δPT_t is: -2.7–6.3%, with the greatest value exceeding 5%, this error takes when t_{dil} = 20–22 °C and $t_{f(max)}$ = 42–44 °C; to eliminate the error δPT_t and improve measurement accuracy PT in the tunnel can control the temperature of the sample.

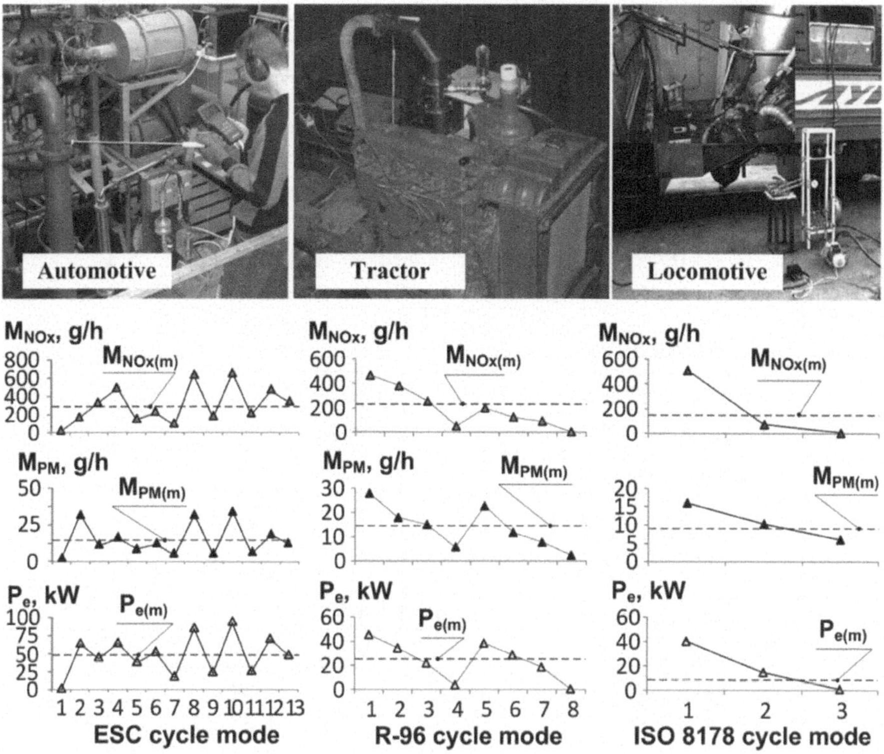

Fig. 3. The results of environmental diagnostics of transport diesel.

Table 1. The results of determining the coefficients KPi, KMgasi and KMpmi, during the tests of the diesel D65M on the cycle R-96.

Coefficient	Coefficient value in cycle mode R-96							
	1	2	3	4	5	6	7	8
KP	1,86	1,41	0,91	0,16	1,57	1,18	0,77	0,03
KMgas	2,19	1,78	1,19	0,25	0,94	0,59	0,44	0,03
KMpm	1,96	1,26	1,05	0,40	1,59	0,82	0,54	0,16

In modes 1, 3 and 8 three repeated measurements were made and the error in reproducibility of the experiment was determined by the method [10], which was: for the K_P – ± 3,7–5,5%; for K_{Mgas} – ± 2,2–4,8%; for K_{Mpm}, – ± 4,0–7,1.

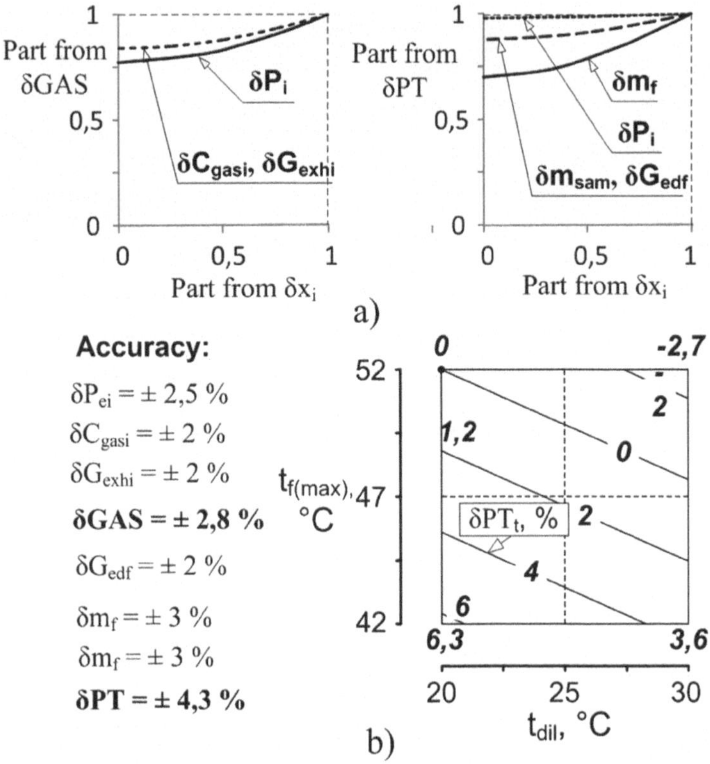

Fig. 4. The accuracy of the system diagnostic.

5 Conclusion

A universal, multifunctional system of environmental diagnostics of HM and BP has was created, allowing to determine indicators characterizing the chemical and physical impact of these objects on the environment: concentrations, mass, specific and average operational emissions of pollutants, noise, thermal pollution, vibration. This measuring system consists of instrumental, testing, demonstration and laboratory modules and, what allow use it like diagnostic tool, a training and test bench and laboratory; and apply it in various fields: transport, energy, environmental and educational fields.

The system diagnostic implements information methods for monitoring and improving the accuracy of measurements of average operational emissions of pollutants: a method for determining the resulting measurement errors of the average operational emissions of gaseous pollutants and PM - δGAS and δPT, which allows one to evaluate the effect on the data of the errors of the measuring equipment of the system diagnostic; a method for increasing the accuracy of measurements of the PT indicator by taking into account the methodological measurement error of this values - δPT_t, due to the influence of the sample temperature in the tunnel on the measured PM emission.

Carried out experimental studies of the system diagnostic and information methods on full-scale objects: transport diesels: automotive - 4CHN 12/14, tractor - D65M, diesel locomotive - diesel train engine DEL-01 and boiler units: gas - DKVR-20/13 and AOGV-100E and solid fuel - KCHM-2 M-4. Transport diesels were tested according to the cycles established by the Regulations R-49, R-96 and the ISO 8178 standard, which allowed us to obtain the following results: the coefficients K_{Pi}, KM_{gasi} and K_{Mpmi} used to calculate the errors δGAS and δPT were established; the values of these errors were determined when using the diagnostic system: δGAS = ± 2,8% and δPT = ± 4,3%; it was found that the greatest influence on the value of δGAS is exerted by the error in measuring engine power, and on the value δPT is the error in measuring PM mass; defined the range of variation of the error δPT_t −2,7− + 6,3%.

In the future, on the basis of the diagnostic system, an educational and scientific center will be created to study the environmental problems of transport and energy.

References

1. Russel, R.: Development of a Miniaturized, Dilution-Based Diesel Engine Particulate Sampling System for Gravimetric Measurement of Particulates. SAE Technical Papers, 931190 (1993). 12 pages
2. Redziuk, A.: With respect to the determination of the mass emissions of pollutants by wheeled vehicle engines. Avtoshlyahovik Ukraine **4**(228), 2–7 (2012). (in Ukrainian)
3. Varlamov, G.: General approaches to creation of methodological bases of energy-ecological analysis of operation of FEC facilities. Energy Saving Energy Audit **10**(116), 2–9 (2013). (in Ukrainian)
4. Hirakouchi, N.: Measurement of Diesel Exhaust Emissions with Mini-Dilution Tunnel. SAE Technical Papers, 890181 (1989). 11 pages
5. Engeljehringer, K.: Meeting ISO 8178 Requirements for the Measurement of Diesel Particulates with Partial-Flow Dilution Systems. SAE Technical Papers, 932466 (1993). 10 pages
6. Nagano, H.: Measurement of Unregulated Exhaust Emissions from Heavy Duty Diesel Engines with Mini-Dilution Tunnel. SAE Technical Papers, 900643 (1990). 10 pages
7. Klimenko, O.: Research and development of a promising system for determining the mass emissions of pollutants in the exhaust gases of engines. Avtoshlyahovik Ukraine **5**(229), 2–8 (2012). (in Ukrainian)
8. Polivyanchuk, A.: Improving the Efficiency of Diesel Particulate Emissions Control Systems. KNAHU, Kharkiv (2015). 220 pages (in Ukrainian)
9. Polivyanchuk, A.: Experimental verification of microtunnel MKT-2 on the brake stand autotractor diesel engine. Ind. Technol. Eng. **2**(15), 11–16 (2015)
10. Novinsky, P., Zograph, I.: Estimation of Errors of Measurement Results. Energoatomizdat, St. Petersburg (1985). 248 pages (in Russian)
11. Littera, D.: Comparison of Particulate Matter Emissions from Different Aftertreatment Technologies in a WindTunnel. SAE Technical Paper 2013-24-0175 (2013). 17 pages
12. Foote, E.: Evaluation of Partial Flow Dilution Methodology for Light Duty Particulate Mass Measurement. SAE Technical Paper 2013-01-1567 (2013). 10 pages
13. Lach, G.: Specific Problems of Sampling and Measuring Diesel Exhaust Emissions. SAE Technical Papers, 881763 (1988). 12 pages
14. Polivyanchuk, A.: The accuracy of the part-flow systems for controlling emission particulate matter with diesel engine gases. Public Utilities of Cities **5**(151), 32–36 (2019)

Evaluation of the Support-Coupling Patency of Self-propelled Transport

Tassybek Bekenov[1(✉)], Zhanibek Nussupbek[2], and Zhandos Tassybekov[1]

[1] L.N.Gumilyov Eurasian National University, Munaitpasov Str. 5, Nur-Sultan, Kazakhstan
tas-bek@mail.ru, zhandos.org@mail.ru

[2] S.Seifullin Kazakh AgroTechnical University, Zhenis Str. 62, Nur-Sultan, Kazakhstan
nussupbek@gmail.com

Abstract. It is established that basically all the known indicators of the evaluation of the patency of machines are single, which allow them to be compared only on one specific indicator. At the same time, the support and coupling indicators for assessing the patency of machines are also evaluated separately. Moreover, it is known that the support and coupling indicators of the permeability of machines have an impact on each other. In this regard, the relationship between the support and coupling indicators of their permeability is not taken into account when assessing the patency of machines. At the same time, the smaller the bearing capacity of the soil base, the more intense it is manifested. Therefore, the calculation equations for the evaluation of interconnected coupling and support indicators of self-propelled machines are proposed. The results of calculations on the proposed equations confirm their adequacy to the practical data obtained during the operation of specific self-propelled machines.

Keywords: Self-propelled transport · Coupling and support patency · Impact assessment

1 Introduction

Self-propelled vehicles and wheeled mechanization vehicles for various purposes occupy leading positions in terms of cargo transportation and other work performed worldwide. Operation of self-propelled transport in various industries often occurs in difficult road conditions, and the planned volumes of work over the years only increase. Therefore, increasing the productivity of technological complexes, including the productivity and quality of transport equipment are of paramount importance.

For existing types and sizes of transport equipment in various industries are relevant issues of their effective use. This requires the determination of their operational and technical qualities, both for timely and effective performance of preventive maintenance and forecasting of overhaul periods, and to improve their parameters in the design and operation. New types and sizes of self-propelled transport in these

E. Ginters et al. (Eds.): ICTE ToL 2019, LNITI, pp. 368–374, 2020.
https://doi.org/10.1007/978-3-030-39688-6_46

industries are necessary depending on the needs of production and the introduction of new equipment and technology or improvement of existing ones.

Thus, in relation to the existing types and sizes of transport equipment, operational issues are relevant, herewith the correct organization of transport and technological processes will ensure the efficiency of vehicles in various road and freight conditions. It is known that effective operation of self-propelled transport is impossible without analysis and calculation of their operational and technical qualities. In particular, when operating such machines in heavy road conditions, their permeability is relevant. Given the fact that the operating conditions of self-propelled vehicles differ in a variety of different road conditions, the compliance of their support-coupling permeability to such conditions will determine the effectiveness of their work.

The use of self-propelled vehicles for operation in severe road conditions requires the solution of a number of problems associated with the problems of improving the patency of machines in plan and profile. In this regard, the assessment and selection of rational types and parameters of self-propelled transport for use in industrial enterprises and the determination of the main indicators of the traction patency is an urgent task. In this paper, the authors consider the possibility of calculating the interrelated parameters of the support-coupling patency of self-propelled transport.

The aim of this work is to briefly review and analysis of existing research in the field of patency of machines, in justification of the equation for assessing the performance of reference-cross coupling of self-propelled transport, as well as in the calculated and experimental estimates of the support-coupling patency of self-propelled transport.

The second section provides an overview and analysis of existing work in the field presented. In the third section a study of the support-coupling patency of self-propelled transport, which includes the rationale for the equation for calculation of the support-coupling patency of self-propelled transport, and their calculation and experimental evaluation. The fourth section presents the conclusion of the work.

2 Review of Existing Researches

The analysis of operating conditions of self-propelled machines shows that they differ in a great variety [1–3]. The development of the theory of patency of self-propelled machines is devoted to the work [4–8] and others, which also proposed a number of indicators for the evaluation of machines. Basically, all these indicators of patency are single, which allows to compare the machines only on one specific indicator. In most cases, it is necessary to resort to the results of experiments, which narrows the scope of their application. Analysis of existing studies of wheeled vehicles shows that the basis in this type of research is the process of interaction of the wheel with the support surface. It is known that supporting and coupling characteristics of the machines are evaluated separately. It is known that the reference and coupling characteristics of machines have an impact on each other. The analysis also shows that in existing studies, the assessment of rolling and slipping losses is carried out separately, although this is a conditional measure. In fact, experimental data show that rolling resistance increases with increasing slippage. In this respect, the evaluation of machines does not take into account the relationship of the support and coupling indicators of their patency [4, 5]. At the same time, the lower the load-bearing capacity of the soil base, the more intense it is manifested.

Consequently, now, there are no methods of analytical evaluation of interrelated support-coupling indicators of self-propelled machines. Respectively, there is no possibility of evaluation and calculation of both single and compared machines by their complex support-coupling indicator. Assessment of the patency of self-propelled wheeled vehicles on typical surfaces of motion is also advisable to make on the basis of the forecast, on the basis of theoretical calculations to eliminate the huge costs of experiments. Therefore, a feasibility study of the interrelated assessment of the support-coupling indicators of patency of self-propelled vehicles and the development of the analytical method of their calculation, accordingly, the rationale for the recommendation on the efficient operation of existing machines and the design of new construction self-propelled cars of the increased passability is an important issue.

3 Theoretical Background for Support-Coupling Patency of Self-propelled Transport

3.1 Justification of Equations for Estimation of Parameters

In this work evaluation of the support-coupling patency of self-propelled transport is made on its most loaded wheel.

The stress state of the ground under the wheel by the coupling weight G_k is determined by the vertical and horizontal loads and the corresponding stresses on the support base. The development of the model of the ultimate support-coupling patency of self-propelled transport was based on the use of the known shear strength condition:

$$[\tau] = \sigma tg\varphi + C \tag{1}$$

where $[\tau]$ – allowable shear stress; σ – normal stress; C – the coupling of soil; φ – the angle of internal friction of soil.

Assuming that the stresses under the contact area are distributed uniformly and using the known transformation formulas, the equations are obtained [9, 10] to determine the maximum possible forces of the machine wheel adhesion to the ground and soil compaction

$$P_c = (G_k \sin\varphi \cos\varphi + CbL\cos^2\varphi)(1 + \sin\varphi)^{-1}, \tag{2}$$

$$P_y = G_k(1 - \sin\varphi) - CbL\cos\varphi, \tag{3}$$

$$P_d = \frac{G_k}{bL}, \tag{4}$$

where P_c и P_y – accordingly the forces of adhesion and compaction of the soil; C and φ – the grip and angle of internal friction; G_k – coupling weight; b and L – the width and length of tire contact with the base; S_k – wheel contact area with ground; ($S_k = bL$); P_d - wheel specific load.

3.2 The Estimation of Indicators

Calculations of the maximum possible forces of adhesion of the most loaded wheel with soil and soil compaction wheel self-propelled transport carried out by Eqs. (2 and 3) at different values of the strength parameters of the soil and the constant contact area of the wheel with the soil are presented graphically in Figs. 1, 2 and 3.

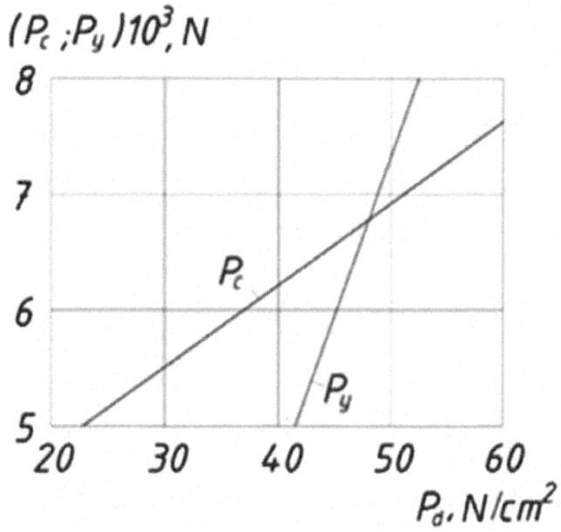

Fig. 1. Dependence of the wheel coupling force with the soil and the soil compaction force by the wheel of self-propelled transport on the specific load (S_k = *325* cm^2; *C* = *15* N/cm^2; φ = *15*°).

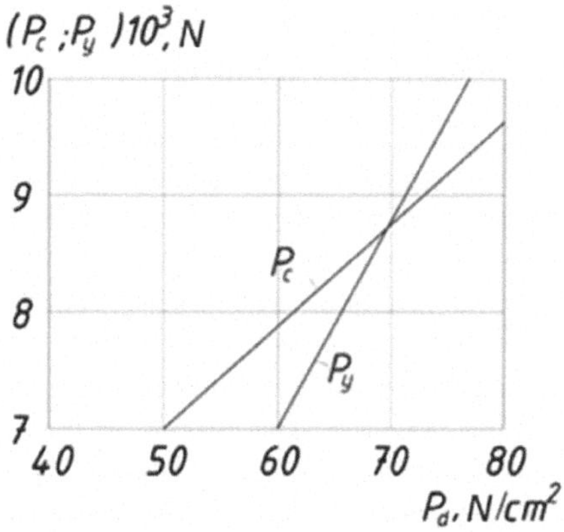

Fig. 2. Dependence of the wheel coupling force with the soil and the soil compaction force by the wheel of self-propelled transport on the specific load (S_k = *325* cm^2; *C* = *15* N/cm^2; φ = 25°).

Analysis of the findings shows that any increase the coupling weight of the wheels is accompanied by increasing both traction wheels with the ground, and the forces of the compaction wheel self-propelled transport, the difference between them decreases as it approaches a unit load bearing capacity of the soil to destruction, as the point of intersection of lines corresponds to the bearing capacity of a given soil [11].

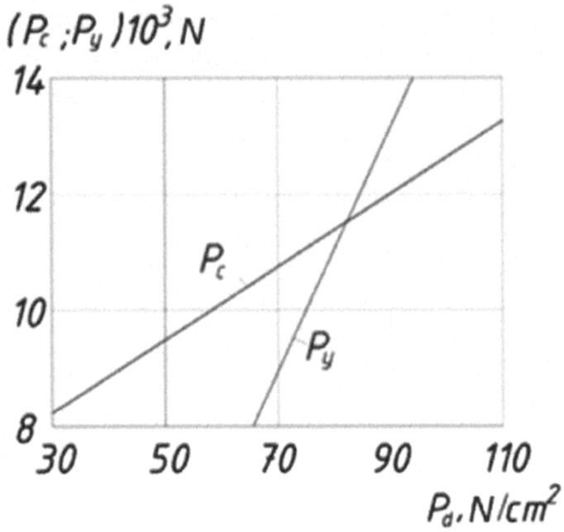

Fig. 3. Dependence of the wheel coupling force with the soil and the soil compaction force by the wheel of self-propelled transport on the specific load (S_k = *325* cm^2; C = *26* N/cm^2; φ = *15*°).

In such cases of increasing the coupling weight of the wheel at a constant area of contact with the ground, the support-coupling permeability of self-propelled transport decreases, since its stock decreases.

3.3 Experimental Evaluation of the Parameters

An experimental test of Eq. (2) to determine the maximum force of adhesion of the wheel to the ground of the base was carried out on a self-propelled machine BC5E (4 × 4) [10]. At the same time with tests measurements of characteristics of soil of the basis were made [10], when the soil moisture fluctuated from 4% to 12%. The coupling weight in the loaded state of the machine was 120000 N. The footprint area of a single tire averaged was 888 cm^2 [10]. According to the formula (2) it turned out the following that at a humidity of 12% and G_k = 120000 N, P_c = 45143 N, but at humidity 4% and G_k = 120000 N, P_c = 90944 N. The average of the calculated P_c within the soil moisture from 4% to 12% was 68043 N. Experiments to determine P_c respecting the temperature within the fluctuations of soil moisture from 4% to 12% [10] allowed to establish the average value equal to 65000 N [10], that confirms adequacy the results of the theoretical data.

4 Conclusion

The review and the analysis of operating conditions of self-propelled transport in various industries and methods of an estimation of their passability on a ground were carried out. The necessity of estimation and calculation of the interconnected support-coupling indexes of pass ability was proved. The equations for determination and estimation of force of coupling of a wheel with soil and force of compaction of soil by a wheel on the most loaded axis were discussed. The calculated estimation of indicators of the support-coupling patency was done. The results of the experimental evaluation of the traction force of wheels with soil and its comparison with the results of the calculated evaluation, which showed sufficient adequacy were presented. The calculations showed that increasing the coupling weight of the wheels is accompanied by increasing both traction wheels with the ground, and the forces of the compaction wheel self-propelled transport, the difference between them decreases as it approaches a unit load bearing capacity of the soil. At the same time calculations showed that in cases of increase of coupling weight of a wheel at the constant area of its contact with soil the support-coupling pass ability decreases as its stock decreases. The research shows that the results do not contradict experimental data and can be recommended for use both in operation and in the design of self-propelled transport.

The approach proposed in the paper allows to assess the patency of self-propelled transport by reference and coupling indicators in their relationship with each other, whereas in existing studies, the support and coupling indicators of patency of transport are evaluated separately.

Further research will be carried out in terms of choosing the optimal parameters of the support-coupling patency of self-propelled transport. Research will also be carried out in terms of improving the design parameters of self-propelled transport, taking into account its support-coupling patency for specific ground conditions.

References

1. Darguzis, A., Ligonnet, F., Pilkauskas, K.: Investigation of vehicle's dynamics in crossing vertical obstacles. In: Transport Means, Proceedings, p. 228 (2010)
2. Kosharniy, N.: Technical and Operational Properties of High-Terrain Vehicles, p. 208. Higher School, Kiev (1981)
3. Bekker, M.: Introduction to the Theory of Control Systems – Machines, p. 520. Machine Building, Moscow (1973)
4. Ageykin, Y.: Permeability of Automobile, p. 231. Machine Building, Moscow (1981)
5. Platonov, S.: Assessment of the patency of vehicles taking into account the operating conditions. In: Proceedings of the Universities, no. 10, pp. 88–91. Machine Building (1987)
6. Perkowski, Y., Shuchman, S.: Fundamentals of the Theory of Wheel Propulsion (Applied Problems of Optimization of Structures), p. 92, Moscow (1996)
7. Volskaya, T.: Assessment of the Patency of a Wheeled Vehicle When Driving on an Uneven Ground Surface, p. 215. Moscow State Industrial University, Moscow (2007)
8. Larin, V.: Methods for Predicting the Reference Patency of Multiaxial Wheeled Vehicles on the Terrain, p. 224. Bauman MSTU, Moscow (2007)

9. Bekenov, T.: Determination of the coupling force of the wheel mover of mining and transport machines with a deformable support surface. Min. J. **11**, 58–59 (1984)
10. Bekenov, T.: Theory and Calculation of Traction and Support-Coupling Patency of Self-propelled Machines and Their Applications, p. 112. Buki Vedi, Moscow (2012)
11. Bekenov, T., Tassybekov, Z.: Investigation of the influence of design parameters on the traction permeability of self-propelled machines. In: Theory and Calculation of Traction and Support-Coupling Patency of Self-propelled Machines and Their Applications, pp. 50–53. Buki Vedi, Moscow (2012)

The Uses of Smart Hydrones in Case of Floods or Tsunami Relief Operations

Mario Arturo Ruiz Estrada(✉)

Mantarraya Negra Smart Platforms Co. Ltd., Kuala Lumpur, Malaysia
marioarturoruiz@gmail.com

Abstract. This research presents three new smart hydrones in case of floods or Tsunami rescue and relief operations anywhere and anytime. Three new models of smart hydrones such as the MAR21, MAR25 and MAR27 smart hydrones are provided. Initially, the MAR21 smart hydrone has a large structure to facilitate the carrying of food, medicines, and water to areas were affected severally by any flood or Tsunami. At the same time, the MAR21 smart hydrone can handle also to do it a deep cleaning service of any area was affected by floods or Tsunami subsequently. Additionally, the MAR21, MAR25 and MAR27 smart hydrones offer to government and humanitarian agencies an alternative natural disasters emergency transportation system to support large floods or Tsunami rescue and relief operations anytime and anywhere.

Keywords: Smart hydrones · Floods and Tsunami rescue · Usability assessment

1 Introduction

This paper proposes the use of three different smart hydrones in case of floods or Tsunami such as MAR21 [1] (see Fig. 1), MAR25 [2] (see Fig. 2) and MAR27 [3] (see Fig. 3) smart hydrones can be put to resourceful use. The areas obviously include the natural disasters and environment issues. Their applicability is dictated by design, size and power and therefore these three smart hydrones (MAR21 [6], MAR25 [7] and MAR27 [8]) can be remarkably different as observed in Figs. 1, 2 and 3. The MAR25 and MAR27 are basically smaller of real high-speed ships, whereas smart hydrones are aimed to formal maritime transportation applications with precise instrumentation of water navigating and uses. The MAR25 and MAR27 smart hydrones have the same concept but designed with smaller or medium main structures with stronger electric engines that enable it navigate longer distances with light cargo for different missions assigned to it. The MAR21 is a large smart hydrone of massive proportions powered by huge five electric engines and engineered with sophisticated software and hardware systems that give it immense navigating and operations capabilities. The MAR21 smart hydrone is controlled by very experienced pilots who operate it using high or short-wave reception remote controls systems or advanced computer systems from control base hundreds of miles away. This article provides a clear distinction of the different MAR21 smart hydrone vs. MAR25 and MAR27 hydrones.

E. Ginters et al. (Eds.): ICTE ToL 2019, LNITI, pp. 375–383, 2020.
https://doi.org/10.1007/978-3-030-39688-6_47

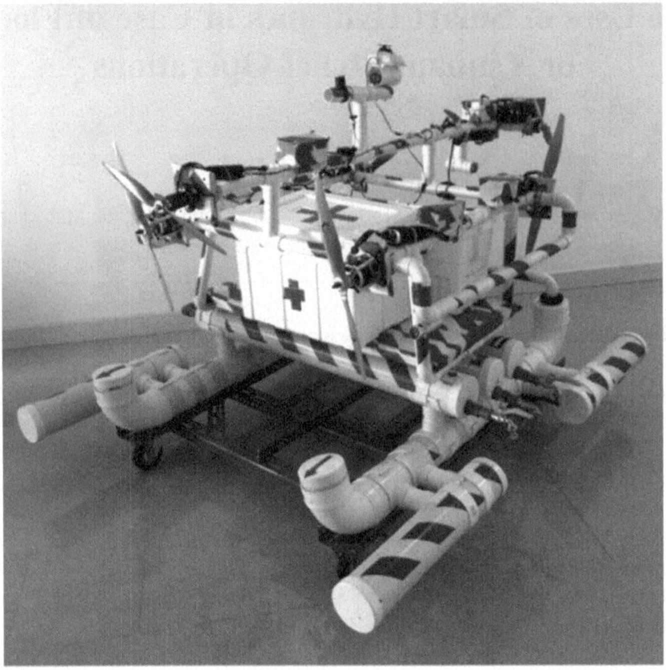

Fig. 1. The MAR21 smart hydrone.

The main difference depends on the structure size and missions, distance of navigation, radio remote control reception systems, GPS navigation, HD camera systems, energy supplier (batteries), large or small electric engines, and endurance respectively [9].

Usually, the MAR21 smart hydrone style systems can be observed mainly in five rotors and stronger PVC plastic structure. The main advantage of MAR21 is the capability to navigate anywhere and anytime, while the disadvantage is mainly the limitation to navigating long distances related to batteries power and ESC resistance particularly when compared to similar the MAR25 and MAR27 smart hydrones. In addition, the limitation of the batteries systems presents many difficulties particularly the delay during departure and arrival to areas with large extension of floods or areas affected by a large Tsunami. In the area of maritime HD video filming and photography, the MAR21, MAR25 and MAR27 smart hydrones can take better videos and photographs according to the weather, waves size, water turbulence, and distance coverage in emergency zones. In contrast, the MAR21, MAR25 and MAR27 hydrones can take pictures and videos under water and water level with high resolution based on its perfect angle, stability and coverage location.

The article further classify the MAR21, MAR25 and MAR27 smart hydrones under study by size and weight namely large-size, medium-size, and small-size. More importantly is the focus of this article on the application of the MAR21, MAR25 and MAR27 smart hydrones in case of floods and Tsunami emergencies. Furthermore, it is important to clarify that the these smart hydrones are aimed to a mix of missions such as the case of the MAR21 smart hydrone can help in rescue and relieve mission. At the same time, cleaning missions in the post-flood or Tsunami disaster event can be combined. Hence, the primary objective of the MAR21, MAR25 and MAR27 smart hydrones is to offer an alternative system to evaluate pre and post-floods, and Tsunami disaster evaluation support with different applications such as monitoring floods or Tsunami impact and rescue missions in all levels.

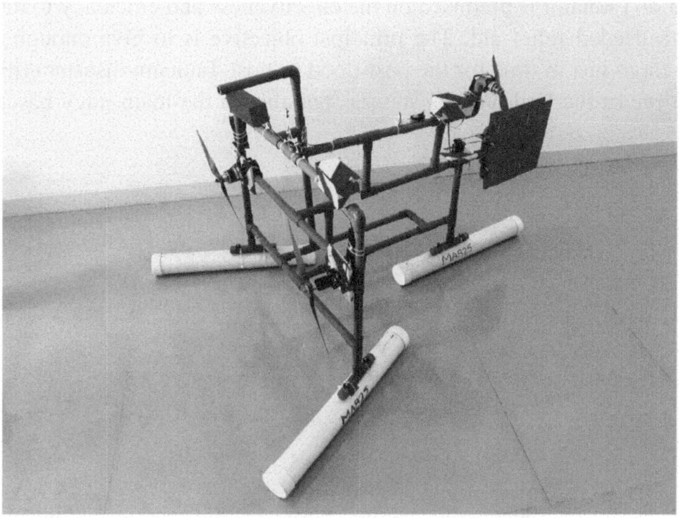

Fig. 2. The MAR25 smart hydrone.

It is suggested that the MAR21, MAR25 and MAR27 smart hydrones are a new natural disaster smart platforms that can supply necessary provisions like water, food, medicine, lights and communication systems (radios) to survivors. To boost these smart hydrones there is need for the construction of large storages where provisions are deposited and easily accessible for supply to survivors. Ultimately for smooth operation, the MAR25 and MAR27 requires the construction of unique light medium and small platforms that can facilitate fast and safe departure and navigating in the event of a flood or Tsunami issue. These hydrones can facilitate the efficient operations that cannot be accessed in case of floods and Tsunami areas affected severally.

In essence, MAR21, MAR25 and MAR27 require to generate an efficient and systematic logistic coordination of their missions. There must be efficient computing systems and excellent maintenance in the case of a huge and destructive floods or Tsunami. The MAR21, MAR25 and MAR27 smart hydrones require five essential elements:

- Availability of large storage deposits;
- Efficient electricity plants and fast charging systems;
- Large stock of parts and accessories;

- The strategic geographical location of large number of hydrones according to the high population density;
- The design of particular routes for navigating in different areas affected by floods and Tsunami anytime and anywhere. Also important is the need for operators to follow needful design structures and communication systems according to the floods and Tsunami dimensions.

The advantages of using the smart hydrones to help survivors in the critical moment of any flood or Tsunami is premised on the effectiveness and efficiency to dispatch and supply much needed relief aid. The principal objective is to give enough support to offload any cargo just in time for the post-flood or post-Tsunami disaster critical timing and the fast return the hydrones as soon as possible to the main navy base.

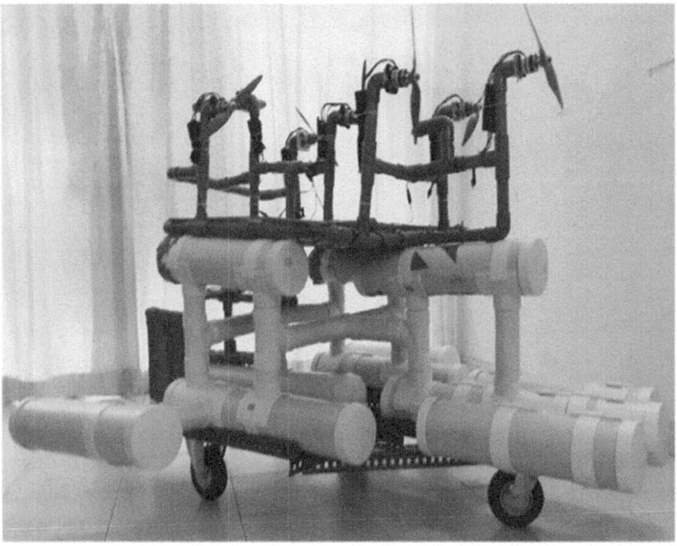

Fig. 3. The MAR27 smart hydrone.

Hence, this research proposes that the smart hydrones have three core missions in any natural disaster mission:

- The aerial monitoring of post-natural disaster damage evaluation;
- The floods or Tsunami logistic and cargo delivery;
- The post-flood or Tsunami maritime assessment.

Firstly, the maritime monitoring post-flood or Tsunami damage evaluation by using the hydrones can help evaluate the real situation and magnitude of the flood or Tsunami damage as soon as possible.

Secondly, the floods or Tsunami logistic and cargo delivery role depends on the hydrones capability. The mission of the hydrones is to convey essential provisions such as water, food, lights, and radios with massive coverage, medicines, and communication systems as primary Internet connectivity as soon as possible.

Thirdly, the post-flood or Tsunami assessment can help evaluate the overall human damage and the magnitude of infrastructural destruction caused by any flood or Tsunami.

These three core missions of the smart hydrones facilitate the quick and efficient floods or Tsunami disaster response and humanitarian relief framework when disasters occur. Finally, the hydrones require experienced maritime pilots located in different main control stations with high radars and antenna to operate them efficiently. In addition, survivors who would engage the hydrones at floods or Tsunami disaster sites must be provided with quick and easy guide on how to offload all provisions from the dispatched hydrones faster and efficiently. With the accelerated rate of technological advancement today hydrones is likely to be transformed into robots in the very near future to help in the informal interaction between survivors and rescue agencies directly under the uses and application of artificial intelligence (AI). When this happens, the hydrone robots are going to facilitate the efficient distribution of necessary provisions to survivors in floods or Tsunami. The efficiency of the hydrone robots particularly after a critical floods or Tsunami depends on the AI based advanced computer systems, large antennas coverage, sophisticated GPS navigation and better mapping systems.

2 The Missions Assessment According to Floods and Tsunami Levels

This section introduces the technical aspect that evaluates the smart hydrones in floods and Tsunami response and humanitarian relief aid. It proceeds in the first section with the technical analyses of the use of hydrones. The second section looks at the different type of missions suitable for the varying the hydrones in a post-floods and Tsunami disaster missions.

2.1 The Smart Hydrones Technical Specifications

To approach the technical evaluation for floods and Tsunami disasters response and humanitarian relief aid for the smart hydrones, first, nine main factors [5] can be mentioned:

- The hydrones design must have enough size (space) and capacity (weight) into consideration and using light materials that will save battery energy and generate a better performance of the electric engines;
- The voltage capacity of batteries which are highly significant to navigate long distances. Many companies today have begun considering and experimenting the use of solar energy on hydrones for longer range of operating hours. Together with efficient charging systems for batteries it will boost floods and Tsunami emergency response in different geographical areas at the same time;
- The size of electric engines and power to generate more speed and carry heavy weights for long distances for an urgent and successful natural disaster mission;

- The antenna coverage and GPS navigation to keep a large reception for better location in the departure or navigating from any hydrone location. At the same time a large stock of spare parts such as a set of propellers, batteries, motors and electronic speed control units;
- The pilot's experience to keep a stable navigating and high precision to deliver any cargo in particular areas affected by any area affected by a flood or Tsunami;
- The geographical location of hydrone-ports (km^2) to generate the perfect logistic coordination in floods or Tsunami areas affected severally;
- The suitable electronic speed control units (ESC-40, ESC-80 and ESC-100) to keep a good connection between the batteries and electric engines all the time mainly in the propulsion system aspect;
- The aerodynamic propellers particularly their design and material quality to generate a better propulsion and stability in the air;
- The HD camera systems with good enough resolution for photo and video recording under water and at water level.

The Table 1 or SHTCE-Table involves the evaluation of these nine technical factors for the MAR21, MAR25 and MAR27 smart hydrones. The SHTCE-Table is based on the uses of a binary system (1 or 0) to facilitate the evaluation of each technical factor, which directly depends on nine different parameters that were established for each technical factor respectively. If the technical factor located in the setting was created before, then the result is equal to 1 (excellent performance) and vice versa (see Table 1). In addition, the SHTCE-Table helps to create an indicator to evaluate the final technical performance of the smart hydrones respectively. The name of indicator or index is the Smart Hydrone Technical Performance θ (1). This indicator could possibly be used by natural disaster rescue and management agencies, governments, and non-governmental organizations (NGOs) to evaluate the smart hydrones technical performance:

$$\theta = \sum_{t=1}^{9} \frac{V_t}{9} \tag{1}$$

where V_t is technical factor in SHTCE-Table. Because $V_t = [0, 1]$ then $0 \leq \theta \leq 1$.

The Smart Hydrone Technical Performance θ differentiate four levels of efficiency: Level-1: 0.00–0.25 (Out of performance); Level-2: 0.26–0.50 (Poor performance); Level-3: 0.51–0.75 (Good performance) and Level-4: 0.76–1.00 (High performance).

The next step is the evaluation of θ (see Table 1). High θ can help in better planning and enforce efficient floods or Tsunami response plans by conveying heavyweight cargo across long distances in the short run. Therefore, the efficiency of the smart hydrones performance directly depends on the technological advances (software and hardware) and the number of pilots able to navigate the hydrones. Given the delicate nature of navigating the MAR21, MAR25 and MAR27 smart hydrones the pilots often begin training from an early age which avails them ability to maneuver and control the hydrones in any flood or Tsunami emergency situation.

Table 1. The smart hydrone technical check evaluation table (SHTCE-Table).

Variable	Parameters	Hydrone
Drone design	Weight: light = 1 or heavy = 0	X
Batteries power	Duration: long = 1 or low = 0	X
The electric engines size and power	Power: high = 1 or low = 0	X
GPS navigation	Coverage: large = 1 or short = 0	X
Pilots experience	Hours: long = 1 or few = 0	X
Droneports location	Distance: short = 1 or long = 0	X
ESC systems	Resistance: high = 1 or low = 0	X
Propellers design	Design: light = 1 or heavy = 0	X
Camera system	Resolution: high = 1 or low = 0	X
Total (θ)		In conformity with (1)

2.2 The Types of Mission Stages

There are three types of mission stages in the evaluation of the MAR21, MAR25 and MAR27 smart hydrones in floods or Tsunami disaster response and humanitarian relief aid:

- Stage-1 is the maritime monitoring pre and post-floods or Tsunami disasters damage magnitude evaluation. The Stage-1 focuses on the uses of the smart hydrones with high-quality resolution HD cameras in real time under water and at water level. The primary objective of mission Stage-1 is to quickly evaluate the preliminary infrastructure devastation after any flood or Tsunami disaster;
- Stage-2 focuses on the channels of distribution or logistic of light cargo by the smart hydrones that can be significantly resourceful in the critical post-floods or Tsunami disaster period. There are five key provisions that are critical to survivors in any flood or Tsunami disaster: water and food in light packaging, essential medicines, essential equipment, Light emission diode (LED) lamps, and radios to establish fast communication in remote areas affected by a natural hazard;

The smart hydrones are of different shapes and levels of cargo capacity. The biggest problem that the hydrones may face would be the reliability test of electric engines that can ultimately run for longer hours in the water and carry heavy cargo for long distances. There is necessary to find the balance between light and durable materials for the smart hydrones design. In addition, there is needful to find suitable electric engines and powerful batteries with high voltage capacity to facilitate long navigating hours. Charging platforms for quick cell charge and fast loading of provisions are equally essential. The smart hydrones require high resolution HD cameras together with large antenna and good GPS coverage as well as high precision for navigating to the droneports for efficient and multiple cargo loading and deliveries.

- Stage-3 focuses on the post-maritime evaluation of any flood or Tsunami disaster. This particular type of stage suggests that each flood or Tsunami disaster require different strategy for the smart hydrones use. In essence, natural disasters such as floods or Tsunami so on demands varying technological features, designs and capabilities that best suit the environmental hazard for efficient emergency response. Importantly, the future of the MAR21, MAR25 and MAR27 smart hydrones effectiveness to support agencies in time of floods or Tsunami is to find the balance between the software and the hardware (structures, battery power and engines) combined [4].

3 Conclusion

The author concludes that the effectiveness of the MAR21, MAR25 and MAR27 smart hydrones in the aftermath floods and Tsunami disasters highly depends on the balance between the technology development of software and hardware combined. The professional training of pilots to implement efficient and successful missions in different stages is highly significant because they are the brains behind every smart hydrones mission. The three types of mission stages analyzed in this article provides a critical insight into the use of different elements that support the smart hydrones efficiency in floods or Tsunami disaster emergency response and humanitarian relief aid. Hence, the smart hydrones response at any stage depends on the fast development of new materials, powerful engines (electric or mechanical), powerful batteries with high voltage capacity or alternative energy resources for electric or mechanical engines (electricity, gasoline, solar), and powerful antennas with advanced GPS navigation respectively. The author strongly accepts the importance and crucial roles the MAR21, MAR25 and MAR27 smart hydrones play in transforming floods or Tsunami disaster emergency response and humanitarian relief aid delivery in the foreseeable future around the world.

References

1. Ruiz Estrada, M.A.: The MAR21 smart hydrone. https://www.researchgate.net/publication/335260876_MAR21_Smart_Hydrone_Food_Medicine_and_Water_Supplier_in_Case_of_Floods_or_Tsunami. Accessed 25 Aug 2019
2. Ruiz Estrada, M.A.: The MAR25 smart hydrone. https://www.researchgate.net/publication/336642021_The_MAR25_Smart_Hydrone. Accessed 25 Oct 2019
3. Ruiz Estrada, M.A.: MAR27 final reforced version. https://www.researchgate.net/publication/335813536_MAR27_FINAL_REFORCED_VERSION. Accessed 25 Sept 2019
4. Ruiz Estrada, M.A.: MAR27-5 M-T, MAR27-6 M-T, MAR27-7 M-T, and MAR21 Smart Hydrones, Project: Mantarraya Negra Smart Platform UAV V-8. https://www.researchgate.net/publication/333972334_MAR27-5M-T_MAR27-6M-T_MAR27-7M-T_and_MAR21_Smart_Hydrones. Accessed 25 May 2019
5. Ruiz Estrada, M.A., Ndoma, I.: The uses of unmanned aerial vehicles UAV's (or drones) in social logistic: natural disasters response and humanitarian relief aid. Procedia Comput. Sci. **149**, 375–383 (2019)

6. MAR21 final testing. https://www.youtube.com/watch?v=IJgeoOkKhX0. Accessed 01 May 2019
7. MAR25 smart hydrone under water and water level cameras. https://www.youtube.com/watch?v=JSwRtBk2Sj4&t=23s. Accessed 01 May 2019
8. MAR27 hydrone. https://www.youtube.com/watch?v=vGKrQam1nAQ. Accessed 18 Oct 2019
9. MAR21, MAR25, and MAR27 smart hydrones by Mantarrya Negra Smart Platforms Group. https://www.youtube.com/watch?v=E0P_79orkAQ&t=2s. Accessed 19 Oct 2019

Author Index

E. Ginters et al. (Eds.): ICTE ToL 2019, LNITI, pp. 385–387, 2020.
https://doi.org/10.1007/978-3-030-39688-6

The manufacturer's authorised representative in the EU is Springer Nature Customer Service Centre GmbH, Europaplatz 3, 69115 Heidelberg, Germany. If you have any concerns regarding our products, please contact ProductSafety@springernature.com

Printed and bound by CPI Group (UK) Ltd, Croydon, CR0 4YY
29/07/2026
02180615-0009